Book of North American
Birds

Book of North American Birds

READER'S DIGEST PROJECT STAFF
Project Editor: James Cassidy
Senior Staff Editor: Richard L. Scheffel
Project Art Director: Gerald Ferguson
Senior Editor: Gayla Visalli
Associate Editor: David Palmer
Art Associate: Colin Joh
Editorial Assistant: Vita Garner

GENERAL CONSULTANTS
John Farrand, Jr.
Editor, The Audubon Society Master Guide to Birding
Past President, Linnaean Society of New York
Elective member, American Orinthologists' Union

Harold F. Mayfield
Past President, American Ornithologists' Union, Cooper Ornithological Society, and Wilson Ornithological Society
Recipient, Brewster Memorial Award and Arthur A. Allen Award

David M. Bird, Ph.D.
Director, Avian Science and Conservation Centre
McGill University, Sainte-Anne-de-Bellevue, Quebec

ART CONSULTANTS
John P. O'Neill
Research Associate, Louisiana State University Museum of Natural Science
Elective Member, American Ornithologists' Union

Kenneth C. Parkes
Senior Curator of Birds
Carnegie Museum of Natural History, Pittsburgh

Robert M. Peck
Fellow of the Academy
Academy of Natural Sciences of Philadelphia

The credits and acknowledgments that appear on page 492
are herby made a part of this copyright page.

First printing in paperback 2005

Visit our website at www.rd.com
In Canada: www.readersdigest.ca

CONTRIBUTORS

Art Assistant: Joe Dyas
Editor: Rita Christopher
Copy Editor: Patricia M. Godfrey
Indexers: May Dikeman, Patricia M. Godfrey

ARTISTS

Paintings:

Raymond Harris Ching
John Dawson
Walter Ferguson
Albert Earl Gilbert
Cynthia J. House
H. Jon Janosik
Ron Jenkins
Lawrence B. McQueen
Garry Moss
John P. O'Neill
Hans Peeters
H. Douglas Pratt
Jim Prutzer
Chuck Ripper
David Simon
John Cameron Yrizarry

Drawings:

Amy Harold
Olena Kassan
Cynthia J. Page
Don Radovich
Dolores R. Santoliquido

WRITERS

Norman M. Barrett
Chuck Bernstein
Robert M. Brown
Jack Connor
Kate Dunham
Pete Dunne
John Farrand, Jr.
David B. Hopes
Kenn Kaufman
Norman Lavers
Mary Leister
Rick Marsi
Wayne R. Petersen
Jan Pierson
Alan Pistorius
Judith Toups

The Library of Congress has catalogued the original edition as follows:
Book of North American Birds.
p. cm.
At the head of title: Reader's Digest
Includes bibliographical references.
ISBN: 978-1-7621-0576-2 (paperback)
ISBN: 978-0-89577-351-7 (hardcover)
I. Birds-North America I. Reader's Digest
II. Reader's Digest
QL681.B657 1990 89-70271
597.297-dc20

Published by World Publications Group, Inc.
140 Laurel Street
East Bridgewater, MA 02333
www.wrldpub.com

ISBN 978-1-4643-0229-9 (paperback)

Printed in China

1 3 5 7 9 10 8 6 4 2

CONTENTS

GALLERY

ABOUT THIS BOOK

More than any other kind of wildlife, birds have an almost magical hold on the human imagination. They are beautiful, vibrantly alive, and everywhere to be seen. They open our eyes to the world of nature. They enrich our spirits with their color, their music, and their wondrous gift of flight.

The BOOK OF NORTH AMERICAN BIRDS celebrates the hundreds of species that spend at least part of the year in the United States or Canada. It includes some 600 species in all—more than 450 of them presented in the Gallery section, which is divided into eight groups according to type and habitat. Each page features a full-color painting, usually of a male in breeding plumage, since in most species the male is more colorful than the female. The portrait is accompanied by text that is meant to be not so much an ornithological profile as a brief essay or narrative—sometimes focusing on a key aspect of the bird's behavior, appearance, or lifestyle, sometimes simply evoking the delight of glimpsing it in the wild.

At the bottom of each page in the Gallery is an information capsule for quick reference, with details on identification, habitat, nesting, and food. In addition, a black-and-white drawing highlights a particular point of interest about the species, and a color-coded range map shows where it can be found. The yellow area indicates a bird's summer range (usually on its breeding grounds, though there are a few exceptions, chiefly among seabirds); blue indicates its winter range; and green indicates areas where it lives year-round. All such maps, of course, can only reflect the best current data on a species' range. Members of any species may wander far outside normal boundaries, and over time the range itself may be altered by habitat and other changes. These maps should thus be viewed as general guides, not as absolutes.

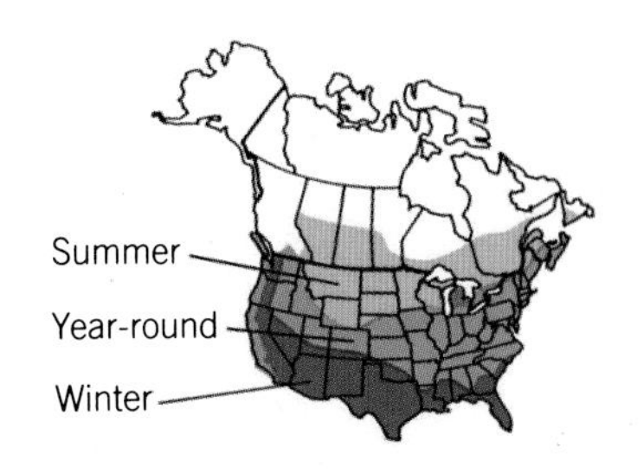

Occasionally, a featured species has a relative that is virtually identical in appearance. In such cases, the second species is named on the same page, along with an extra range map and relevant additions to the information capsule. For an example, see the Eastern Screech-Owl on page 38.

Following the book's main section is a Special Collection of more than 100 species that are rare or have limited ranges in North America. Presented three to a page, these birds have been painted with the same concern for both artistry and accuracy as those in the Gallery.

Finally, the Traveler's Guide will be of practical value to anyone looking for good birding sites anywhere in the United States or Canada. Arranged alphabetically by state and province, the Guide describes more than 350 prime locations for seeing America's birds at their best.

— THE EDITORS

BIRDS OF PREY

They are the warlords of the air—the hawks, falcons, owls, eagles, and other birds referred to as raptors. Swift, fearless, and powerful, they are superbly equipped for their lives as predators—almost too well, for they have also suffered at human hands for the seeming cruelty of their behavior. But such a view mistakes an act for a motive: these fierce hunters, like all wild creatures, are only carrying out the roles given them by nature.

American Swallow-tailed Kite

Turkey Vulture

Cathartes aura

Perch-bound by rain or fog, the turkey vulture is a sullen-looking creature. But there must be magic in the touch of daylight. When the sun burns through the morning mist, the vulture is transformed: quitting the earth to glide above the countryside, wings cocked at a shallow angle, the bird looks for all the world like some feathered truant balanced on an invisible rail fence in the sky. For hours at a time it patrols the air; when conditions are right, it may never need to flap its wings at all. But when the day is spent, or clouds roll in, the magic that holds the vulture aloft slips through its slotted feathers and it sinks to earth, becoming a sullen perch-pumpkin once more.

This master soarer has another specialty: sanitation. It seldom kills its own prey; so the vulture depends on carrion, the flesh of dead animals—a road-killed rabbit, a stillborn lamb, or a deer shot by hunters and never found. Even long-dead food is welcome; indeed, decay is a vulture's ally, softening the bodies of large animals so that they can be picked clean. But an old carcass can be a messy meal, and so it is that the vulture's head is naked, unfeathered—not pretty, perhaps, but beautifully suited to the way this hardworking bird makes its living.

Before it begins a day's soaring, a turkey vulture often basks in the warm morning sun.

Recognition. 26–32 in. long; wingspan 6 ft. Black, with long narrow wings and tail; underside of flight feathers silvery; head naked and small, red at close range. Soars with wings held slightly above horizontal; flaps infrequently.
Habitat. Open country, farmland, and forests.
Nesting. Eggs 1–3, white with reddish-brown spots and blotches; laid on bare ground or in old building, hollow log, or cave. Incubation 39–40 days, by both sexes. Young downy; leave nest at 11 weeks.
Food. Carrion; some small mammals and birds.

Black Vulture

Coragyps atratus

Among the turkey vultures wheeling over an open field in Pennsylvania, there is a smaller, stockier bird. When the group lines out, the little fellow trails behind. Flapping in short, choppy bursts, a black vulture flying with turkey vultures looks like the runt of the litter trying to catch up to its athletic siblings. But farther south, where black vultures are more common, the two species seem not to associate at all—certainly not in a neighborly way.

Black vultures are noticeably more aggressive than turkey vultures and at times take on a predatory manner. Their smaller size does not prevent them from driving turkey vultures away from a carcass or a contested nest site—not that they are overly choosy when setting up house. Black vultures nest in caves, hollow logs, or on sheltered ledges; or, as sometimes happens, they may simply deposit their eggs on the ground. In the tropics, black vultures are familiar residents of cities and towns, where they earn their keep by foraging on assorted human waste. But in more natural settings their principal food is carrion—the riper, the better. Black vultures will often let a fresh carcass go untouched, waiting until the ripening is well advanced, and the pickings are easy.

Unlike turkey vultures, black vultures often soar in tight circles high above a carcass.

Recognition. 23–27 in. long; wingspan 5 ft. Black and stocky; wings short and rounded, with white flash near tip; tail short and fan-shaped; head naked and black. Soars with wings held horizontally; flaps frequently.

Habitat. Open areas, deserts, farmland, and garbage dumps.

Nesting. Eggs 1–2, pale green sparsely marked with brown; laid on bare ground or in shelter of ledge, rock, or shallow cave. Incubation 39–41 days. Young downy; leave nest at 10–11 weeks.

Food. Carrion and garbage; occasionally small mammals and birds.

California Condor

Gymnogyps californianus

April 19, 1987, was a somber day for North America's greatest soaring bird, and for conservationists everywhere. On that date the last free-flying California condor was captured and taken to join the 26 other birds already in captivity, objects of an all-or-nothing effort to keep the species from vanishing into extinction.

A living fossil, surviving from an age when mastodons and giant bison roamed the continent, this bird with the nine-foot wingspan once ranged widely over the coastal mountains of California. But its numbers declined as civilization advanced, and by the 1980's North America's largest soaring bird was making its last stand in the mountains north and west of Los Angeles. Condors are normally long-lived, but they also reproduce slowly. Pairs lay only a single egg every two years, not enough to break even in the face of higher and higher mortality. Condors died from eating the poisoned carcasses set out by ranchers to kill coyotes. They died from lead poisoning after eating deer that had been shot by hunters—or worse, from gunshot wounds that should never have been inflicted.

Scientists concluded that the only hope was to put all the remaining birds in captivity. So it is that the world's last California condors are confined now to zoos in San Diego and Los Angeles. And only if captive breeding efforts—delicate, difficult undertakings—prove successful will the nine-foot wings ever again soar over America.

The blunt claws of a condor reveal it to be a carrion-eater, not a true predator.

Recognition. 45–55 in. long; wingspan 9 ft. Very large and black, with bold white patches on wing linings. Head naked, orange or yellow. Soars with wings horizontal; flaps infrequently.
Habitat. Rugged mountains and nearby agricultural land.
Nesting. Single white egg laid every second February on floor of cave in cliff face or behind boulder. Incubation 42–50 days, by both sexes. Young downy; leave nest at about 5 months, but remain dependent on parents until following summer.
Food. Carrion.

American Swallow-tailed Kite

also known as

Swallow-tailed Kite

Elanoides forficatus

The thunderheads build over Florida every afternoon, offering some relief from the summer heat. Just as the first gusts shake the treetops, a black and white dream of a bird appears. It circles, demonstrating its mastery of the air and its contempt for storms. Then, as slick as an otter in a stream, it glides on, moments before the downpour.

The swallow-tailed kite is everyone's candidate for "loveliest aerialist in North America." If talent and poise are given equal standing with beauty, the swallow-tail makes a strong bid for loveliest *bird* in North America. Dressed in white-tie finery, tails streaming, it maneuvers with the grace of a figure skater and the precision of a gymnast. Grasshoppers, dragonflies, winged beetles, and the like are picked off in midair. Lizards, frogs, and snakes that have gone out on a limb are snatched in passing.

Once the swallow-tailed kite ranged over most of the eastern and central United States. But as farming changed the landscape, the bird withdrew to the swamps and river bottoms of the South. Swallow-tailed kites return to their nest sites early in the spring, usually by March, and the journey to their tropical winter quarters begins in August. Some birds take the land route through Texas and Mexico. Others, it appears, cross the Straits of Florida to Cuba and then navigate on to South America. Many raptors would shun a water crossing of so many miles. But then, few raptors fly with the mastery of a swallow-tailed kite.

A skilled aerial hunter, the swallow-tail can easily snatch dragonflies in midair.

Recognition. 20–25 in. long. Black above, with white head and underparts; tail black, long and deeply forked.
Habitat. Open woods, swamps, and bayous.
Nesting. Nest is a shallow cup 50–100 ft. above ground in tree near water; built of dead twigs plucked with feet by adults in flight. Eggs 2–4, white blotched with reddish brown. Incubation 28 days, by both sexes. Young downy; leave nest 5–6 weeks after hatching. Several pairs may nest close together.
Food. Large insects caught in flight; also snakes, frogs, and lizards.

Black-shouldered Kite *Elanus caeruleus*

also known as

White-tailed Kite

Elanus leucurus

At first glance, that bird seen cruising the freeway median might be mistaken for a ring-billed gull foraging for cold French fries or sesame-seed buns jettisoned by motorists. But gulls don't hover—at least, they don't hover with the style of a black-shouldered kite. Swooping low along the coastal hillsides, sweeping up to gain a better view, the bird suddenly stops in midair. Facing into the wind, the pumping wings must react to changes in wind velocity if the bird is to remain stationary and aloft. So when the coastal winds gust, wingbeats quicken; when wind velocity drops, wingbeats slacken. But let one field mouse stir, let a lizard scuttle, and the fluttering wings beat a different tune. They stop! The kite parachutes to earth, back arched, head down, talons extended, and wings held aloft, signaling "V" for victory. Prey secured, the kite retires to a nearby perch to celebrate its success with a meal.

Though common in Central and South America, the black-shouldered kite long seemed destined to disappear from North America, hanging on only in California and Texas. Now, suddenly, the bird is reclaiming much of its historic range, nesting once more in Louisiana, Mississippi, and Florida. The versatile kite has taken a liking to interstate highways, whose medians and shoulder strips provide ideal habitat for this predator of open, grassy plains. Even the fences and overpasses make welcome additions, offering all the privacy and protection a bird could ask for.

Briefly glimpsed, a gracefully soaring black-shouldered kite could be mistaken for a gull.

Recognition. 15–17 in. long. Pale gray above with white head, underparts, and tail; black patch on shoulder. Immature similar, but tinged and spotted with brown. Soars and hovers.

Habitat. Open plains and farmland with scattered trees.

Nesting. Nest is a mass of sticks 15–60 ft. above ground in tree, usually near water. Eggs 3–6, white, heavily marked with brown. Incubation 30 days, by female only. Male brings food but downy young fed by female. Young leave nest at 35–40 days.

Food. Voles and other small rodents, birds, lizards, and large insects.

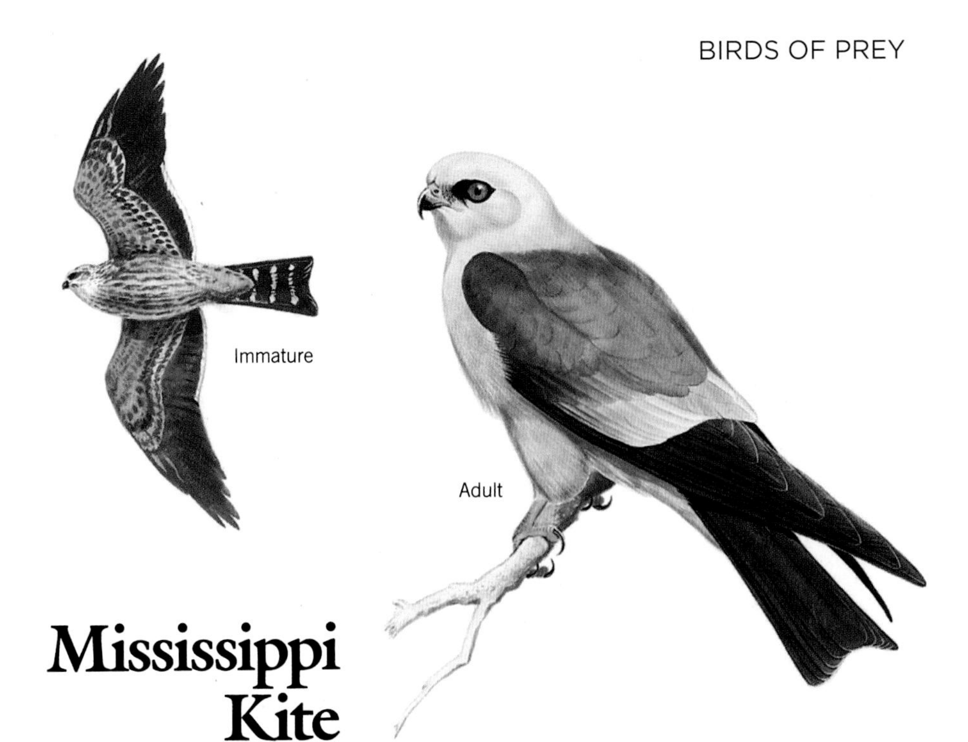

Mississippi Kite

Ictinia mississippiensis

This creature was made for the air. No other bird, not even the elegant swallow-tailed kite, seems so much at home aloft. Before the morning mist has dissipated, flocks of these pale-headed kites are airborne, weaving through the treetops, waiting for the warming air that will carry them skyward. Sometimes they soar on outstretched wings, their tails fully fanned. Sometimes they draw in their sails and glide, tails moving like rudders in the current. But when the mood strikes, Mississippi kites can fly like birds possessed, twisting and turning, parrying each changing gust of wind, then streaking down on unsuspecting dragonflies to snatch them out of thin air.

A lesser hunter might have to carry its prey to a perch, but not the Mississippi kite. Head down, feet extended, these sociable birds delight in feeding aloft, and 20 or 30 of them wheeling overhead can make dragonfly wings fall like confetti. On occasions, the feeding flocks hunt low—at treetop level, even at grasstop level. But more often in the superheated air of a southern summer, Mississippi kites feed so high that they are all but invisible to human eyes. Even when nesting, the birds seem loath to surrender the sky. Treetop nests may be a hundred feet off the ground. Only two powers of nature can defeat the wings of a Mississippi kite. One is rain, the other darkness. As for landing on the ground, a star would be more likely to fall to earth than a Mississippi kite.

Mississippi kites are highly social birds, gathering in large numbers at nighttime roosts.

Recognition. 14–15 in. long. Slender, with pointed wings. Mainly gray, darkest on back, palest on head; tail black; wings pointed. Immature with heavy brown streaks on underparts, faint bars across tail.
Habitat. Open country, groves, brushy areas, and streamside thickets.

Nesting. Nest is a mass of sticks in tree or tall shrub, 4–100 ft. above ground. Eggs 1–3, pale blue. Incubation about 32 days, by both sexes. Young downy; leave nest 32 days after hatching.
Food. Mainly insects; occasionally small snakes and frogs.

Snail Kite

Rostrhamus sociabilis

The female snail kite, unlike other kite females, looks markedly different from her all-dark mate.

As aptly named as any bird, the snail kite feeds exclusively on apple snails, fist-size mollusks that are common throughout the tropical regions of the New World. Lodging around saw-grass marshes, this floppy-winged bird has little need for stealth or lightning reflexes. Apple snails are not particularly wary creatures, and wherever an adequate supply of them can be found, the kite's hearty appetite is certain to be satisfied.

But such specialization in feeding is a double-edged sword. So long as apple snails are abundant, the birds thrive. But let the number of snails dwindle, and the kites—too specialized to eat anything else—helplessly decline. Over the decades, the ditching, diking, and draining of Florida greatly reduced that state's great "River of Grass" and all that lived in it. Forty years ago only a handful of snail kites remained in Florida. Since then a major campaign to save them has achieved some success, securing healthy pockets of the birds in portions of the Everglades and elsewhere. But the water table in Florida continues to fall. The great River of Grass is dying. The snail kite is linked to the snail. The snail is linked to the marshes. And neither seems linked to Florida's future.

Recognition. 16–18 in. long. Wings rounded; tail fan-shaped with large white patch at base; bill slender and strongly hooked. Male slate-black; bill orange or red; feet red. Female dark brown with heavily streaked underparts; feet orange or yellow.
Habitat. Freshwater marshes.

Nesting. Nest is a shallow cup of sticks in marsh grass or shrub. Eggs 3–4, white, boldly marked with brown. Incubation about 30 days, by both sexes. Young downy; leave nest about 4 weeks after hatching.
Food. Exclusively apple snails (*Pomacea*).

Northern Harrier

Circus cyaneus

The silver-backed hawk calls once and begins his descent. Red-winged blackbirds form an angry cloud around him, but the hawk's determination is unswerving. He calls again. From a featureless marsh, a bird rises to meet him—a tawny-colored bird with a white spot on her rump. As they pass, the silver one tosses a furry bundle to his mate. Turning on her back, she reaches out with taloned feet to snatch it, then drops lightly into the marsh.

The male returns to the hunt and the female will return to her nest, but not yet. An open marsh is filled with eyes, some belonging to creatures that would devour her young in a moment if they could. The female waits, flies half the length of a football field, then drops to the marsh again. But still she is not home. With the top of her head barely visible above the grass, she studies the horizon and sky. Satisfied at last, she takes wing again, flying this time to the nest.

As with many other raptors, the male harrier—once called the marsh hawk—is smaller than his mate. Why? A possibility is that during nesting season, the male's quickness and agility make him a more effective hunter, while at other times the pair can avoid direct competition by seeking out prey suited to their different sizes.

Harriers spend the night on roosting platforms that may be shared with short-eared owls.

Recognition. 17–24 in. long. Slender, with wings and tail long and narrow; rump white. Adult male pale gray. Female brown above, streaked with brown below. Soars and glides with wings held slightly above horizontal.
Habitat. Marshes, fields, and grasslands.
Nesting. Nest is a platform of sticks, reeds, and weed stems on ground or in shrub. Eggs 3–9, pale blue, rarely spotted with brown. Incubation about 32 days, by female. Male brings food but downy young are fed by female. Young leave nest about 5 weeks after hatching.
Food. Mainly voles; also other small rodents, frogs, reptiles, insects, and small birds.

Golden Eagle

Aquila chrysaetos

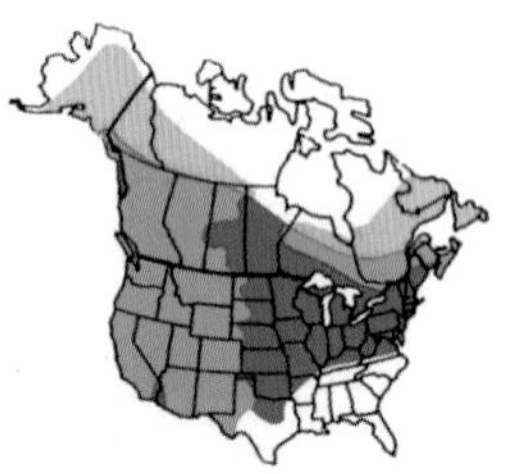

The King of Birds, it has been called. It has ridden on the banners of victorious armies, stood on the fists of emperors, and figured in the religious life of many cultures. The poet Lord Tennyson wrote of the lordly golden eagle:

He clasps the crag with crooked hands:
Close to the sun in lonely lands,
Ring'd with the azure world, he stands.

But human regard for this splendid bird is tainted with ambivalence. The golden eagle has also been a symbol of malice—accused of stealing infants from cribs and of killing domestic animals. The charges of infanticide are pure fiction. The damage golden eagles inflict on lambs is real, but far less common than rumor would have it.

Though they are powerful enough to attack larger prey—even snowbound deer and antelope on rare occasions—golden eagles feed mainly on marmots, jackrabbits, and prairie dogs. What they crave most is solitude. They thrive where the terrain is rugged and people are few. When land becomes populated and tamed, these birds disappear, as they have in the eastern United States. So to the list of the golden eagle's symbolic attributes, we can add one more. It is a true emblem of America's wilderness: as the wilderness goes, so goes the golden eagle.

Most Sioux war bonnets were fashioned from the tail feathers of young golden eagles.

Recognition. 30–41 in. long; wingspan 7 ft. Very large. Adult dark brown with golden buff feathers on back of head, all-dark tail. Immature has white at base of tail, white patches on wings.
Habitat. Rugged mountains and dry mesas.
Nesting. Nest is a huge mass of sticks and brush on cliff or in tall tree. Eggs 1–4, white speckled and blotched with brown. Incubation 43–45 days, by both sexes. Young downy; leave nest 9–10 weeks after hatching.
Food. Mainly small mammals; also birds, and occasionally large mammals up to size of deer.

Bald Eagle

Haliaeetus leucocephalus

On June 20, 1782, the Continental Congress adopted what was called the American eagle as our national emblem, though not without dissent. Benjamin Franklin, for one, opposed the choice. So too, a century later, did the ornithologist Arthur Cleveland Bent, who lamented that its carrion-eating habits and its piratical attacks on the smaller, weaker osprey "hardly inspire respect and certainly do not exemplify the best in American character." To these reproaches one might as well add the charge "lazy." Eagles are amazingly sedentary creatures, often remaining on the same perch for hours at a time.

But by what right are human standards applied? Nature has her own yardstick, and in nature's eyes the bald eagle is blameless. What we perceive as laziness is actually competence. Inept birds waste energy searching for food; adult eagles are free to sit and conserve energy precisely because they *can* secure food at will. True, eagles do eat carrion; but they are also skilled hunters, able to capture waterfowl in flight and rabbits on the run. And, yes, the bald eagle is an accomplished food thief, as many an osprey has learned. But only the osprey has standing to make that accusation—for if the bald eagle is to be judged, it is surely entitled, like all Americans, to a jury of its peers.

Renovated and enlarged each year, the bald eagle's nest may grow to an enormous size.

Recognition. 35–40 in. long; wingspan to 8 ft. Very large. Adult blackish brown with head and tail white, bill yellow. Immature dark brown with dark bill, varying amounts of white in plumage.
Habitat. Rivers, lakes, and seacoasts.
Nesting. Nest is a very large heap of sticks 10–150 ft. above ground in tree, usually near water. Eggs 1–3, dull white. Incubation about 35 days, by both sexes. Young downy; leave nest about 10 weeks after hatching.
Food. Mainly fish, often stolen from osprey; also muskrats, other small mammals, water birds, and carrion.

Osprey

Pandion haliaetus

From a distance, the soaring bird might be taken for a gull. But the wings seem too broad, the head is too small, and . . . wait a minute. Gulls don't dive! No, but the osprey does, with breathtaking mastery. From heights of 100 feet or more it searches for fish swimming near the surface of a bay, a lake, or an ocean—then folds its wings and dives headfirst like an incoming missile. The target is not quite where it seems—refraction in the water distorts the picture—but the osprey knows this and takes it into account during the dive. Just before hitting the water, the bird throws its taloned feet forward. If its aim is true, the hunter will emerge with a fish firmly in its clutches.

The nest to which the "fish hawk" returns is generally a bulky affair. An osprey pair will use the same nest season after season, adding material until the overburdened structure collapses in a storm. House-hunting birds keep an open mind when selecting new nest sites. A dead pine beside some inland lake is ideal; along the coast, a cedar is the preferred setting. But telephone poles, duck blinds, channel markers, and lighthouses serve nicely, too. Sticks are the main building materials, but all sorts of things find their way into osprey nests: conch shells, muskrat skulls, discarded toys, plastic webbing from deck chairs—even, superfluously enough, fishing lines and lures.

In a power dive for a fish, an osprey hits the water feet first with talons spread.

Recognition. 21–25 in. long; wingspan 6 ft. Large and long-winged. Head white with broad dark band through eye and across cheek; back and wings brown. Flies with distinct crook in wing; often hovers over water.

Habitat. Rivers, lakes, and seacoasts.

Nesting. Nest is a structure of sticks in tree, on bush or nesting platform, or, rarely, on ground. Eggs 2–4, white, pinkish, or cinnamon, heavily marked with brown. Incubation 32 days, usually by female. Young downy; leave nest 8 weeks after hatching.

Food. Almost exclusively fish caught in dives from air; occasionally small rodents and birds.

Sharp-shinned Hawk

Accipiter striatus

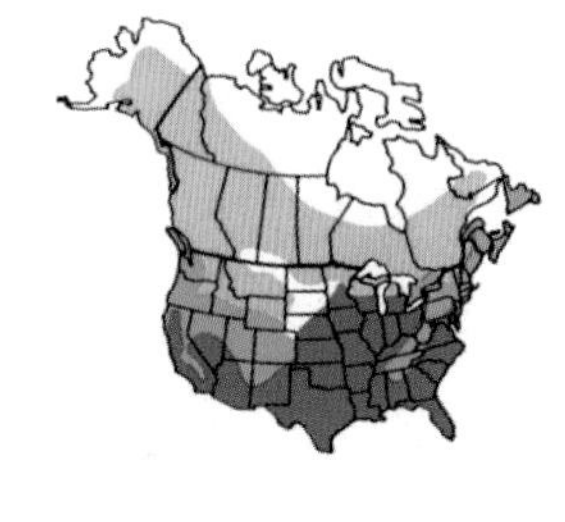

A jay screams an alarm, and the backyard erupts in a flurry of wings. But the warning comes a fraction too late. Beneath the bird feeder, a junco lies pinned to the snow by a small blue-backed hawk. What should be done? The answer, easy to give but difficult to accept, is nothing. The natural world works its own patterns, and human sympathies have little place there. The sharp-shinned hawk is, after all, just one more bird coming to the feeder. The only difference is that this little hawk doesn't eat seeds. It eats the birds that eat the seeds.

The sharp-shin is an accipiter, a woodland raptor, skilled at capturing birds on the wing. Its short, rounded wings permit it to snake through brushy confines. Its long, narrow tail serves as a rudder. But a sharp-shin's speed and reflexes alone are no match for those of a healthy chickadee or junco. So the sharp-shin relies on two tactical edges. One is surprise: a songbird caught off guard forfeits precious split seconds. The other is infirmity: birds weakened by injury or disease are easier to catch than healthy ones. And the sharp-shin actually performs an important service as it meets its own needs. Backyard feeders concentrate birds unnaturally, increasing the spread of infection or disease. By culling the sick birds from the flock, sharp-shinned hawks reduce the risk to all.

Young female sharp-shins begin nesting before they have acquired their adult plumage.

Recognition. 10–14 in. long. Wings rounded; tail long, notched or square-tipped. Adult blue-gray above; white barred with rusty below. Immature brown with pale spots above, streaked and barred with brown below.

Habitat. Forests and woodlands, especially mixed coniferous and deciduous.

Nesting. Nest is a mass of sticks and twigs 10–60 ft. above ground in conifer. Eggs 4–5, white with brown blotches. Incubation 34–35 days, by both sexes. Young downy; leave nest about 23 days after hatching.

Food. Small birds; also rodents and insects.

Cooper's Hawk

Accipiter cooperii

So much like the sharp-shinned hawk that even experts are often fooled, the Cooper's hawk has just one real distinction: it is larger, more powerful, and able to kill larger prey. As luck would have it, the common barnyard chicken of the 19th century fell within the Cooper's prey range. The "chicken hawk" thus became an outlaw, to be shot on sight. Since few people troubled to distinguish between the Cooper's and other species, all hawks by definition became chicken hawks. For years, over much of North America, they were slaughtered by the thousands.

Fortunately, most people have come to understand the role that predators play in nature, and hawks are protected by federal law. But many are still inadvertent victims of man. Flying hawks strike roadside wires; some die after eating gophers or other animals that have been poisoned to control their numbers. Perhaps the greatest threat, though, comes from an unseen foe—the plate glass window. Cooper's hawks are woodland birds; they know nothing about reflecting surfaces. When Cooper's hawks see a window, they see whatever the glass reflects, be it sky or trees. They think they can just fly through it. Sadly, they sometimes even succeed, but the price of success is still a broken neck.

Small songbirds make up a large percentage of the diet of a Cooper's hawk.

Recognition. 15–20 in. long. Similar to sharp-shinned hawk but larger. Wings rounded; tail long and rounded. Adult dark blue-gray above; white barred with rusty below. Immature brown with blackish streaks above, finely streaked with dark brown below.
Habitat. Deciduous forests and woodlands.

Nesting. Nest is a mass of sticks 10–60 ft. above ground in tree. Eggs 3–6, white or greenish, sometimes spotted with brown. Incubation 24 days, by female. Food brought by male, but downy young fed by female. Young leave nest 30–34 days after hatching.
Food. Chiefly small mammals and birds.

Northern Goshawk

Accipiter gentilis

Gray like the color of gunsmoke, eyes that glow like living embers, wings sheathed in silence, and gripping feet tipped with daggers—these are the elements of a nightmare. And if ruffed grouse or snowshoe hares dream, the image of the great northern goshawk haunts their nights just as the living bird hunts them by day.

This large, powerful raptor of northern forests combines the size of a buteo with the killing skills of an accipiter. An adult goshawk usually stays in the north during even the coldest winters—so long as prey populations are sufficient. But nature is rarely static; once every decade or so, snowshoe hare and grouse populations crash simultaneously. Faced with starvation, goshawks flee by the tens of thousands, heading south in what are called invasions. At one site in Minnesota, over a thousand of the migrating gray ghosts have been counted in a single day.

Immature goshawks, by contrast, wander south almost every winter. By spring, their brown juvenile feathers are replaced by the gray plumage of an adult. The youthful yellow eyes begin to glow orange, then red-orange. It will take more than a year before all the brown feathers give way to gray and the streaky bib is exchanged for the adult's fine gray vest. And for several more years their eyes will continue to darken, until they turn the color of old blood . . . the stuff that nightmares are made of.

A hiker who unwittingly comes too close may get a dramatic warning from a nesting female goshawk.

Recognition. 21–26 in. long. Large, with wings rounded, tail long and banded. Adult slate-gray above with black crown and bold white eyebrow; whitish below with fine gray bars. Immature brownish above, whitish with bold streaks below.
Habitat. Coniferous and mixed forests.
Nesting. Nest is a bulky mass of twigs 20–60 ft. above ground in tree. Eggs 2–5, pale blue. Incubation 36–38 days, by female. Food brought by male, but downy young fed by female. Young first fly 45 days after hatching.
Food. Hares, rabbits, other mammals, birds up to size of grouse, and some insects.

Red-shouldered Hawk

Buteo lineatus

It's mud time, and the wooded New England hillsides ring with the cries of courting red-shouldered hawks. A male spirals skyward, climbing high over the treetops. His mate circles below. Suddenly, he pitches forward—a screaming projectile whose looping flight traces a roller coaster route through the sky. Blue jays scream back insults and mimicked cries. New England farmers, who have seen this pageant many times, stop their plowing to watch the display. Spring has returned to the land.

The fidelity of red-shouldered hawks to their nest sites, and often to a specific nest, is well known. The stick nests are usually fitted into the crotch of a mature hardwood, or sometimes a pine. They are sturdy structures, made to last several seasons. But each spring brings alterations and new materials—fresh leaves, mosses, shredded bark, and a freshly trimmed sprig, apparently to proclaim occupancy.

The diet of this woodland raptor is a hearty and varied one. Its hunting skills are most often focused on the cold-blooded creatures of the forest floor—lizards, toads, frogs, and particularly snakes. The red-shoudered is also adept at catching birds, as evidenced by the remains that surround its nest, and mammals up to the size of rabbits and squirrels appear on the menu too. Only because the red-shouldered's talons are small is larger prey safely out of reach.

Red-shouldered hawks usually decorate their nests with sprigs of fresh green foliage.

Recognition. 17–24 in. long. Adult brownish and streaked above with rusty shoulder patch, barred with rusty below; tail with several light bands. Immature mottled brown above, streaked and blotched below.

Habitat. Moist deciduous forests and swamps.

Nesting. Nest is a deep cup of sticks and twigs 20–60 ft. above ground in tree. Nest often decorated with green sprigs. Eggs 2–6, white with brown blotches. Incubation 28 days, by both sexes. Young downy; leave nest 5–6 weeks after hatching.

Food. Reptiles, frogs, small mammals, birds, and large insects.

Broad-winged Hawk

Buteo platypterus

One minute, the September sky is empty except for a few fair-weather clouds. Then it happens—a ribbon of birds materializes. They are so high they look like grains of pepper sprinkled across the sky. The ribbon reaches the bottom of a cloud and frays into filaments, looping and knotting itself into a tangle. In some places they call this swirling mass of broad-wings a kettle; elsewhere it's a boil. By whatever name, the annual exodus of the broad-winged hawk is one of the great spectacles in nature.

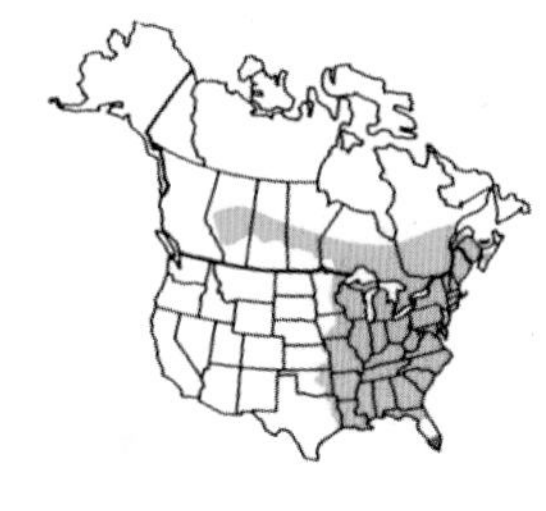

Most broad-winged flocks number in the hundreds. But when the mid-September push is on, kettles run into the thousands. Migrating flocks cross the continent, then funnel down the land bridge of Central America to the tropical forests. Come March, the clouds of birds will once again boil up out of the Mexican plains and swirl over the Rio Grande, tracing their path northward.

This shy, almost docile raptor is hardly larger than a crow and specializes in cold-blooded prey (which is why it must vacate northern forests so early in the fall). From a favorite hunting perch near a stream or woodland pond, the bird watches and waits. Then, on silent wings, it angles swiftly down on its target: a careless frog, a water snake, or perhaps a crayfish that ambles into the shallows—and the broad-wing's waiting clutches.

The broad-wing's swift, decisive attack gives its prey no warning and little hope for escape.

Recognition. 13½–19 in. long. Adult plain brown above, barred with rusty below; tail with 2 light bands. Immature brown with pale bars above, whitish below with brown blotches and fine streaks.
Habitat. Deciduous forest.
Nesting. Nest is a mass of twigs 15–50 ft. above ground in tree. Eggs 2–3, white, marked with brown and purple. Incubation about 28 days, by both sexes. Young downy; leave nest 6 weeks after hatching.
Food. Frogs, toads, reptiles; also small mammals, small birds, and insects.

Swainson's Hawk

Buteo swainsoni

The farmer directs the mower through his field. Crickets and grasshoppers spring for safety as he passes. In the wake of the tractor, where the blades have passed and the grass is cropped short, a dozen russet-bibbed birds scramble and dance. Are they injured? Hardly. The birds are reaping the bounty of insects and small rodents exposed by the mower. The farmer does the work. The Swainson's hawk repays him for his kindness by consuming the creatures that consume his crops.

Among the world's long-distance champions, Swainson's hawks nest as far north as Alaska and winter on the plains of Argentina. In March, Swainson's hawks by the thousands pour across the Rio Grande from Mexico into Texas. When evening comes, the great flocks descend. In the morning, when the sun begins drawing thermals aloft, the birds rise up: dozens, scores, hundreds, thousands of Swainson's funnel skyward. The swirling tornadoes of birds sweep north and fan out across the western states. As the miles and the days fall behind, the great migratory flocks diminish. One by one, the birds reach their nest sites across the western United States and Canada. The business of raising young consumes the summer; by August the task is done, and the birds turn southward. Another exodus of these endurance champions has begun.

With few trees available on the open plains, Swainson's hawks often nest on rocky ledges.

Recognition. 19–22 in. long. Adult dark brown above; breast brown, belly pale. Immature heavily streaked with brown. Wings longer than tail when perched. All-dark birds are common. Soars with wings held above horizontal; often seen in flocks.
Habitat. Grasslands and deserts.

Nesting. Nest is a large mass of twigs and grasses on ground, ledge, or in tree. Eggs 2–4, white, sometimes faintly marked with brown. Incubation 28 days, by both sexes. Young downy; fledge at about 4 weeks.
Food. Mainly grasshoppers and crickets; also small mammals and birds.

Red-tailed Hawk

Buteo jamaicensis

From a forest snag, a sandstone cliff, a tall cactus, a fence post, or the skies over a midwestern farm, the red-tailed hawk throws its challenge to the world.

Keee-yrrrrr . . . Kee-yrrrrrr!

All across North America, people know this large, handsome raptor with the chestnut-colored tail. Farmers welcome it because, like the Swainson's hawk, it eats the rodents that eat their crops. People who feed songbirds in winter have no need to worry: most birds that come to feeders are small and agile enough to evade a burly buteo.

Red-tailed hawks boast a range of plumages tailored to the peculiarities of different regions. Western red-tails tend to be darker than eastern birds. The underparts of western birds are warm and rufous, stained as if by western sunsets. Arizona red-tails are pale, their underparts bleached by desert sun, and prairie red-tails are paler still.

Almost as distinctive as the red tail of a typical red-tailed hawk is the "bellyband"—the dark swath of feathers that runs across the bird's underparts. Aside from its usefulness as a field mark, the bellyband provides at least a modicum of camouflage. Without one to break up the bird's outline, a perched red-tail would stand out like a beacon.

In the West, some dark-phase red-tailed hawks are a solid dark brown except for the rusty tail.

Recognition. 19–26 in. long; wingspan 4½ ft. Typical adult dark brown above; breast white; sides streaked; tail brick-red. Immature similar, but tail brown and banded. All-dark birds with reddish tails common in West.

Habitat. Plains, farmlands, deserts, and open woodlands.

Nesting. Nest is a solid mass of sticks and twigs 15–70 ft. above ground in tree. Eggs 1–5, white, sparsely marked with brown. Incubation 28–32 days. Young downy; leave nest about 45 days after hatching.

Food. Mainly voles and other mammals; also birds and large insects.

Ferruginous Hawk

Buteo regalis

Regal! That describes this western raptor. Standing beside its large stick nest, or perched high atop some sandstone butte, the bird has all the bearing of an eagle. Unlike many birds of prey, the ferruginous hawk is not shy about sitting on the ground. It may alight on a hillside even when elevated perches are handy. Takeoffs from the ground are lumbering procedures; direct flight is a pushing, powerful motion. But when the morning sun draws warm air upward, a ferruginous hawk spreads its long, broad wings and soars. The bird may soar so high that the pale underparts disappear and only the russet leggings that form a dark V remain visible against the sky.

Ground squirrels and prairie dogs are its favorite prey, although animals the size of jackrabbits are not too large. Despite its strength and size, surprise is still this hawk's best ally. Flying low, navigating the contours of the land, the "ferrug" strives to catch prairie dogs away from the safety of their burrows. If it does, the race is on, as hunter and hunted speed for the dark opening that means safety for one, failure for the other. If the hawk is alone, it's an even contest. But if two are working together—a common trick—the smart money is on the hunters every time.

Ferruginous hawks usually pick nest sites that command a wide view of their surroundings.

Recognition. 22–27 in. long. Adult brown above with rusty streaks, white below with rusty thighs; tail whitish with rusty tinge near tip. Immatures and dark-phase adults have whitish tail.
Habitat. Plains and dry mesas.
Nesting. Nest is a mass of sticks, twigs, and bones in top of tall tree when available, otherwise on cliff or pinnacle. Nests reused; often become very large. Eggs 2–6, white or bluish, heavily spotted with brown. Incubation 28 days, by both sexes. Young downy; leave nest after 44–48 days.
Food. Rodents, rabbits, and hares.

Rough-legged Hawk

Buteo lagopus

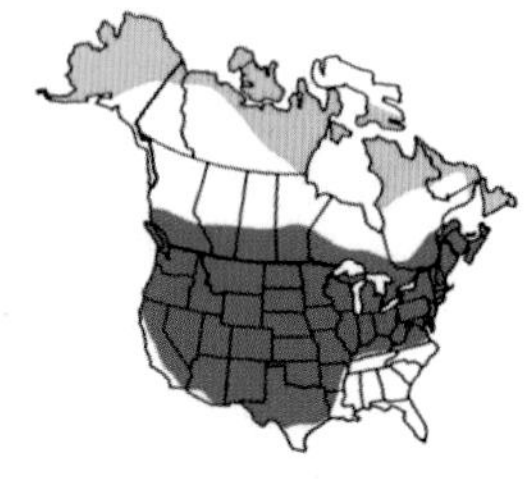

During the long days of an Arctic summer, the rough-legged hawk hunts continually, its rowing wingbeats propelling it over the treeless land like some Viking oarsman of the sky. At every movement on the tundra, it stops and hovers. Broad buteo wings arch up and down, holding the bird in place. The long neck moves from side to side. Suddenly the hawk plunges straight down, feathered legs extended. In a moment the lemming is secured, and the hunter returns to its nest, where ravenous young will fight for their share—and more, if they can get it.

Rough-legs are circumpolar birds, breeding in the Arctic regions of Europe, Asia, and North America. They are also relatively nomadic, nesting wherever mice and lemmings abound. The same opportunism applies in winter, when rough-legs descend on marshes, meadows, and farmlands to the south. Sometimes scores of them mingle with other hawks and owls to feast on a surplus of mice or voles. But it is easy to pick one out of a crowd: just look for the hovering bird with the long, arching wings. Perched rough-legs are even easier to find. They tend to sit on the springy tips of trees—a difficult trick for most buteos, but no problem at all for this agile hunter from the Far North.

In years when lemmings are abundant, rough-legs can raise as many as six or seven young.

Recognition. 19–24 in. long. Dark brown above; head and breast sandy and streaked; belly blackish; tail with white base. Underside of wing has black "wrist" patch. Flies with wings slightly above horizontal; hovers frequently.

Habitat. Open areas and marshes; nests on Arctic tundra.

Nesting. Nest is a mass of twigs on ledge. Eggs 2–7, white, variably blotched with brown. Incubation 28–31 days. Young downy; leave nest about 41 days after hatching.

Food. Rodents and other small animals.

Harris' Hawk

Parabuteo unicinctus

On her nest, a mat of sticks, roots, and grass held aloft by a saguaro cactus, a female Harris' hawk broods her young. Chestnut wing patches flashing, her mate arrives, a snake clutched in his talons. Then a second bird arrives—another male, bearing a desert cottontail. While the female tends to the snake and her hungry nestlings, the male with the rabbit sits nearby, patiently waiting his turn.

Two males at one nest? Yes, polyandry is fairly common in the Harris' hawk, an amazing bird with an amazing array of traits. These beautiful hawks of the desert are social raptors, gathering in small flocks in winter, hunting in pairs, and even sharing the intimate duties of nesting. There are other species of raptors that have a social side, but such close cooperation is rare in hunting birds—and the Harris' hawk is a ferocious hunter.

Why Harris' hawks are so communal is a subject of lively curiosity among experts. Team hunting offers a clear advantage when the prey is agile or too large for one bird to handle. But then the prize must be shared, an obvious disadvantage. A nest with two male providers will produce more young than one tended by a male and female alone—but not as many as there would be if each male had his own nest. Where is the advantage? That is still anyone's guess.

After the young have fledged, families of Harris' hawks stay together for several months.

Recognition. 18–23 in. long. Adult blackish with rusty shoulders, wing linings, and thighs; rump and base of tail white. Immature similar, but streaked with reddish brown.

Habitat. Deserts, desert scrub, and riverside woodlands.

Nesting. Nest a mat of sticks, twigs, and roots 5–30 ft. above ground in desert tree or saguaro cactus. Eggs 2–4, white. Incubation 33–36 days, by both sexes. Young downy; leave nest about 40 days after hatching. A second male sometimes helps tend young.

Food. Mainly small mammals; also birds and reptiles.

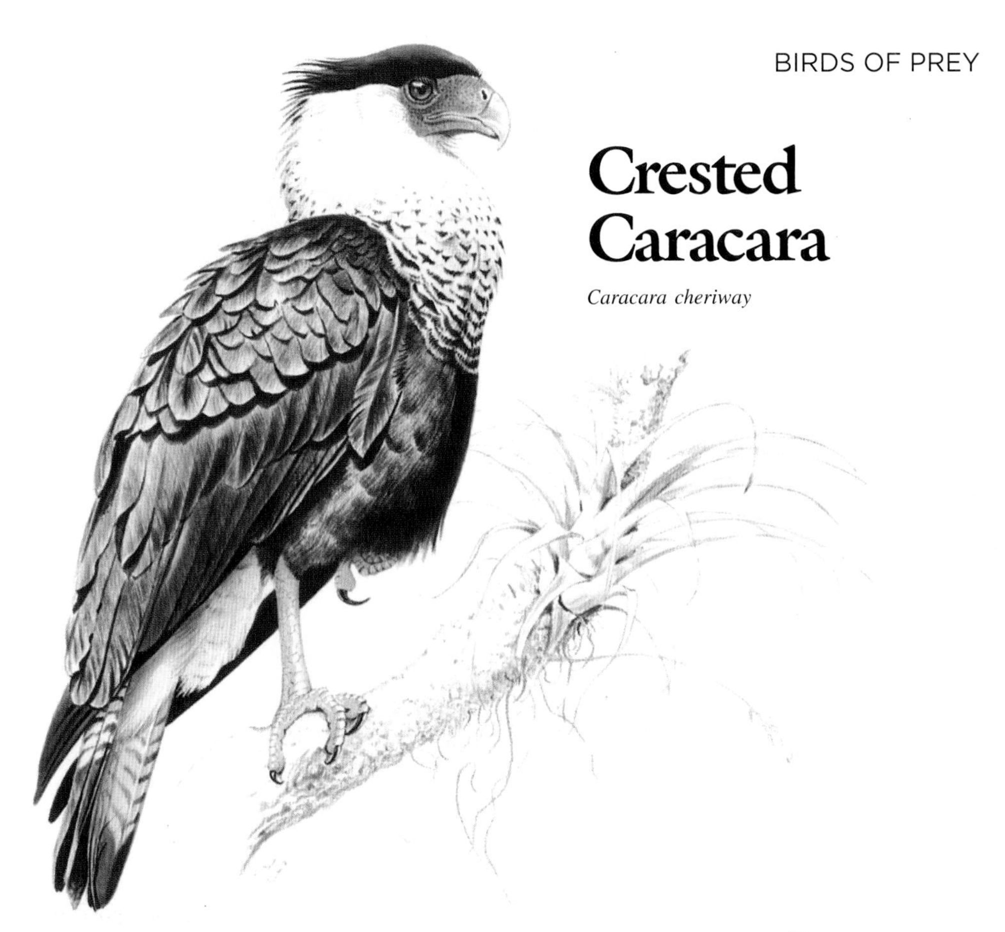

Crested Caracara

Caracara cheriway

From its palmetto perch, the caracara surveys its realm. The shaggy crest gives it an air of royalty. The bill, heavy and hooked, commands respect. The raiment, black and white and boldly patterned, makes a lordly impression. It can only disillusion, then, to recall that this bird, this "Mexican eagle," cavorts with the likes of vultures.

The crested caracara is indeed a triumph of contradictions. Behaviorally it is lumped with vultures; structurally it is closer to falcons. A capable flier, it nevertheless spends much of the time earthbound. Even its range is disjointed, with separate populations in the Florida prairies, the plains of south Texas, and the deserts of Arizona.

Like the vulture, the caracara is a carrion-eating bird, but with a grimly practical approach: it routinely patrols roads in search of animals killed by cars. Pilfering dinner is another old standby. A harrier with a mouse, a pelican with a fish, even a vulture with a full crop is fair game; the caracara simply harasses the burdened owner until it finally drops its catch. Still, if all else fails, the caracara shows itself to be a perfectly able hunter, and when opportunity beckons, snakes, birds, insects, and small mammals all fall prey to this odd, opportunistic raptor.

The caracara's long legs allow it to pursue snakes and other small prey on the ground.

Recognition. 21–25 in. long; wingspan 4 ft. Head and neck white with black crown and crest; upperparts blackish; underparts barred; wings long, with white patch near tip; tail white with black band near tip; legs long; face naked and reddish.
Habitat. Grasslands and brushy country.

Nesting. Nest is a bulky platform of stems and twigs 8–50 ft. above ground in palmetto, tree, or cactus. Eggs 2–4, white or pinkish, marked with reddish brown. Incubation 28 days. Nestling period unknown.
Food. Mainly insects and carrion; also small mammals, birds, and reptiles.

American Kestrel

Falco sparverius

Most falcons seize their prey in flight, but kestrels often hover and scan the ground for food.

Commuter traffic on Route 3 is at a standstill eastbound from New Jersey into the Lincoln Tunnel. Westbound traffic, the radio says, is backed up for two miles because of rubbernecking delays. But the tiny falcon, poised on a sapling overlooking the highway, is oblivious. Its hungry eyes see nothing but the fresh snow. The stalled vehicles might as well be rocks for all it cares.

If any of the frustrated commuters happen to notice, they will be startled by such a sight in this unlikely setting. Rust-colored above, wings a gunmetal blue, the robin-sized bird seems too beautiful and much too small to be a raptor. But the mice digging their tunnels beneath the snow, and the house sparrows huddled in a nearby hedge, are not deceived. The sharp, hooked bill, the large, talon-tipped feet, and the intensity with which the bird surveys the world mark it for what it is: a hunter.

Suddenly the kestrel's head stops turning. The black falcon eyes study the snow even more intently. Leaning forward, head bobbing with excitement, the bird launches itself with a series of rapid wingbeats. Ninety feet out, the bird pulls up, hovering like a tiny helicopter over its target. Then it dives, feet first, into the snow. The struggle is short and furious—and one-sided. A hungry kestrel is a consummate mouser. Prey held firmly in its talons, the bird returns to its perch and proceeds to feed. When traffic finally begins to move, the raptor doesn't even notice.

Recognition. 9–12 in. long. Wings pointed. Male rusty, with blue-gray wings, black spots on back and sides, 2 vertical black stripes on cheek. Female rusty with black bars on back and tail. Hovers; pumps tail when perched.
Habitat. Open country, farmland, deserts, towns, and cities.

Nesting. Eggs 3–7, white or pinkish, heavily blotched with brown, laid in woopecker hole or crevice in building. Incubation 29–30 days, by both sexes. Young downy; leave nest 30–31 days after hatching.
Food. Mainly insects in summer; mainly small birds and mammals in winter.

Merlin

Falco columbarius

Birds of prey are specialists. Some hunt from perches; others search for victims while soaring overhead. Falcons, in turn, have been specially crafted for expertise at intercepting prey in open flight. And, ounce for ounce, few of them can rival the spunky little merlin, so swift and agile that it can outmaneuver even such masters of flight as dragonflies and tree swallows.

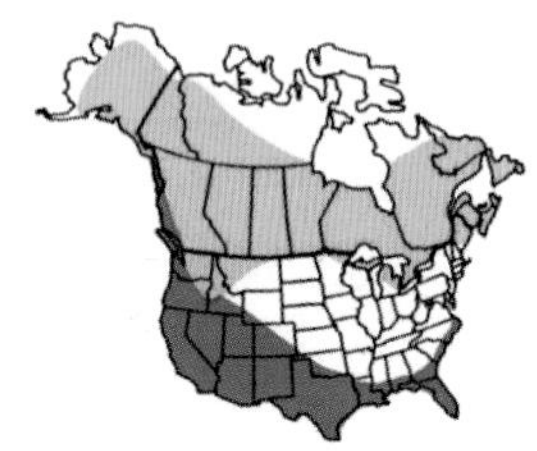

The merlin's wings—long and sharply tapered like those of jet fighters—beat with rapid downward-flicking power strokes. Its nostrils, like those of other falcons, are specially modified to allow it to breathe freely even when it flies at very high speeds. The merlin's principal weapons, however, are its feet. Though they appear large and ungainly, in fact each is a four-pronged arsenal of talons that forms a fearsome trap. From these sharp snares, skillfully deployed at high speed, few creatures marked as prey escape.

Over much of its northern breeding range, in clearings and muskegs amid vast coniferous forests, the merlin is difficult to find. But in Saskatoon, Saskatchewan, this daunting hunter has grown so accustomed to humans that it makes its home in abandoned magpie nests located in shade trees along placid residential streets.

Small birds are the mainstay of the merlin's diet.

Recognition. 10–14 in. long. Wings pointed. Male gray above, streaked and barred with brown below; tail dark, with light gray bands. Female and immature similar, but brownish above, usually with buff tail bands.
Habitat. Open woodlands and groves.
Nesting. Nest built of sticks 15–35 ft. above ground in tree or in natural cavity. Eggs 2–7, buff, speckled with brown and purple. Incubation 28–32 days, by both sexes. Young downy; leave nest 25–30 days after hatching.
Food. Mainly small birds caught in flight; also large insects and small mammals.

Peregrine Falcon

Falco peregrinus

Cliff-dwelling peregrines also nest high on city skyscrapers.

From ledges high on rocky cliffs, the peregrine falcon once fell like judgment upon its prey, primarily smaller birds caught in flight. With meteoric stoops, clocked in excess of 200 miles per hour, the falcon—one of our fastest birds—flew with a mastery that made the flight of lesser birds seem a feeble imitation.

Over much of its former nesting range, however, the bird is gone now, the victim of a chemical, DDT. Shortly after World War II the insecticide was widely applied to crops and mosquito-infested areas. Carried up the food chain to peregrines and other birds of prey, the poison caused eggshells to thin. Some eggs crushed under the weight of incubating birds; others failed to hatch. By the mid-1960's no peregrines were breeding east of the Mississippi River, and they were in severe decline throughout much of the West.

Now, thanks to captive-breeding programs, the peregrine population is slowly recovering. Reintroduced into the wild, captive-bred falcons show signs of successful reproduction. Though many of their ancestral ledge-top aeries are still empty, perhaps in time the cliffs will once again echo with the cries of the peregrine falcon.

Recognition. 15–20 in. long. Large, with pointed wings, broad black stripe below eye. Adult slate-gray above, whitish or buff barred with black below. Immature are brownish above, heavily streaked below.
Habitat. Cliffs in open country or near water; also on buildings in cities.

Nesting. Eggs 2–6, buff marked with reddish brown, laid on bare rocky ledge or outcrop, sometimes on ledge of building in city. Incubation 28 days, by both sexes. Young are downy, leave nest 5–6 weeks after hatching.
Food. Mainly birds caught in flight; occasionally mammals and large insects.

Prairie Falcon

Falco mexicanus

Looking like a pallid version of its near relative, the peregrine falcon, the prairie falcon is a bird of wide-open spaces—the dry country of the West, where rolling plains stretch toward distant, dim horizons. And its hunting strategies are finely attuned for success on such terrain. In contrast to the peregrine, which plummets in steep power dives to nab flying birds in midair, the prairie falcon takes most of its prey on or near the ground, relying on a breathtaking combination of maneuverability and speed.

If it can find a raised perch, the prairie falcon sits there motionless, surveying its surroundings. When it spies its quarry—an unwary ground squirrel, perhaps—the falcon launches into the air. Accelerating in seconds from a standstill to a blur of speed, it strikes with one taloned foot before the rodent has time to react.

The prairie falcon also hunts by flying low over the grass. When a small bird flushes ahead of it, the falcon shifts into high gear. Streaking forward like an arrow, then twisting and turning in zigzag pursuit, it easily outmaneuvers even the swiftest of smaller birds. In surroundings where the dive-bomber tactics of the peregrine might fail, the stunt-plane agility of the prairie falcon scores again and again.

In flight, the prairie falcon shows dark patches on its underwings.

Recognition. 17–20 in. long. Similar to peregrine falcon but slimmer and paler. Sandy brown and faintly barred above, with narrow dark stripe below eye; buff or cream and lightly streaked and spotted below. Dark wing linings are clearly visible in flight.
Habitat. Dry plains and desert cliffs.

Nesting. Eggs 3–6, white or pinkish, coarsely spotted with brown and purple; laid in depression in soil or gravel, on ledge, or in old nest of large bird. Incubation 30 days. Young are downy, leave nest in about 40 days.
Food. Mainly birds caught in flight; also small mammals; rarely insects and reptiles.

Barn Owl

Tyto alba

More tolerant of human presence than most owls, the barn owl earned its name from its willingness to nest in barns, belfries, and other buildings as well as in hollow trees and caves. And it mingles with the people of many nations: found in both North and South America, it also ranges across parts of Europe, Asia, Africa, and even far-off Australia.

Described by Geoffrey Chaucer in the 14th century as a "prophet of woe and mischance," the barn owl has probably given rise to tales of haunted houses on every continent it inhabits. For owls around the world—creatures of the night, swift, silent hunters given to hooting spookily—have been perceived in all cultures as birds of ill omen and harbingers of death.

Yet such is the ambivalence of the human mind that owls have also long been revered as symbols of sagacity. The ancient Greeks associated owls with Athena, goddess of wisdom, and embellished their coins with an image of that deity on one side and an owl on the reverse. Certainly the bird's solemn stare conveys at least a suggestion of superiority and hidden knowledge. And so it is hardly surprising that, even in the space age, we continue to cherish "the wise old owl."

Seen from below, the barn owl looks almost pure white.

Recognition. 13–19 in. long. Large, with long legs and round or heart-shaped facial disc. Golden brown above; white below and on wing linings.
Habitat. Grasslands, marshes, and deserts; also residential and urban areas.
Nesting. Eggs 3–11, white, laid on bare floor of cavity in tree, building, or cliff. Incubation 32–34 days, by female. Young are downy; leave nest 2 months after hatching; become independent at 3 months. Adults may raise a second brood.
Food. Mainly mammals up to size of jackrabbit; occasionally birds.

Flammulated Owl

Otus flammeolus

On moonlit nights in spring, the voice of the flammulated owl permeates the pine forests of the West. It is a pleasing sound, a soft, mellow, almost musical *boop* or *bu-boop.* But its source is devilishly difficult to pinpoint, for this charming gnome is a ventriloquist who cannot risk revealing its location to a predator. Its tiny feet and bill are designed for catching moths and similar small prey, not for self-defense.

Even in the beam of a flashlight, the owl is hard to spot. Not that it lacks pattern or color—it is intricately marked with black on gray, along with shades of brown and rusty accents that account for the name *flammulated,* or flame-shaped. The pattern, in fact, is a perfect match for the bark on pine trees—an ideal camouflage.

By day, the flammulated owl normally roosts in a cavity in a tree. But if it must roost in the open, it nestles up against a tree trunk. Huddling motionless, with its eyes closed, the owl looks exactly like a gnarled branch; and the illusion persists even if the bird opens its eyes. Although most owls' eyes are yellow, those of the flammulated owl are brown, lending it a gentle expression that matches its ways, while also contributing to its disguise.

The flammulated owl catches its prey in mid-flight.

Recognition. 6–7 in. long, slightly larger than a sparrow. Similar to western screech-owl, but with dark eyes and shorter ear tufts. Gray-brown, with irregular streaks and rusty mottling; white spots on shoulders. Facial discs have rusty margins. Females larger than males.
Habitat. Mountain pine forests.

Nesting. Eggs 3–4, white, laid in abandoned woodpecker hole in dead pine or aspen. Lengths of incubation and nestling periods unknown.
Food. Mainly large moths, beetles, and spiders; also grasshoppers, small birds, and rodents.

Eastern Screech-Owl

Eastern Screech-Owl *Otus asio*
Western Screech-Owl *Otus kennicottii*

Screech-owls are an example of birds that are not all "of a feather," for they occur in two distinct color phases—red and gray-brown. These differences in plumage have nothing to do with age, sex, or season: once a red-phase owl, always a red-phase owl. Most of the western species are gray-brown, although the red phase is occasionally encountered in the Pacific Northwest. But eastern screech-owls in both colors are common. Indeed, reds and grays can be found in the same brood.

As varied as their coloration are the birds' eating habits. In typical owl fashion, screech-owls prey heavily on mammals and birds, swooping in for the kill on silent wings. Despite their modest size, they have no trouble taking creatures as large as rats and ruffed grouse. But they also make do with such disparate fare as insects and earthworms, snakes and snails. Sometimes they catch insects in flight, flycatcher-style, snatching them with their beaks in midair. And not surprisingly for birds that seem to enjoy a nightly dip—including visits to backyard birdbaths—the screech-owls are talented anglers, plunging right into streams to catch fish and crayfish.

Recognition. 7–10 in. long. Gray or gray-brown with black streaks, yellow eyes, and ear tufts. Red-phase are rusty above, barred with rust color below; common in eastern species, rare in western. Two species best separated by range and voice: eastern gives long whinny, western a series of notes speeding up at end.

Habitat. Woodlands, orchards, and gardens; western also in deserts.
Nesting. Eggs 2–8, white, laid on bare floor of cavity in tree. Incubation 26 days, by female. Young downy; leave nest 4 weeks after hatching.
Food. Mainly mice, shrews, and large insects.

Great Horned Owl

Bubo virginianus

Once described as "the tiger of the woods," the great horned owl is indeed a superbly adapted nighttime hunter. Keen of eye, acute of ear, and silent on the wing, it is equipped with powerful talons for striking and gripping prey and a sharp hooked beak for tearing flesh from its kill. And thus arrayed, it seems to know no fear. Squirrels, rabbits, skunks, songbirds, geese, hawks, even porcupines—all are fair game for satisfying the horned owl's enormous appetite. On rare occasions, the bird has even been known to swoop in and attack people wearing fur hats, apparently mistaking the pelts for living prey.

The great horned owl normally carries its catch to a feeding roost—its "dining room"—where it tears the creature into bite-size pieces. The ground beneath the roost is characteristically littered with bones, feathers, and other leftovers from the feast, along with an array of telltale owl pellets. Like hawks, which also swallow prey whole or in large chunks, the owls regularly regurgitate felty wads of indigestible animal remains—fur, feathers, bones, and the like. Essential for the well-being of the birds, regurgitated pellets are a boon to biologists as well: analysis of their contents has yielded invaluable information on the feeding habits of all the birds of prey.

The great horned owl uses its beak to tear larger prey into bite-size pieces.

Recognition. 18–24 in. long. Very large, with widely spaced ear tufts, white throat, and finely barred underparts. Usually calls with four hoots.
Habitat. Deep forests, open country, deserts, and wilder city parks.
Nesting. Eggs 1–6, white, laid as early as January in old nest of crow, eagle, or hawk; on rocky ledge; or on ground in treeless areas. Incubation up to 35 days, by female only. Young downy; fed by both parents; leave nest 9–10 weeks after hatching.
Food. Mainly mammals up to the size of porcupines; also birds, reptiles, and frogs.

Snowy Owl

Nyctea scandiaca

Most owls are creatures of the night. But not all. Above the Arctic Circle, where summer sunlight shines around the clock, the snowy owl, like all living things of the Far North, adapts to days that are not bracketed by hours of darkness.

To judge by the range of prey taken by North America's heftiest owl, hunting without the concealing cover of darkness is no great handicap. The remains of animals as varied as ptarmigan and arctic hare have been found in and around the owl's tundra nests. The bird's principal food, however, is the lemming, a small rodent. During years when lemming numbers are at their peak, the snowy owl may hunt nothing else.

In winter, when darkness reigns for nearly 24 hours a day, the snowy owl moves southward from the upper limits of its breeding range. The prairie provinces of Canada are a traditional winter stronghold, and there the bird is a common fixture on fence posts and rooftops. During years when prey within its normal wintering range is in short supply, however, the great white owl of the Arctic occasionally wanders as far south as Alabama—a surprising sight to behold in such balmy climes.

The snowy owl commonly nests on low Arctic hillocks.

Recognition. 19–25 in. long. Large and round-headed, with no ear tufts and yellow eyes. Mainly white, with scattering of black bars and spots. Markings more numerous on younger birds. Active during day and night.
Habitat. Open areas, dunes, and marshes; nests on Arctic tundra.

Nesting. Eggs 5–8, up to 11 in good lemming years, are white, laid in shallow depression in ground on tundra. Incubation about 32 days, by female only. Young downy; at first fed only by female; leave nest 6–8 weeks after hatching.
Food. Various mammals and birds; also carrion.

Northern Pygmy-Owl

Glaucidium gnoma

The knowing stare—a compelling characteristic of the northern pygmy-owl, as it is of all the owls—has inspired much of the folklore surrounding these creatures of the night. But it also reminds us that keen vision, especially in dim light, is vital to the owls' way of life.

Several factors account for their exceptional eyesight. Like all birds, owls have unusually large eyes—on some of the bigger species they are nearly the size of human eyes. And they have correspondingly large pupils, which permit as much light as possible to enter the eyes. Their retinas, furthermore, are especially rich in rods, the light receptors most sensitive to low-intensity light.

Unlike those of most birds, owls' eyes face forward. The result: a narrower field of vision but also the incalculable advantage of binocular eyesight. Like humans, owls see a three-dimensional world—which greatly helps them to judge distance, especially when homing in on prey.

Even their narrow visual field poses no problem for owls. Because their eyes are fixed in their sockets, the birds cannot glance from side to side. Instead, they simply turn their entire heads. Endowed with incredibly mobile necks, owls can swivel their heads in an arc of a full 270 degrees.

The northern pygmy-owl has two black "eye" spots on its neck.

Recognition. 6–7 in. long. Small, with no ear tufts. Gray-brown spotted with white above; 2 black spots on nape; white with black streaks below; sides brownish, with fine white spotting. Tail narrow and banded.

Habitat. Mountain pine and pine-oak forests; also in woodlands along streams.

Nesting. Eggs 2–7, white, laid on bare floor of old woodpecker hole or other natural tree cavity. Incubation about 28 days, by female only. Downy young are fed by both parents; leave nest in 4 weeks.

Food. Chiefly mice and large insects; occasionally small birds.

Elf Owl

Micrathene whitneyi

Elf owls are tiny. A mere 5 to 6 inches in length and weighing only about 1½ ounces, these sparrow-size birds are North America's smallest owls and one of the smallest owls in the world. Delightful sprites, they are among the most abundant birds in our southwestern deserts.

Like most of their kin, elf owls are nocturnal hunters. But because of their diminutive stature, they are scarcely able to swoop down and fly off with the rodents, rabbits, and similar creatures that make up the menu of most of their kind. Instead, they specialize in insects, sometimes snatching them in midair with their feet, sometimes snagging them off the ground. Even such small creatures cannot be dealt with carelessly, however. Should they catch a desert scorpion, the elf owls are prudent enough to remove its poisonous stinger before swallowing their prey.

Elf owls nest in abandoned woodpecker holes in trees and saguaro cacti. There, too, they roost by day, snug and sheltered from the oppressive desert heat. To spot one of these feathered mites as it peers, wide-eyed, from the comfort of its nest hole high in a giant saguaro is to garner a memory that will last a lifetime.

The tiny elf owl is hardly larger than a saguaro cactus blossom.

Recognition. 5–6 in. long. Sparrow-size, with stubby tail, yellow eyes, blurry streaks on breast and belly. No ear tufts.
Habitat. Deserts, dry woodlands, and streamside thickets.
Nesting. Eggs 1–5, white, laid on bare floor of old woodpecker hole in tree or saguaro cactus, rarely in other natural cavity. Incubation 24 days, by female only. Male brings food to female until downy young are half grown, then feeds them directly. Young leave nest about 33 days after hatching.
Food. Moths, beetles, grasshoppers, and scorpions; also small reptiles.

Burrowing Owl

Athene cunicularia

Owls are shameless opportunists when it comes to nesting. Some, such as the great horned owl, regularly take over old hawks' nests. The elf owl of the Southwest happily adopts abandoned woodpecker nest holes as its own. And the snowy owl of the treeless Arctic tundra makes do with a simple moss- and grass-lined hollow scooped into bare ground.

The burrowing owl of the western plains and southern Florida has found a totally different solution to providing safety for its eggs and young. True to its name, it nests in burrows—usually those of prairie dogs, ground squirrels, and other rodents, but also in the renovated dens of armadillos and even gopher tortoises. Often returning to the same nest year after year, the birds remodel them annually by scraping and digging with their beaks, legs, and wings. Frequently the nesting chamber is littered with shredded, dried cattle droppings or horse manure.

Sociable creatures, burrowing owls sometimes nest in small colonies. Tales of owls, prairie dogs, and rattlesnakes all living peaceably together in the same burrows, however, are nothing more than fanciful folklore. The birds nest only in abandoned burrows.

Active by day, burrowing owls are frequently seen perched atop fence posts.

Recognition. 8–10½ in. long. A ground dweller with long legs, short tail, and no ear tufts. Sandy brown above, blackish bars below. Active both day and night; bobs up and down when alarmed or curious.
Habitat. Grasslands and deserts.
Nesting. Eggs 5–11, white, laid on bare soil in old burrow of mammal (or gopher tortoise in Florida); eggs quickly become stained by soil. Incubation about 28 days, probably by both sexes. Young downy; soon emerge from burrow, but nestling period unknown.
Food. Mostly large insects and small rodents; occasionally birds.

Spotted Owl

Strix occidentalis

The spotted owl, one of our rarest species, can be a noisy bird. Its rich baritone whoops and catlike shrieks can fill the forest with sound—fair warning to other spotted owls not to trespass on its territory. But it can also be a silent bird—decidedly a virtue in a hunter. Ranging through the forest on wings that span more than three feet, it can swoop in for a landing, ghostlike, without a sound.

Unlike other large birds, whose wingbeats are clearly audible at close range, most owls are specially designed for silence. Their wings, for one thing, are comparatively large, the better to carry them through the air without need for frantic flapping. Their body feathers are extremely soft and fluffy. And, most remarkable, the leading edges of the foremost flight feathers on their wings are finely serrated, which muffles the sound of the wings stroking the air.

Not all owls are silent in flight. The fishing owls of Africa and Asia, which swoop down to capture fish at the water's surface, lack the softened edges on the wing feathers; since the fish cannot hear them, they have no need for stealth. But the spotted owl preys on small creatures of the forest floor—shy nocturnal mammals with acute hearing. The only way the owl can take them by surprise is to swoop like a shadow in the moonlight, as silent as the night.

Spotted owls sometimes take over abandoned hawks' nests.

Recognition. 15–19 in. long. Large and round-headed, with dark eyes and no ear tufts. Dark brown above spotted with white or whitish; barred with dark brown below. Hard to locate, but very tame.

Habitat. Deep, moist coniferous forests and wooded canyons.

Nesting. Eggs 1–4, white, laid on bare floor of tree cavity or cave, or in old nest of large bird. Incubation 32 days, by female only. Young downy; fed by both parents; leave nest about 35 days after hatching.

Food. Mainly small rodents; also birds (including smaller owls) and insects.

Barred Owl

Strix varia

Who-cooks-for-you, who-cooks-for-you-all? is the usual transcription for the most commonly heard call of the barred owl. A far cry from the simple hoot we tend to think of as the typical owl's call, it is a reminder that owls as a group are capable of a varied range of vocalizations.

The barred owl is one of the virtuosos of the lot, with a raucous repertoire that includes all sorts of squeaks and squawks, screeches, barks, and yowls. Launching into a maniacal mixture of demonic cries, it can produce a chorus that is truly spine-tingling—especially when heard late at night in the deep, swampy forests that are among its preferred haunts. Small wonder that the bird is known locally in some areas as the crazy owl.

Owl watchers, of course, have long known how to use the hoots to their own advantage. By imitating some of the birds' simpler calls, birders can usually lure any nearby owls out of hiding, curious to learn the identity of the interloper who has invaded their territory. If the bird thus summoned turns out to be a barred owl, the human hooter will find himself confronted by a genuine hooter with a memorable dark-eyed stare.

Roosting by day, barred owls may be mobbed by smaller birds.

Recognition. 16–23 in. long. Large and round-headed, with dark eyes and no ear tufts. Dark gray-brown above barred with whitish; underparts with barring across neck and upper breast, vertical streaks on belly.
Habitat. Wooded swamps and lowland forests; also drier upland forests.

Nesting. Eggs 2–4, white, laid on bare floor of natural tree cavity or in old nest of crow, hawk, or squirrel. Incubation 28 days, mostly by female. Young downy; leave nest 6 weeks after hatching.
Food. Mostly mice; also other mammals, birds, frogs, snakes, insects, and crayfish.

Great Gray Owl

Strix nebulosa

With an overall length of up to 32 inches and a wingspan approaching five feet, the great gray owl is considered to be North America's largest owl. But overall dimensions are misleading, for the bird in fact is a fluffy fraud. Cloaked in thick down and insulating feathers that protect it from the numbing cold of its northern haunts, an adult weighs only two to three pounds—barely half the weight of its more formidable cousin the snowy owl.

Preying mainly on mice, squirrels, and other small mammals, the great gray owl is a year-round resident of remote wilderness areas. Even when winter locks northern forests in deep snow, it seldom ventures farther south or east. On the rare occasions when it does, it is driven not by severe weather but by scarcity of food. Only twice in the 19th century did the bird reach the northeastern United States in large numbers. A third migration in 1978–79 brought hundreds of great gray owls across New England and even south to Pennsylvania.

In general, however, this fluffy, well-insulated impostor remains secluded in its isolated wilderness, where it has little contact with humans. As a result, the bird can be so closely approached that experienced birders have even captured the great gray owl by hand.

Gliding on wings that span nearly five feet, the great gray owl is a graceful flier.

Recognition. 24–32 in. long. Very large and round-headed, with huge facial discs, yellow eyes, and no ear tufts. Mottled and barred gray-brown, with 2 bold white crescents on throat.
Habitat. Coniferous forests in Far North; mountains in West.
Nesting. Eggs 2–5, white, laid in abandoned nest of hawk or eagle or on top of stump. Incubation 30 days, by female only. Young downy; leave nest 5 weeks after hatching but depend on parents for several weeks longer.
Food. Primarily small mammals; also crows and smaller birds.

Long-eared Owl

Asio otus

True to its name, the long-eared owl is crowned with prominent ear tufts. But, like similar adornments on many other owls, the tufts are simply feathers. They have nothing to do with hearing.

As with most birds, the owls' ear openings are hidden beneath feathers on the sides of the head. Proportionally larger than on other birds, in some species they also are asymmetrical—in shape, in size, and in position. The resultant differences in sound received by the two ears, though minute, seem to be sufficient to enable an owl to pinpoint the source.

The arrangement of feathers on owls' faces, moreover, serves to funnel sound into the ears. As a further refinement, many species have feathered flaps in front of the ear openings. Raised or lowered at will, they function much like a hand cupped behind the human ear. But in the case of owls, they help detect sounds originating *behind* the bird. And owls tend to be exceptionally sensitive to high-frequency sounds—such as the squeaking of a mouse.

Thus equipped, owls enjoy uncommonly acute hearing. Their eyesight is excellent too, as befits nocturnal hunters. But experiments have proven that owls can catch prey in absolute darkness, homing in with uncanny accuracy on the basis of sound alone.

With feathers compressed, a roosting long-eared owl looks like a broken branch.

Recognition. 12–15 in. long. Similar to great horned owl but smaller and more slender, with longer, more closely spaced ear tufts. Gray-brown above; whitish with dark streaks and bars below; buff or rusty facial discs.

Habitat. Dense forests; rarely in orchards and streamside woodlands.

Nesting. Eggs 3–10, white, laid in old nest of hawk, crow, magpie, or squirrel, less often in natural tree cavity. Incubation 26–28 days, usually by female only. Young downy; leave nest about 5 weeks after hatching.

Food. Mainly small rodents, shrews, and rabbits caught at night; occasionally birds.

Short-eared Owl

Asio flammeus

On mating flights, male short-eared owls make a clapping sound with their wingtips.

Unlike the inconspicuous ear tufts responsible for its name, the short-eared owl itself is one of the most readily observed of all the owls. It is, for one thing, often active during the daylight hours, beginning its nightly hunt for rodents late in the afternoon. Unlike most of its kin, moreover, the short-eared owl is a bird of open country, frequenting fields and marshes instead of forests. And it is one of the more cosmopolitan owls; ranging across virtually all of North America, it also flourishes in Europe, Asia, and South America.

The short-eared owl is most often seen on the hunt, flying low with fluttering, mothlike wingbeats and pausing frequently to hover in midair. (When mice are especially abundant, several owls can sometimes be seen at once, patrolling the same area.) But the birds can also surprise observers with more unusual behavior. The males, for instance, are noted for their courtship flights; diving earthward, they flap their wingtips together as if to applaud their own performances. And if disturbed at the nest—a mere depression in the ground—the adults feign injury, killdeer-style, squealing pitifully and flapping their wings in an effort to lure intruders away from their downy young.

Recognition. 12–16 in. long. Round-headed, with ear tufts very short and usually not visible. Sandy brown or tawny, streaked above and below. In flight, shows dark marks at "wrist" and wingtip, and rich buff patch near end of wing. Active both day and night.

Habitat. Fields, plains, marshes, and dunes.

Nesting. Eggs 4–9, white, laid in shallow, grass- and feather-lined depression in ground. Incubation about 3 weeks, by female. Young downy; leave nest in about 6 weeks. Several pairs may nest close together.

Food. Mainly voles; also other small rodents, insects, and small birds.

Northern Saw-whet Owl

Aegolius acadicus

Named for one of its calls—a raspy note that resembles the sound of a saw blade being sharpened with a file—the northern saw-whet owl is more often heard than seen. Yet even its voice hardly ever reveals its existence, for the little bird is mostly silent except during its late-winter, early-spring mating season.

Shy, retiring, and strictly nocturnal, the smallest owl in the eastern states thus goes mostly undetected in its preferred haunts—dense, usually evergreen forests and swamps. Far from rare, it simply is rarely seen.

Once spotted, however, the saw-whet owl is unlikely to be forgotten. And there is more than its elfin size and appealing appearance to etch this tiny hunter in the memory of anyone lucky enough to see it: one of the saw-whet's most striking characteristics is what Roger Tory Peterson described as its "absurdly tame" behavior.

Perhaps overly confident in its typically owlish concealing coloration, the bird not only allows a close approach, it can even be caught by hand and held without putting up a struggle. When flushed from its nest—most commonly an abandoned woodpecker's hole—it simply perches, unruffled, on a nearby branch and has even been known to settle on the head of a human onlooker.

Young northern saw-whet owls show a distinctive white between their eyes.

Recognition. 7–9 in. long. Small and round-headed, with large facial discs and no ear tufts. Adult dark brown above; whitish below, with heavy brown streaks. Immature are dark brown above, with buff belly and bold white eyebrows.

Habitat. Dense forests in summer; often winters in thickets.

Nesting. Eggs 4–7, white, laid on bare floor of old woodpecker hole, natural tree cavity, or birdhouse. Incubation about 4 weeks, mainly by female. Young downy; leave nest 4–5 weeks after hatching.

Food. Mostly insects; also rodents and birds.

Large Land Birds

Varied in their habits and habitats, the large land birds range from well-camouflaged grouse to speedy roadrunners, brazen crows, and dignified pheasants. Each is a master of its own way of life—the sage grouse male, enacting his ages-old courtship ritual; the bold raven, shattering the silence with its strange, haunting calls; the turkey hen, patiently incubating the eggs in her well-concealed nest. All are integral elements of the American landscape, irreplaceable gems in our wildlife heritage.

Black-billed Magpie

Ruffed Grouse

Bonasa umbellus

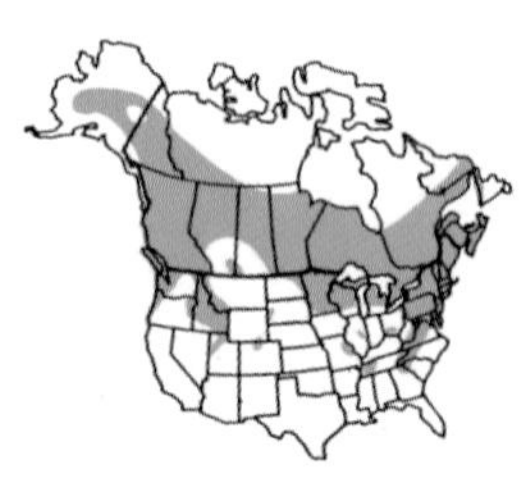

The ruffed grouse takes its name from the ruff of dark feathers on its neck.

A distant thumping, like a muffled drum, sets the spring air to throbbing in a secluded woodland. Beginning with a steady, pounding rhythm, it speeds faster-faster-faster until, suddenly, it stops. And then begins again. The drumming continues for hours upon hours throughout the day and, during the spring, is likely to go on for half the night as well.

The drummer—patient and repetitive—is a male ruffed grouse making his statement to the world. Standing on a fallen log, with breast feathers ruffled, head up, and tail braced against the log's rough surface, he cups his wings and brings them forcefully forward and up. With each wingbeat, the resultant compression of the air produces a low thump that can be heard half a mile away.

The bird's message, in the spring, is threefold. He is staking claim to his territory; he is warning off all other male grouse in the vicinity; and he is inviting all females in the area to come visit him at his log. His drumming, in fact, serves exactly the same purposes as does singing among songbirds. And, like them, he does most of his communicating in the spring. However, the ruffed grouse may burst into drumming at any time of year. Especially noticeable—and unexplained—is a bout of drumming that goes on for a few weeks in the middle of the fall.

Recognition. 16–19 in. long. Chickenlike. Brown, reddish brown, or gray-brown with barred sides; tail fan-shaped, with black band near tip. Black ruffs on sides of neck more visible on male. When surprised, flushes suddenly from ground and flies away.
Habitat. Deciduous forests with undergrowth.

Nesting. Eggs 9–12, buff, often spotted with brown; laid in depression lined with leaves and placed near log, rock, stump, or under dense bush. Incubation 24 days, by female. Chicks leave nest at once; fly in 1 week; stay with female about 12 weeks.
Food. Insects, seeds, berries, and buds.

Blue Grouse

Dendragapus obscurus

In contrast to most mountain wildlife, which summers in the highlands and moves downslope as temperatures begin to fall, blue grouse move *up*slope for the bitter months of winter. Their yearly cycle begins in the warming months of spring, when blue grouse gather on the lower mountain slopes and in the foothills for their annual mating rituals. The males perform elaborate courtship maneuvers, and the hens choose mates from among them.

As soon as breeding is over and the females have settled down to incubate their eggs, the males move back up the mountainside to spend the remainder of the year in coniferous forests at higher elevations. In late summer the hens and half-grown young follow behind; by the end of September, all the birds are back on the peaks, experiencing the first fine snowfalls of approaching winter.

Forsaking their summer diet of berries, insects, leaves, and flowers, in winter the blue grouse feed mainly on the needles of conifers. When winter storms are raging, they bury themselves in deep snowdrifts for warmth and for shelter from buffeting winds. And then, with the return of spring, they resume their upside-down migration, retreating once again to lowland breeding grounds.

Blue grouse often spend winter nights buried in snowdrifts.

Recognition. 15–21 in. long. Chickenlike. Male blue-gray, mottled above, gray below, with red or orange combs over eyes. Tail black, usually with gray tip. Female brown, barred with black. Displaying male exposes bare yellow or red air sacs on sides of neck.

Habitat. Open forests and shrubby woodlands.

Nesting. Eggs 7–16, buff speckled with brown, laid in depression near log or base of tree. Incubation 26 days, by female. Chicks stay with female 12 weeks.

Food. Berries, insects, leaves, and flowers in summer, needles and buds in winter.

White-tailed Ptarmigan

Lagopus leucurus

When molting, the white-tailed ptarmigan is a mixture of browns and white.

As still as the ancient stone that surrounds it, a white-tailed ptarmigan warms itself in the summer sun on a western mountaintop. Except for the glint of sunlight in its eye, this master of disguise is all but invisible, so perfectly does its mottled plumage blend with the jumble of barren rocks on which it crouches. The bird itself seems well aware of the efficacy of its camouflage, for it is reluctant to flush even when closely approached. In fact, people have nearly stepped on ptarmigans without seeing them.

Remarkably, this smallest of North American grouse remains in its high-mountain habitat throughout the year, changing its disguise to suit the seasons. Molting begins in mid-autumn, when the browns of the ptarmigan's summer garb are replaced by new winter feathers of pure, snowy white. Feathers also grow on the bird's legs and feet, scales fringe its toes, and its claws become much longer—resulting in a pair of natural snowshoes that have four times the walking surface of the ptarmigan's summertime feet. Burying itself in snowdrifts for protection from raging storms or foraging for food in windswept clearings, the white-tailed ptarmigan is a steadfast and adaptable mountaineer, who endures nature's adversity as indomitably as the aged summits themselves.

Recognition. 12–13 in. long. Chickenlike. Barred and mottled with brown, with white wings, tail, and belly during most of year. (Barred feathers often hide white tail.) Entirely white in winter.

Habitat. Rocky tundra in high mountains; willow and alder thickets in winter.

Nesting. Eggs 4–16, buff marked with brown, are laid in depression among rocks above timberline. Incubation, by female, about 23 days. Chicks stay with female for nearly 1 year.

Food. Leaves, flowers, buds, insects, and seeds; catkins, needles, and buds in winter.

Sage Grouse

also known as

Greater Sage Grouse

Centrocercus urophasianus

The show begins at dawn on mornings in spring: on sagebrush plains across the western states, male sage grouse gather in flocks of 20 to several hundred birds to perform their communal courtship dance. With tail feathers fanned and air sacs on their breasts inflated, the dancers transform acres of land into exuberant parade grounds as each cock, fiercely protective of his own small plot, struts and bobs, all the while producing loud popping sounds by releasing air from the inflated sacs.

Such communal courtship is typical of several species of grouse, as well as certain other birds, including some of the sandpipers and birds of paradise. In each case the basic pattern is the same. The display ground, called an arena or lek (from the Swedish word for "play"), is a traditional meeting place where the birds assemble year after year. Females enter the lek to select a male for mating, but no permanent pairing takes place; the mated hens depart to nest and rear their young on their own.

The system serves the species well, for it results in selection of the fittest males to father the next generation. Indeed, among the sage grouse it is not uncommon for a single dominant cock to mate with some 75 percent of the hens that enter the lek.

Camouflaged by mottled plumage, sage grouse nest safely on western plains.

Recognition. 22–30 in. long. Large and chickenlike. Male gray-brown above with white breast, black bib, black on belly, and long, pointed tail. Female much smaller, barred with brown; belly black. Displaying males inflate air sacs on breast and fan tail feathers.
Habitat. Sagebrush plains.

Nesting. Eggs 7–15, pale green marked with brown, are laid in depression under sagebrush. Incubation 25–27 days, by female. Length of time chicks stay with female not known.
Food. Leaves and shoots of sagebrush and some other plants.

Greater Prairie-Chicken

Tympanuchus cupido

The sea of grass that once spanned mid-America was home to vast numbers of greater prairie-chickens. With the arrival of the plow, however, and the replacement of the grass with grain, those numbers dropped drastically. The birds now survive only in scattered pockets that have escaped the plow, places where the native grasses and wildflowers still flourish.

Like their ancestors, the prairie-chickens that remain today depend on the richness and variety of the original prairie vegetation. There they find their food—mostly insects in summer and seeds and fruits in winter. They hide their nests in places where the weeds and grass grow tall and dense. When the eggs hatch, the parents lead their tiny chicks to areas of shorter, sparser grass.

They also assemble for their courtship rituals in settings where the grasses are short enough for the hens to watch the springtime spectacle. Competing with each other in the misty light of sunrise, the males drum their feet in the stylized rhythms of their dance. All the while their air sacs resonate with a booming that can be heard a mile or more away, an echoing reminder of a time when the sound was a prairie-wide harbinger of spring.

Courtship competition sometimes leads to open fighting between male prairie-chickens.

Recognition. 17–18 in. long. Chickenlike. Boldly barred brown and buff, with short, dark brown, fan-shaped tail. Displaying male raises narrow black plumes on neck to expose inflated orange air sacs.
Habitat. Moist grasslands and prairies.
Nesting. Eggs 7–17, whitish, buff, or olive, spotted with dark brown; laid in shallow depression hidden in vegetation. Incubation 23–26 days, by female. Chicks stay with female 6–8 weeks.
Food. Insects, especially grasshoppers, in summer; leaves, seeds, and berries in winter.

Sharp-tailed Grouse

Tympanuchus phasianellus

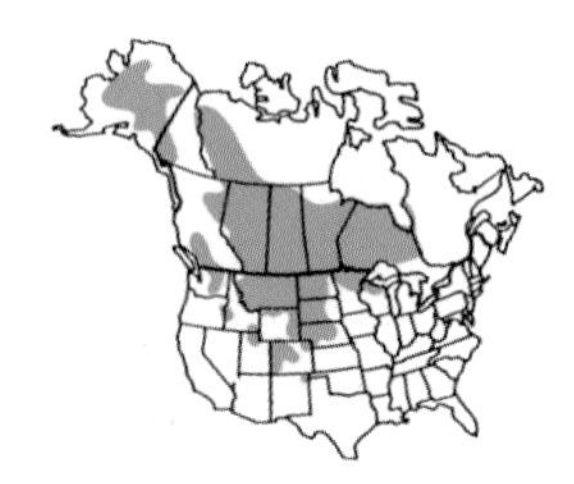

Adaptable denizens of the plains and brushlands, sharp-tailed grouse vary their lifestyle with the changing seasons. Although they are among the strongest fliers of all the grouse—swift enough to escape from hawks and owls—sharp-tails prefer to keep their feet planted firmly on the ground. Ambling through grasslands and brushy areas, heads bobbing lightly, they nibble on dandelions, rose hips, berries, buds, and insects. Except for the long, pointed tail feathers from which they take their name, they resemble the prairie-chicken in appearance and habits.

But as winter's chill sweeps the plains, stripping the leaves from poplar, willow, and cottonwood trees and dusting the amber grasses with a cover of snow, the versatile sharp-tails take to the trees. There they feed on buds by day and seem as much at home in this arboreal world as they were on foot amid the summer grasses. Only at night do they descend from the branches to sleep buried in snowdrifts, insulated by the fluffy cover against the bitter prairie cold.

With the coming of spring, the sharp-tails change their habits once again, returning to the ground and their more earthbound warm-weather way of life.

A grassland forager in summer, the sharp-tailed grouse takes to the trees in winter.

Recognition. 14–20 in. long. Chickenlike. Gray-brown, with whitish bars and spots above; tail pointed, with white edges. Male has purple air sacs on neck, inflated during displays.
Habitat. Prairies, cleared areas, brushlands, thickets, and bogs.
Nesting. Eggs 5–17, olive, buff, or light brown, lightly speckled with dark brown; laid in shallow depression lined with grasses and hidden under vegetation. Incubation about 24 days, by female. Chicks stay with female 6–8 weeks.
Food. Berries, buds, flowers, and insects.

Northern Bobwhite

Colinus virginianus

When European settlers first arrived in eastern North America, they encountered a chunky little game bird reminiscent of the quail and partridge of their former homelands. In spring, from brushy forest edges and open woodlands, it jauntily proclaimed its name with clear whistled tones: *Bob-white . . . Bob-white . . . Bob-white.* The bird, still common over much of the East, has since been introduced in several western states and in Europe as well.

With the onset of the nesting season, aggressive males face off against each other to secure a mate, but after breeding, both sexes gather into coveys of 25 to 30 birds for the fall and winter. Relying on their concealing coloration, bobwhites characteristically freeze at any sign of danger. If approached too closely, however, the entire covey explodes into the air with a loud whirring sound produced by their rapidly beating wings.

In the evening the birds gather to roost under the shelter of a shrub or brush pile. There they form a feathered circle with all heads pointing outward. Not only does this help conserve warmth through a cold winter night, it also assures each bird a clear flight path to escape should danger threaten.

Bobwhites roost in a ring, with all birds facing outward.

Recognition. 8–10 in. long. Small and chicken-like. Reddish brown, with white streaks on sides. Male has black and white stripes on face; female has black and buff stripes. Easily identified by male's whistled *Bob-white* call.

Habitat. Open woodlands, brushy fields, meadows, and pastures.

Nesting. Eggs 7–18 (more when two females share nest), white or cream-colored, are laid in depression hidden among plants. Incubation about 24 days, by both sexes. Chicks stay with adults until following spring.

Food. Seeds, berries, leaves, and insects.

Scaled Quail

Callipepla squamata

Common in the arid Southwest, scaled quail frequent a landscape of mesquite and creosote bushes, scattered cacti, and cottonwood bottomlands. And though they are able to thrive far from moisture in this dry region, they can often be seen near the dwellings of humans. There the "cotton-tops," so called for their white-tipped topknots, seek grain left over from the feed of ranch animals and take advantage of daily opportunities to drink at man-made water holes.

Scaled quail are social birds, sometimes gathering in coveys of 100 or more. Their daily routine, like that of most other desert dwellers, is busiest in the cooler hours of the early morning and late afternoon, while midday is spent loafing in the shadow of a cactus.

Scaled quail time their nesting to coincide with the rainy months of June and July. Insects, new green leaves, and tender shoots are in good supply then, and the chicks can drink from temporary puddles close at hand. Should the rainy season be too short or perhaps not come at all, however, scaled quail sometimes forgo nesting altogether and wait until the next year for the arrival of the desert's most precious resource.

Scaled quail make themselves at home on farms and ranches.

Recognition. 10–12 in. long. Small and chickenlike. Mainly brown, with dark-edged, gray scalelike feathers on neck, upper back, and breast. Short, white-tipped tuft on top of head. Usually seen running in small groups.
Habitat. Scrubby deserts and dry grasslands.
Nesting. Eggs 9–16, white, sometimes speckled with brown, laid in grass-lined depression under bush or tuft of grass. Incubation about 22 days, mainly by female. Young remain with parents until the following spring. Adults sometimes rear a second brood.
Food. Insects and seeds of weeds and shrubs.

Gambel's Quail

Callipepla gambelii

Residents of the parched and wild deserts of the Southwest, Gambel's quail easily rise to the challenge of survival in that harsh terrain. For a nest, they simply scratch a hollow in the soil beneath a creosote bush or cactus and line it with a few sticks, a feather, a leaf or two—whatever materials they find near at hand. When their dozen or so chicks have hatched, the parents take them off at once to explore the food sources of their home. The father leads the way, while the mother acts as rear guard, tagging along behind the little band.

The chicks eat more insects than their parents do. With their boundless curiosity—and need for protein—they eagerly snatch up lively grasshoppers and ants. Soon they also learn to recognize the dry seeds of weeds and grasses and, more importantly, of mesquite and other legumes, along with a variety of fruits and berries.

During daylight hours, in family groups and in larger coveys, the quail forage in the desert, but by nightfall they must have water. Gathering in great numbers at the edges of reliable water holes, they drink their fill before roosting for the night in nearby trees and shrubs. Then, early the next morning, they drop back to the ground and drink once more before setting off to cope for another day in the arid, unforgiving desert.

Mesquite pods are a favorite food of the Gambel's quail.

Recognition. 10–12 in. long. Small and chickenlike. Male gray above; sides chestnut with white stripes; face and chin black; crown chestnut, with curved black plume; belly whitish with black patch. Female duller; plume shorter.

Habitat. Desert thickets, cactus scrublands, and streamside woodlands.

Nesting. Eggs 9–14, buff marked with brown, laid in shallow scrape under bush or in tall grass. Incubation 21–23 days, by female. Chicks stay with parents for 3 months.

Food. Seeds, fruits, leaves, and insects.

California Quail

Callipepla californica

Elegant, graceful, and among the liveliest of all North American game birds, the California quail seems never to be still for one moment of the daylight hours. It is constantly on the move and, when it does pause to rest on a stump or fence post, its curly-plumed head moves continually from side to side, its bright eyes never missing the slightest flicker of a shadow.

This sprightliness is soon apparent, for the instant the downy, daintily striped chick is free of its shell, it is on its feet and ready to go. In fact, chicks have been seen to flee from danger with pieces of eggshell still sticking to them.

Like the young of domestic chickens, baby quail are termed precocial: they are covered with down at hatching, their eyes are open, and their strong little legs are able to run almost as soon as their down is dry. Their opposites, known as altricial—baby robins, for instance—are born naked, blind, and utterly helpless.

How much care precocial chicks require after hatching varies with the species. In the case of California quail, although the little ones are accomplished fliers at just 14 days, they continue to be brooded for warmth for a full four weeks after leaving the nest, at which time they begin to roost in trees with the adults.

California quail chicks can fly just two weeks after hatching.

Recognition. 9–11 in. long. Small and chicken-like. Male gray above; sides chestnut striped with white; face and chin black; crown brown, with long, curved, black plume; belly whitish, scaled with black. Female similar but duller, with shorter plume.

Habitat. Woodlands, gardens, and chaparral.

Nesting. Eggs 12–16, buff or white, marked with brown, laid in grass-lined scrape under bush or brush pile, or in crevice in rocks. Incubation 21–23 days, by female. Chicks may form flocks when only a few weeks old.

Food. Seeds, berries, leaves, and insects.

Mountain Quail

Oreortyx pictus

Very much creatures of the western mountains, these largest of our North American quail are noted for their seasonal movements up and down the slopes. Their journeys are not long—just 20 to 40 miles or so. But, surprisingly, the birds travel every inch of that distance on foot.

Far from being a forced march, the trek more closely resembles a leisurely jaunt. In spring and summer, mountain quail nest and rear their young in undergrowth along the foaming streams of high mountain glens at elevations ranging from 1,500 to 10,000 feet. Then, in late summer and early fall, the birds move gradually down the slopes to lower valleys and canyons, where they escape the extremes of high-altitude winters. When spring returns, they go up the mountains again, striding along on sturdy legs, feeding and conversing, resting and roosting along the way.

Mountain quail can fly, of course. Like other quail, they have short, rounded wings that are arched and strong, designed for quick takeoffs—from the ground into full flight in just an instant. But for the most part, whether seeking safety beneath the almost impenetrable brushy cover in which they live or traveling long distances, they seem to prefer to walk or run.

The mountain quail relies on dense underbrush for cover.

Recognition. 11–12 in. long. Small and chickenlike. Brown above, with gray head, neck, and breast; throat chestnut; sides chestnut, with white bars; black plume on crown long and straight. Usually timid and hard to see.

Habitat. Dense brush in mountain forests; also mountain meadows and cleared areas.

Nesting. Eggs 5–15, buff or pinkish, laid in depression under bush, tuft of grass, or shrub, always near water. Incubation about 25 days, mainly by female. Chicks can fly in 2 weeks.

Food. Leaves, buds, and insects in summer; seeds, berries, and nuts in winter.

Wild Turkey

Meleagris gallopavo

Wild turkeys may gobble at any time of year, but it is in spring that secluded woodlands ring most often with the mellow vibrato gobblings of dominant males. Each tom reigns over his own small clearing. And there, with breast outthrust, head drawn back, wingtips dragging, and tail feathers spread in a magnificent fan, he struts from left to right and back again, in a timeless ritual designed to lure female turkeys into his private harem.

And they come. Out of the half-light under the leafing trees, the willing hens walk in—sometimes 10 or even more to a single tom. When egg-laying time approaches, the wary females depart to scrape depressions in the ground and build their hidden nests. Laying an egg a day for 8 to 20 days, the hens return to the mating ground each morning for the courtship ritual and then slip away to guard their nests.

The chicks—dainty bundles of fluff—are able to run soon after hatching and can make short flights when just two weeks old. Cared for entirely by the hen, the young birds remain with her through the following winter, then set out to live adult lives on their own.

Although they feed on the ground, wild turkeys roost in trees at night.

Recognition. 36–48 in. long. Very large and chickenlike. Metallic bronze with black barring; head and neck naked; tail broad and fan-shaped, opened when male displays. Female smaller, more slender, and paler than male.
Habitat. Open woodlands, brushy forest edges, and wooded swamps.

Nesting. Eggs 8–20, white or buff marked with brown or red, laid in leaf-lined depression on ground in woods, often near edge of clearing. Incubation about 28 days, by female. Chicks stay with female until next spring. Male takes no part in nesting.
Food. Seeds, nuts, berries, and insects.

Ring-necked Pheasant

Phasianus colchicus

Resplendent in exotic plumage, the male ring-necked pheasant struts through northern fields and meadows with all the pomp and dignity of an oriental potentate. Such ostentation may seem misplaced in a leading American game bird, but in fact the pheasant was a Chinese import who arrived in this country in the late 1800's.

Benjamin Franklin's son-in-law, Richard Bache, was the first person to try to establish pheasants in America, but the 1760 experiment at his New Jersey plantation failed. In 1881 Judge O. N. Denny, the U.S. consul general in Shanghai, China, sent 30 birds back home to Oregon's Willamette Valley, where they flourished with such ease that he soon supplemented them with a second flock. By the early 1890's pheasants had also become established on the Atlantic Coast, first at Rutherford Stuyvesant's estate in New Jersey and later in Massachusetts.

Through organized stocking programs and the pheasant's own adaptiveness in the wild, this colorful immigrant (now South Dakota's state bird) has spread across most of the northern states. Unmistakable when flushed from its weedy haunts, it can vanish back into its surroundings in an instant, completely at one with its adopted homeland.

Recognition. 22–35 in. long. Chickenlike. Male has glossy green head, red face wattles, long, pointed tail. Many males have white neck ring. Female smaller, sandy brown, mottled with darker brown; tail shorter than male's.
Habitat. Fields, pastures, croplands, marshes, and brushy areas.

Nesting. Eggs 7–15, buff or olive-brown, laid in grass-lined scrape in ground. Often several nests close together. Incubation 23–25 days, by female. Chicks fly in 2 weeks. Male usually plays little part in nesting.
Food. Seeds, nuts, berries, and insects.

Greater Roadrunner

Geococcyx californianus

According to legends of the old Southwest, the roadrunner had a foolproof way of outfoxing its enemy, the deadly rattlesnake. Whenever it found a rattler sleeping, the roadrunner would fence it in with pieces of cactus, then awaken the snake with a quick jab. Alarmed and thrashing, the rattler would be stabbed to death by cactus spines.

More recently, cartoons have portrayed the roadrunner as a spunky character that calls *beep-beep* as it races to escape its sworn foe, the coyote. But it never is in real danger: besides being quick-witted, the roadrunner is also portrayed as the quickest-footed critter in the West.

The legends and the cartoon traits are greatly exaggerated, though rooted in fact. The bird may not build corrals around dozing reptiles, but it does eat snakes—even fair-sized rattlers. And while it is no speed demon, it is agile enough to dodge coyotes at close range and take flight when necessary.

The stories, no doubt, were inspired by the roadrunner's sprightly ways. Approaching us fearlessly, regarding the world with a curious eye, and raising its crest to a jaunty angle, the bird attracts attention wherever it goes. Small wonder that Mexicans call it *paisano,* or "countryman." Like many another wild creature, the roadrunner deserves to be considered a *paisano* on both sides of the border.

Sleek and quick, the roadrunner darts about at speeds of up to 15 miles per hour.

Recognition. 20–24 in. long. Pheasantlike but slender, with bushy crest, long, mobile tail, sturdy legs, and streaked plumage. Runs rapidly, and seldom flies. Raises and lowers crest and tail when curious or alarmed.

Habitat. Grasslands, brushy areas, deserts, and open woodlands.

Nesting. Nest is a cup of sticks lined with leaves and grasses, in cactus, thicket, or small tree. Eggs 2–6, white. Incubation 18–20 days, probably by female alone. Young fed by both parents; leave nest after 16–19 days.

Food. Snakes, lizards, insects, and small birds.

Clark's Nutcracker

Nucifraga columbiana

An unwitting conservationist, the Clark's nutcracker plays a significant role in maintaining the pine forests on mountains in the West. Every year, this industrious creature spends late summer and early fall pulling pine seeds from their cones, then carrying them away to bury them in the ground. This is no small task: a single nutcracker, in a single season, may transport more than 30,000 seeds.

The bird begins its work by prying the cones open with its bill and extracting the seeds. It eats some on the spot after crushing them in its beak, but most are slipped into an expandable pouch of skin under the bird's tongue. Once the pouch is full (and it can hold as many as 70 seeds), the nutcracker flaps heavily away to an area of open ground. There it digs a series of holes with its bill and deposits a few seeds in each one.

Of course, the bird is not making a conscious effort to plant a forest; it is storing the seeds for future use. Showing an uncanny knack for finding the stashes many months later, it dines on pine seeds for most of the winter and spring, even feeding them to its young when the nesting season arrives. But the nutcracker does not always retrieve them all. Some remain in the ground and grow into the pines that feed future generations of nutcrackers.

Clark's nutcrackers have elastic pouches under their tongues for storing pine seeds.

Recognition. 12–14 in. long. Stocky and pale gray; bill long and pointed; wings and tail black, with bold white patches on rear edge of wings and at sides of tail.
Habitat. Barren rocky areas in high mountains; sometimes visits lowlands in winter.
Nesting. Bulky nest on branch of pine or juniper is built of twigs and bark and lined with grasses. Eggs 2–6, pale green dotted with brown. Incubation about 18 days, by both sexes. Young leave nest in about 4 weeks.
Food. Mostly pine seeds and juniper berries; also insects in summer. May steal eggs and nestlings of other birds.

Black-billed Magpie

Black-billed Magpie *Pica hudsonia*
Yellow-billed Magpie *Pica nuttalli*

Pesky rogues, magpies are among the playful bad boys of the West. Although they normally eat carrion and insects, both the black-billed magpie and its nearly identical cousin, the yellow-billed, are notorious for stealing the eggs and young of other birds. Members of the Lewis and Clark expedition, who discovered the black-billed species while exploring the West in 1804–06, reported that these daredevils even snatched food from inside tents where men were eating. Often the birds flee with other stolen treasures—buttons, pins, and the like—and bury them just as a pirate buries his loot.

Like all good thieves, magpies know how to defend themselves against rival marauders. Nesting in loose colonies among brushy thickets, they top their homes with protective domes of stout, thorny twigs.

Though they are normally about two feet across, nests up to seven feet wide have been found. Whatever the size, the interior is comfortably lined with rootlets and animal hair. Not only other birds but even mammals sometimes move in after the builders have moved on. Reportedly, four young house cats once lived in a deserted magpie nest, and a gray fox took up residence in another.

Black-billed Magpie

Yellow-billed Magpie

Recognition. 17–22 in. long. Large and long-tailed, with bold black and white pattern. White wing patches conspicuous in flight. Black-billed has black bill; yellow-billed has yellow bill, is found only in California.
Habitat. Open country with scattered trees; croplands and brushy areas; also towns.

Nesting. Eggs 7–13, greenish or buff, spotted or speckled with brown. Nest is bulky mass of sticks in tree. Incubation 16–18 days. Nestling period not known.
Food. Insects, carrion, fruit, and nestlings of other birds.

American Crow

Corvus brachyrhynchos

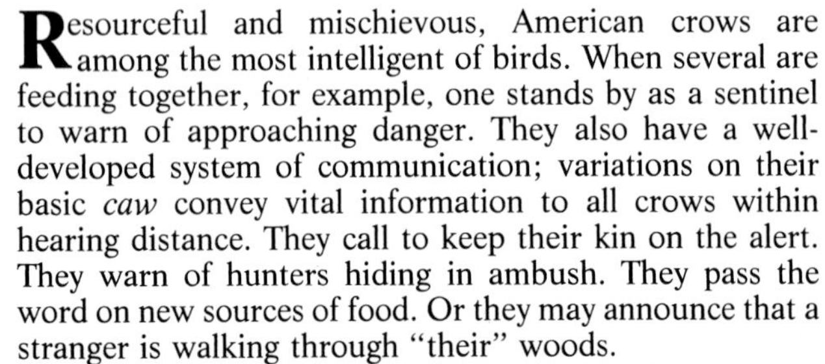

Resourceful and mischievous, American crows are among the most intelligent of birds. When several are feeding together, for example, one stands by as a sentinel to warn of approaching danger. They also have a well-developed system of communication; variations on their basic *caw* convey vital information to all crows within hearing distance. They call to keep their kin on the alert. They warn of hunters hiding in ambush. They pass the word on new sources of food. Or they may announce that a stranger is walking through "their" woods.

Indeed, noise seems to delight these brigands, so long as it is of their own making. Near the nest, however, the crows turn suddenly silent. Then they slip furtively through the woods, drifting like shadows among the trees, wary even beyond the wariness typical in their everyday lives.

These wonderfully resilient birds have adapted to a man-changed environment, moving from forests to farmlands, then to suburbs and city parks. In the process, humans have slaughtered them, poisoned them, and made them the objects of organized hunts. And still the crows flourish—sportive, audacious, triumphant, they remain among the most widespread of American birds.

The male crow feeds his mate while she incubates their eggs.

Recognition. 17–21 in. long. All black, with fan-shaped tail. Call note a distinctive, harsh *caw.* Often travels in pairs or small groups.
Habitat. Woodlands, plains, farmlands, orchards, towns and cities.
Nesting. Eggs 3–7, greenish, blotched with brown. Nest is a large mass of sticks, usually 18–60 ft. above ground in tree, occasionally on utility pole, rarely on ground. Incubation 18 days. Young leave nest in 5 weeks.
Food. Almost omnivorous: insects, small reptiles, eggs and nestlings of other birds, clams, carrion, fruit, and crops such as corn.

Fish Crow

Corvus ossifragus

Looking like a smaller version of the American crow, the fish crow is easily distinguished by its voice: its nasal falsetto is quite different from the cawing of the larger bird. Its lifestyle stands out too. While it may venture inland with the American crow, the fish crow is more at home in coastal marshes and on beaches, and in the broad, watery swamps of the southern states.

An opportunist, the fish crow dines on the remains of animals that wash ashore or lie afloat in shallow water. But it also hunts actively for prey. Hovering above a school of fish teeming near the surface, it drops and grasps one in its claws. If the catch is small, the crow eats it on the spot. Larger fish are carried off to a perch, where they are torn into bite-size shreds.

Scouring the shallows and the shoreline mud, the fish crow catches such other tasty fare as fiddler crabs, shrimp, and crayfish. When it uncovers a clam, the crow breaks the shell with its beak, then pulls the flesh out with the aid of its claws. And when it wants to drink, it flies low over a pond, cutting a silver slash across the water with the tip of its lower bill as, with mouth wide open, it skims the glistening surface.

Its bill extended, a fish crow skims over the water for a thirst-quenching drink.

Recognition. 16–20 in. long. Similar to American crow but smaller, glossier, and usually found in groups near water. Also has more rapid wingbeats. Call note a nasal *ca-hah,* or *ca-uk.*
Habitat. In tidewater areas along coast and along larger rivers in lower Mississippi Valley.
Nesting. Nest built of sticks and twigs in tree along river or at edge of marsh. Eggs 4–5, greenish, spotted with brown. Incubation 16–18 days. Young leave nest in 3 weeks. Often nests in loose colonies.
Food. Fish, carrion, shellfish, and eggs of water birds; some insects and berries.

Common Raven

Corvus corax

Ravens often build their deep, bulky nests high on a cliff.

Few birds have captured man's imagination for so long or been credited with such conflicting attributes as the raven. To some it has been a trusted messenger: in the Bible, Noah first sends out a raven to test the floodwaters, and the ancient Norse god Odin relied on ravens perched on his shoulders to serve as his eyes and ears. Like the early Nordic tribes, Indians of the American Northwest also associated the raven with divinity.

Elsewhere, the raven was a messenger of another sort. A medieval Irish treatise on magic lists more than two dozen prophecies that can be based on a raven's behavior. And in Edgar Allan Poe's most famous poem, a young man mourning the death of his beloved is visited one night by a raven that can speak but one word: "Nevermore." Feverishly he asks if he will meet his lost love in an afterlife, only to sink deeper into despair with each response of the raven, which comes to embody his own dark torment.

In addition to its reputation for divine or magical powers, the raven has long been noted for a variety of more earthly traits. Its cunning is legendary, but just as evident is the raven's deep concern for its life partner—as is a streak of playful exuberance that verges at times on pure hilarity.

Recognition. 22–27 in. long. Large and all black, with shaggy throat, wedge-shaped tail, and long wings with feathers that spread out like fingers in flight. Distinctive call note a hoarse, ringing *crock* or *corruk.*

Habitat. Remote forests, deserts, sea cliffs, mountains, and Arctic and alpine tundra.

Nesting. Eggs 3–7, greenish, heavily spotted with brown. Nest is a bulky, deep cup of branches and sticks, situated in tree or on cliff. Incubation 18–20 days, by female. Young leave nest in 5–6 weeks.

Food. Carrion, crustaceans, bird eggs and young, insects, berries, and small mammals.

Chihuahuan Raven

Corvus cryptoleucus

While the widespread common raven ranges from Arctic tundra to coastal cliffs and mountain forests in the temperate zone, its smaller cousin, the Chihuahuan raven, makes its home in arid grasslands and mesquite thickets near the Mexican border. The two species have much in common, though the Chihuahuan is by far the more gregarious of the two: after the breeding season it sometimes gathers into flocks of as many as 100 birds. Soaring high on currents of warm air rising from the desert floor, the birds indulge in the aerial antics for which all ravens are famous. But in the case of the Chihuahuan raven, such displays of swooping, diving, and tumbling through space are all the more breathtaking because of the sheer numbers of airborne acrobats at play.

As impressive as its agility in the air is the Chihuahuan raven's uncanny intelligence—a trait it shares with all members of the crow family. The bird seems to be able to learn new behavior even from other kinds of animals. Observers in the Southwest claim that it has developed a knack for killing snakes—far from typical raven fare—by watching roadrunners at work. And it has discovered how to get nectar from desert flowers by imitating the feeding habits of nectar-eating bats.

Road-killed animals are one food source for the opportunistic Chihuahuan raven.

Recognition. 19–21 in. long. Similar to common raven but wingbeats are faster, call note is a flat *kraak* or *quark*. Seen in flocks more often than common raven.
Habitat. Desert flats and arid grasslands.
Nesting. Eggs 3–8, pale green, unmarked or blotched with brown. Nest a spherical mass of thorny twigs and sometimes pieces of barbed wire, 4–40 ft. above ground in tree or on utility pole. Incubation 21 days, by both sexes. Nestling period unknown.
Food. Carrion, rodents, eggs and young of other birds, insects, seeds, and fruit.

Smaller Open Country Birds

To small birds such as robins, finches, and swallows, America's vast expanses of open country offer a variety of habitats—undulating prairies with sheltering groves along their rivers; deserts with scattered thickets of brush; meadows bedecked with wildflowers; even man-made parks and croplands. Each of these environments is a special place, with its own distinctive set of living conditions. And each is home to a host of wild birds.

74 Rock Dove
75 Band-tailed Pigeon
76 White-winged Dove
77 Mourning Dove
78 Inca Dove
79 Common Ground-Dove
80 Common Nighthawk
81 Common Poorwill
82 Black Swift
83 Chimney Swift
84 White-throated Swift
85 Black-chinned Hummingbird
86 Anna's Hummingbird
87 Costa's Hummingbird
88 Allen's Hummingbird
89 Belted Kingfisher
90 Green Kingfisher
91 Gila Woodpecker
92 Black Phoebe
93 Eastern Phoebe
94 Say's Phoebe
95 Vermilion Flycatcher
96 Ash-throated Flycatcher
97 Cassin's Kingbird
98 Western Kingbird
99 Eastern Kingbird
100 Gray Kingbird
101 Scissor-tailed Flycatcher
102 Horned Lark
103 Purple Martin
104 Tree Swallow
105 Violet-green Swallow
106 Northern Rough-winged Swallow
107 Bank Swallow
108 Cliff Swallow
109 Barn Swallow
110 Pinyon Jay
111 Cactus Wren
112 Rock Wren
113 Canyon Wren
114 Sedge Wren
115 Marsh Wren
116 Eastern Bluebird
117 Western Bluebird
118 Mountain Bluebird
119 American Robin
120 Northern Mockingbird
121 Sage Thrasher
122 Bendire's Thrasher
123 Curve-billed Thrasher
124 California Thrasher
125 American Pipit
126 Sprague's Pipit
127 Phainopepla
128 Northern Shrike
129 Loggerhead Shrike
130 European Starling
131 Lucy's Warbler
132 Prairie Warbler
133 Palm Warbler

Horned Lark

Rock Dove

Columba livia

More commonly known as pigeons, rock doves vie with domestic chickens for status as the world's most familiar birds. They were introduced into North America from Europe long ago and are conspicuous in cities and villages throughout much of the world.

Pigeons, in fact, have been associated with humans for several thousand years. Believed to have been the first domesticated birds, they were raised for meat as far back as the time of the ancient Egyptians.

Because of their powers of flight and their remarkable homing ability, moreover, pigeons have played important roles in history. A domestic pigeon taken from its home loft and released many miles away almost invariably returns. And if a message is tied to the bird's leg, the result is a kind of airmail—a fact that humans learned to exploit many centuries ago. When Julius Caesar marched against Gaul, the news of his victories was carried back to Rome by a network of carrier pigeons. Other pigeons carried messages for Alexander the Great and for Hannibal. In modern times opposing armies in both World War I and World War II made use of thousands of carrier pigeons. And today pigeons are still bred for their homing ability.

Domesticated rock doves have long been valued for their speed and stamina as carrier pigeons.

Recognition. 11–14 in. long. Typically gray, with 2 black bars across wings, white rump, and dark tail tip. Color variants include blackish, rusty, or mostly white. Very tame; often begs for food in parks and city streets.
Habitat. Cities, towns, and farms.
Nesting. Nest is a shallow platform of twigs and grass, placed on ledge of building or on cliff. Eggs 1–2, white. Incubation 17–19 days, by both sexes. Young leave nest about 35 days after hatching. Usually several broods each season; may nest year-round.
Food. Grain, seeds, and scraps of bread.

Band-tailed Pigeon

Columba fasciata

When storms build over the Rockies and other western mountains, black clouds loom above the ridges and lightning crackles in the air. As peals of thunder echo down the slopes and gusts of wind whip the tops of the trees, most birds quite sensibly take shelter. But not the band-tailed pigeons. At such times flocks of them can often be seen flying in great circles across the slopes, skimming the treetops and then swinging wide over the valleys, defying the violence of the weather.

At first glance, this native of the wilder reaches of the far western states does not look like a bird designed for such powerful flight. Walking on the ground, it seems a bit portly and somewhat slow. But first impressions can be misleading, for the bird is heavy with muscle, not fat. Like most of its kin, the band-tailed pigeon is an excellent flier. Domestic pigeons, for instance, have been timed at speeds of up to 80 miles an hour. And many of the pigeons are not only fast-moving but also superb long-distance fliers, as exemplified by homing pigeons, which return to their roosts even when released hundreds of miles away.

Large flocks of band-tailed pigeons remain aloft even in severe thunderstorms.

Recognition. 14–16 in. long. Gray or blue-gray, with broad pale band on tail, white crescent across back of neck; bill yellow with black tip. Often shy and easy to overlook, but perches in bare treetops in winter.

Habitat. Forests of pine or oak.

Nesting. Nest is a shallow platform of twigs, usually 8–40 ft. above ground in tree or shrub, often at edge of clearing. Single egg is white. Incubation 18–20 days, by both sexes. Young leave nest about 30 days after hatching. Sometimes nests in loose colonies; may raise more than one chick per season.

Food. Acorns, hazelnuts, berries, and insects.

White-winged Dove

Zenaida asiatica

Desert birds, like birds everywhere, must have water. Some get all they need from the food they eat. But others, including white-winged doves, have to drink regularly and will fly miles, if necessary, to do so.

The doves usually arrive at their water holes in early morning and late evening. But no matter how thirsty they are, they seldom fly directly to the water's edge: landing in nearby trees or shrubs, they survey the scene for several minutes before approaching the water.

Most birds drink by dipping the bill into the water, taking in a mouthful at a time, and then tipping the head back to let the water flow down the throat. White-wings have a faster method: like other doves and pigeons, they submerge the bill and take in the water in a continuous draft. One or two long drinks are usually enough, and the white-wings fly away.

The ability to drink quickly is advantageous for a desert bird, since water holes attract not only songbirds and doves, but hawks as well—predators who may be more interested in eating than in drinking. But because white-winged doves are wary in their approach and prompt in their departure, they usually manage to drink in safety.

In flight, the white-winged dove displays its distinctive wing and tail patches.

Recognition. 11–12 in. long. Gray-brown with large white patches on wings; tail long and fan-shaped, with white patches at corners. Often seen in flocks.
Habitat. Deserts, farmland, open woodlands, and residential areas.
Nesting. Nest is a flimsy saucer of twigs 4–25 ft. above ground in shrub or small tree. Lays 2 buff, white, or cream-colored eggs. Incubation about 14 days, by both sexes. Young leave nest 15 days after hatching. Often nests in colonies; raises several broods each year.
Food. Seeds of weeds and shrubs, acorns and nuts, and occasionally insects.

Mourning Dove

Zenaida macroura

Mournful though the sound may seem, the slow, low-pitched cooing of the mourning dove marks a joyous season. It is the sound of birds claiming a home territory, courting a mate, and preparing to raise young.

The nest the mourning dove builds to cradle its offspring is a surprisingly flimsy structure—little more than a haphazard pile of twigs balanced precariously on a branch. The platform is so loosely built, in fact, that the eggs are often visible from below through the latticework of branchlets. When an incubating adult is frightened from the nest, the fluttering panic of its departure can easily knock one of the eggs to the ground.

But if the eggs remain in the nest long enough to hatch, the young birds get a special treat: doves and pigeons produce a unique food, called pigeon milk. Rich in fat and protein, this substance (which is not milk at all) is produced by glands in the crop of the adult bird. Trying to pour this liquid into the mouth of a clumsy infant could be tricky—so the dove has a better way. The parent opens its mouth wide, permitting the nestling to stick its head deep inside to gorge on the thick, nutritious food. And the chicks obviously thrive on their unusual diet: the mourning dove is one of the most numerous and widespread birds in North America.

"Pigeon milk" for the young is produced in the adult's gullet.

Recognition. 10–12 in. long. Slender and small-headed. Gray-brown; tail long and pointed, with white tips on feathers. Wings make whistling sound in flight.
Habitat. All habitats with the exception of wetlands and dense forests.
Nesting. Nest is a flimsy platform of twigs and sticks, usually 5–25 ft. above ground in tree or on building ledge, rarely on ground. Eggs 2, white. Incubation about 15 days, by both sexes (male by day, female by night). Young leave nest about 15 days after hatching. At least 2 broods per year, nesting from early spring to autumn.
Food. Seeds and occasional insects.

Inca Dove

Columbina inca

Like most doves, Inca doves are perfectly at home near people.

When the first pioneer naturalists explored the territory that would become Arizona, they found a wilderness teeming with wildlife—parrots in the mountains, wolves on the plains, even jaguars roaming the unspoiled terrain. But they found no Inca doves. From the doves' point of view, apparently, Arizona was not yet civilized enough. For Inca doves are birds that prosper most when living side by side with people.

No one knows exactly when we humans were adopted by Inca doves. It probably happened centuries ago, as the birds ventured into the villages of the ancient peoples of Mexico and found conditions to their liking: waste grain and weed seeds were plentiful, water was always available, and wild predators were kept at bay. Tiny and quiet, the birds were so unobtrusive that no one bothered them, and they flourished in the settlements.

The doves did not move north into Arizona until about 1870, after forts, missions, and towns had been well established; today, these delightful little birds are as common on the lawns and golf courses of Phoenix as they are in the parks of downtown Mexico City. And in wilder terrain, in places unaltered by mankind, Inca doves are still almost impossible to find.

Recognition. 7–8 in. long. Small and slender; tail long and narrow, with white borders. Pale gray with black scalelike markings; contrasting rusty patch on wings.
Habitat. Towns, farms, and brushy deserts.
Nesting. Nest is a poorly made saucer of twigs and weed stems 4–25 ft. above ground on horizontal branch of tree or shrub; sometimes on ledge of building. Occasionally breeds in old nest of another species. Eggs 2, white. Incubation about 14 days, by both sexes. Young leave nest about 12 days after hatching. At least 2 broods a year, sometimes 5.
Food. Weed seeds and grain.

Common Ground-Dove

Columbina passerina

Resembling miniature mourning doves, sparrow-size common ground-doves are the smallest doves in North America. As their name suggests, they spend most of their time on the ground, heads nodding as they walk about in search of seeds and other bits of food.

They usually nest on the ground as well, producing two young per brood. Believed to mate for life, common ground-doves are notable for the length of their annual breeding season. Unlike most North American birds, which nest only in the summer months, ground-doves have been known to nest at any time from February through October, with some pairs raising as many as three or four broods in a single year. A few other birds—barn-owls, for example—may lay their eggs in virtually any month of the year. But they do not nest continuously throughout such a long period.

Common ground-doves range from the southernmost rim of the United States to tropical South America, and their nesting habits probably reflect their tropical origins. Whereas most northern birds reproduce at a time of peak food supply, many tropical species, enjoying a more stable year-round climate and an abundance of food, are able to breed throughout the year.

Head bowed and feathers puffed out, a strutting male seeks to attract a mate.

Recognition. 5–6½ in. long. Small and slender; gray-brown with rusty patch on wings; tail is short, fan-shaped, dark with white corners. Walks with quick steps, nodding its head. Looks like tiny, short-tailed mourning dove.
Habitat. Woodlands, roadsides, and farmlands.
Nesting. Nest is a slight depression in ground, lined with grass, rootlets, and feathers. Lays 2 white eggs. Incubation about 14 days, by both sexes. Young leave nest 11 days after hatching.
Food. Weed seeds and grain; occasionally small insects and berries.

Common Nighthawk

Common Nighthawk *Chordeiles minor*
Lesser Nighthawk *Chordeiles acutipennis*

Lesser Nighthawk

Wonderfully evocative names have been attached to the common nighthawk. Like all members of its family, it is known as a goatsucker, a bizarre term based on European folklore. Because its Old World cousin, the nightjar, has a wide, gaping mouth and a fondness for pastures and other open places, peasants long ago concluded that the mysterious night-flier sucked milk from their animals.

Two popular names for our common nighthawk—pork-and-beans and bull-bat—derive from characteristic sounds it makes. *Beans* resembles the nasal note the bird repeats emphatically during its herky-jerky insect-hunting flights. *Bull* refers to an odd nonvocal noise. During courtship, the male frequently interrupts its foraging flights with dramatic vertical plunges, swinging back upward just before hitting the ground. As its wings sweep into braking position, air rushing through the feathers produces a kind of sonic boom that has been likened to everything from a bass trumpet note to a whirring spinning wheel. *Bull,* a shorthand reference to the booming of bullfrogs, is imaginatively suggestive of the sound.

The very similar lesser nighthawk—a denizen of southwestern scrubland and desert—says neither *beans* nor *bull.* Its characteristic call is a purring trill.

Recognition. 8–10 in. long. Long, pointed wings and long, square-tipped or notched tail. Gray-brown mottled with black; white patch on wings. Usually seen in flight. Lesser has shallower wingbeats; white patch closer to wingtips than on common nighthawk.
Habitat. Towns, cities, parks, and open country. Lesser nighthawk prefers dry flats and deserts.
Nesting. Eggs 2, pale buff, pinkish, or greenish, speckled with brown and gray; laid on bare ground or flat roof. Incubation 18–20 days, mainly by female. Young are downy; fed by both parents; leave nest about 21 days after hatching.
Food. Flying insects, usually caught high in the air.

Common Poorwill

Phalaenoptilus nuttallii

Of all the common poorwills in the world, one alone occupies a special niche in the history of science. Discovered by biologist Edmund C. Jaeger and two of his students on December 29, 1946, the bird was literally "holed up" in a hollow on a canyon wall in the mountains of southern California. Apparently in a state of suspended animation, it was cold to the touch and showed no signs of heartbeat or respiration. While being handled on a later visit, the bird opened an eye, squeaked like a mouse, and yawned. During a third visit it roused and flew up the canyon wall. The bird returned to the same dormitory for three successive winters.

What Jaeger and his students had found, to the astonishment of science, was a hibernating bird. While some common poorwills fly south for the winter, others survive by lapsing into an energy-saving state of torpor, their body temperature dropping nearly to the level of the air around them. We now know that several species of birds are capable of short-term dormancy, but the common poorwill is the only one known to practice long-term hibernation.

The discovery that *any* bird utilizes such a survival strategy amazed everyone—except, perhaps, the Hopi Indians. Their traditional name for the common poorwill is *holchko,* "the sleeping one."

The poorwill's nest is little more than a shallow depression in bare ground.

Recognition. 7–8 in. long. Brown, with fine black and silvery markings; tail short and fan-shaped, with white corners. Nocturnal; most often seen in headlights, sitting on road; flies like large moth. Best distinguished by its call, a mellow *poor-will.*

Habitat. Dry, brushy country and rocky slopes.

Nesting. Eggs 2, white or cream-colored, laid in shallow depression in gravel on ground, or on bare rock. Incubation by both sexes; period unknown. Young downy at hatching. Nestling period unknown.

Food. Insects caught in flight; also insects picked up from ground.

Black Swift

Cypseloides niger

The largest and least common of our native swifts is also the highest flier—so lofty an aerialist that the black swift went undiscovered by ornithologists until 1857. Spiraling upward to disappear from view at heights of several thousand feet, the "cloud swift" earned its nickname from its habit of hawking insects at the leading edge of advancing storm clouds in the Far West.

Feeding entirely on the wing, black swifts come down to earth only to roost at night or to visit their nests, cups of moss and mud built on rock faces and overhangs, often near waterfalls. Sometimes they even nest *behind* the falls, only inches from the tumbling cascade.

Studying birds that spend most of their lives high in the sky presents a challenge to scientists, and to this day many questions remain unanswered. Black swifts are known to travel to the tropics, for example, but we still have only a vague idea of their winter haunts—or of their courtship, mating, and many other habits. For most of us, perhaps it is enough simply to marvel at the aerobatics of creatures that are as much at home in the sky as we are on land.

With backswept wings, the black swift is scimitar-shaped in flight.

Recognition. 7–7½ in. long. Larger than other swifts or swallows. Dull black with very long, narrow wings; tail slightly forked.
Habitat. Cliffs in mountains and along seacoast.
Nesting. Nest is a deep, solid cup of moss and mud attached to ledge under sheltering overhang or in cave, often behind waterfall. Single egg is white. Incubation and nestling periods unknown. Young are fed by regurgitation; can survive for days without food. Nests are usually in small colonies.
Food. Insects and spiders caught in flight.

Chimney Swift

Just as the woods and fields of eastern North America are filled with songbirds' calls in summer, the sky is filled with a gentle tinkling. The sounds are made by chimney swifts, tiny bundles of boundless energy that sweep the air for flying insects.

In colonial times, the birds could just as well have been called "tree swifts," from their habit of roosting for the night in hollow trees. But they have since adapted remarkably well to the ways of humans, using not only chimneys but barns and other structures as resting and nesting sites. Vaux's swifts, the western counterparts of chimney swifts, also roost occasionally in chimneys, though they most often still prefer trees.

In late summer and fall, when chimney swifts begin their southward migration (their wintering ground in South American rain forests was discovered only in recent times), huge numbers of them—in one case, 10,000 birds—gather at dusk to enter favorite chimneys. Several swifts per second funnel into the opening; by the time all have entered, the birds are packed shoulder to shoulder, clinging tightly to the vertical walls.

Except when nesting or roosting, chimney swifts never perch anywhere, for they are true denizens of the air. It has been estimated that long-lived individuals may cover more than a million miles before they die.

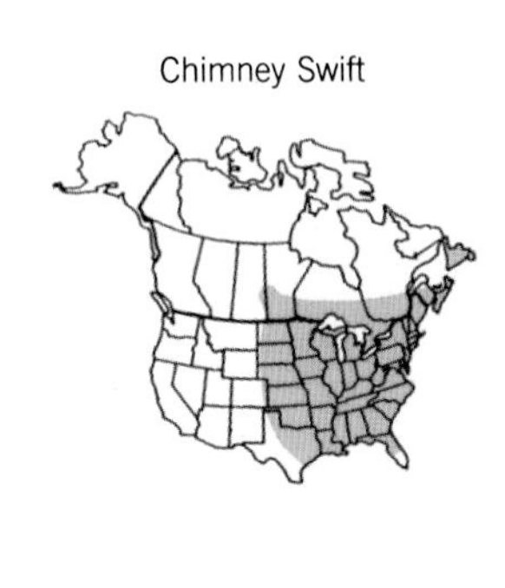

Recognition. 4–5 in. long. Uniform dull brown, palest on throat; wings long and swept back; tail short. Usually seen in flight. Flight rapid and erratic. Vaux's strictly western; has paler throat.
Habitat. Towns and cities; Vaux's also in mountain forests.
Nesting. Nest is a thin semicircle of twigs, cemented together with saliva and attached to inside of chimney or hollow tree. Eggs 2–7, white. Incubation 19–21 days, by both sexes. Young leave nest about 4 weeks after hatching.
Food. Insects and spiders caught in flight.

White-throated Swift

Aeronautes saxatalis

Swifts are aptly named for their rapid flight. But even among this remarkable group, white-throated swifts are exceptional. With an estimated top speed of more than 200 miles per hour, they are said to be the fastest of all North American birds.

Also known as rock swifts, they favor the coastal bluffs and cliff-lined canyons of the West, where their pleasant, twittering flight calls fill the air. Mating takes place mostly on the wing, with two birds tumbling downward together for several hundred feet before parting. Their nests, usually built in cliff crevices high above the ground, are shallow cups of feathers and grasses glued together with a secretion from the birds' large salivary glands. (A similar glue from an Asian swiftlet forms the basis for a well-known oriental dish, bird's nest soup.)

Like others of their kind, white-throated swifts feed on insects caught in rapid and erratic pursuit. As their prey dwindles with the onset of cold weather in the northern part of their range, the birds begin to move south. But some remain behind and, like only a few other birds, are able to enter a sleeplike torpid state during cold spells. On warmer days they become active again, patterning the skies once more with their fluttering flight.

Saliva provides strong cement for the white-throated swift's nest.

Recognition. 6–7 in. long. Wings long and narrow; tail long and slightly forked. Black, boldly patterned with clearly visible areas of white on underparts and flanks.
Habitat. Canyons and dry, rocky cliffs.
Nesting. Nest is a frail, rounded cup of grass and feathers, cemented together with saliva and attached to wall of inaccessible crevice high on cliff face or occasionally in old building. Eggs 3–6, white or cream-colored. Incubation and nestling periods unknown.
Food. Insects and spiders caught in flight.

Black-chinned Hummingbird

Archilochus alexandri

A denizen of woodlands and gardens, the black-chinned hummingbird is the western equivalent of the well-known eastern ruby-throat. So similar are the two, in fact, that even experts have difficulty distinguishing the females of the two species.

Far easier to identify are the two birds' nests, which are built exclusively by the females. While both birds share a fondness for saddling them on limbs that span watercourses, black-chinned hummingbirds disdain the lichen-plastered bowls favored by ruby-throats. Instead, they build their tiny, cuplike bowls with woven plant down and bond them lavishly with spider webs.

The building materials give the nests an elastic quality, providing the fast-growing young with a little extra room in a cup that measures a scant inch and a half in diameter. "Its rim yields to their little strugglings," as one observer put it, and it "opens like a flower bud."

Particularly good nesting sites are frequently reused, with the new nest built right on top of the foundation of one or even two nests that were built in previous years. That they survive so long is impressive testimony to the durability of these seemingly fragile structures.

The black-chinned hummingbird binds its nest with spider webs.

Recognition. 3–3¾ in. long. Male has black chin bordered below by iridescent purple; collar white; tail dark. Female green above, whitish below; tail has white corners. Males make thin dry buzz with wings in flight.
Habitat. Oak woods, gardens, and brushy areas.
Nesting. Nest is a small, round cup of plant down coated with spider web; often built 4–8 ft. above ground on limb, sometimes over water. Eggs 2–3, white. Incubation 16 days, by female. Young leave nest 3 weeks after hatching.
Food. Nectar and small insects taken at flowers.

Anna's Hummingbird

Calypte anna

The hummingbird's long, needle-like beak is perfectly adapted for siphoning the nectar of flowers.

Anna, duchess of Rivoli, could hardly have asked for a more stunning namesake. Cloaked in iridescent feathers that flash green in the sunlight, the male Anna's hummingbird also has a throat and crown embellished with a lustrous, rosy hue unique to the species.

Found mainly in California, the bird is as much at home in parks and gardens as it is in chaparral and open woodlands. Indeed, many homeowners deliberately try to lure it into their yards by planting red flowers, since that color seems particularly attractive to hummingbirds.

Californians who set up hummingbird feeders can even enjoy this charming neighbor all year round, for Anna's hummingbird is unique in its tendency to remain through the winter. (Some nonbreeding birds wander north to British Columbia in spring; wintering birds seldom migrate beyond southern Arizona.) Feeder care is minimal. All that is needed is an inexpensive solution of four parts water to one part sugar, a mixture that approximates the concentration of sugar in floral nectar. Feeders should be emptied and refilled daily and care taken to ensure that supplies do not run out in cold weather. Some people add a bit of red food coloring to the sugar water as a further attraction. Their reward is a delightful parade of flashing feathers as hummers come by for a sip.

Recognition. 3–4 in. long. Stocky for a hummingbird. Male mainly green above and on flanks; crown and throat iridescent rose-red; tail black. Female has red flecks on throat; flanks grayer; tail tipped with white.
Habitat. Gardens and brushy open woods.
Nesting. Nest is a small cup of plant down coated with lichen; placed 2–30 ft. above ground on limb, sometimes on utility wire. Eggs 2, white. Incubation 14–18 days, by female. Young leave nest 25 days after hatching.
Food. Nectar and small insects taken at flowers.

Costa's Hummingbird

Calypte costae

Hovering before a cactus bloom with his back to the sun, a male Costa's hummingbird is unremarkable to behold, his head and throat patch appearing flat and black. But when this little desert hummer turns, catching the light just so, the black ignites into a stunning burst of violet.

Brilliant beyond words, the colors of all hummingbirds have long been a source of wonder. The Aztecs of ancient Mexico, for instance, adorned their finery with the skins of hummingbirds. It is easy to understand why, arrayed in feathered robes that shattered sunlight into a rainbow of color, the Aztec leaders were regarded as divine. Nor were they alone in coveting the beauty of nature's feathered gems. Hundreds of thousands of hummingbird skins were used to adorn the hats of fashionable women during the Victorian era, a mere century ago.

Like the Aztec nobles and Victorian grandes dames, the Costa's hummingbird uses his brilliant plumage to impress. During his spirited aerial display—a series of swoops from high in the air—the male is careful to maintain a proper angle to the sun so that prospective mates receive a full view of his glistening purple throat. For additional effect, each full-throttled dive passes within inches of the perched female and is accompanied by a harsh, metallic shriek.

Artificial feeders can be as tempting as nature's blossoms.

Recognition. 3–3½ in. long. Male mainly green above and on flanks; crown and throat iridescent violet, with throat patch elongated at sides; tail black. Female gray-green above, whitish below; tail black with white tips.

Habitat. Arid brushy areas and nearby gardens.

Nesting. Nest is a small cup of plant down coated with lichen, 2–30 ft. above ground on limb or on yucca stalk. Eggs 2, white. Incubation 15–18 days, by female. Young leave nest 20–23 days after hatching.

Food. Nectar and small insects taken at flowers.

Allen's Hummingbird

Selasphorus sasin

Perhaps the reddish colors of the male Allen's hummingbird were meant to serve as a warning to would-be transgressors: diminutive but hot-tempered, he quickly banishes all rival males who enter his territory. Woe, too, to hummingbirds of any other species that stray too close to a feeder marked by an Allen's as his own. Even red-tailed hawks and American kestrels have been put to rout for no greater offense than entering the airspace of an Allen's hummingbird.

The aggressiveness of Allen's and other territorial hummingbirds is understandable. Since their rate of metabolism is almost 100 times faster than our own, these tiny feathered dynamos require a constant source of fuel. Their breeding success, and sometimes their very survival, depend upon the flowers or feeders they defend.

Their main food is flower nectar, a liquid rich in sugar, which converts quickly into energy. Even with this power-packed fare, hummingbirds still must feed every 10 to 15 minutes—they visit perhaps 1,000 blossoms in the course of a day. They also eat insects, which provide protein and other essential nutrients unavailable in nectar. But the sweet gift of the flowers remains the staple of their diet, and they return the favor by distributing the plants' pollen, a light dusting of which can often be seen atop their heads.

Fiercely territorial, hummers will attack much larger birds.

Recognition. 3–3½ in. long. Male green above, with rusty flanks, rump, and tail; throat patch iridescent orange-red. Female lacks throat patch; rusty coloring only on tail; tail feathers have white tips.
Habitat. Brushy woods, gardens, and flower-filled mountain meadows.

Nesting. Nest is a cup of plant down and weed stems coated with lichen, ranging from 1–90 ft. above ground on limb or weed stalk. Two white eggs. Incubation 15–17 days, by female. Young leave nest about 3 weeks after hatching.
Food. Nectar and small insects taken at flowers.

Belted Kingfisher

Ceryle alcyon

Aggressive loners for most of the year, belted kingfishers tolerate others of their kind only during the summer nesting season. But then, for a few weeks, the mated pairs work together as well-disciplined teams.

They begin by taking turns digging a tunnel into an earthen bank, preferably close to their fishing territory. As they carve out a passageway with their bills, they push the debris out behind them with their feet. At the end of the tunnel, they excavate a nesting chamber where, in total darkness, the female lays her clutch of pure white eggs.

For the next 24 days or so, the two birds share in the incubating chores and spend their free time fishing. The chicks, stark naked when they break from their shells, at first are brooded by the female only, while the male feeds the entire family. Once the little ones have begun to sprout feathers all over their bodies, the female joins the male in providing food.

The chicks finally leave the nest—fully feathered and able to fly—when they are about five weeks old. Within another week or two they learn to fish for themselves. The family then breaks up, and each member goes off to defend a territory of its own—and to live as an aggressive loner until the onset of the next nesting season.

Quick and deadly, the belted kingfisher dives in pursuit of an unwary fish.

Recognition. 11–14 in. long. Blue-gray above and on head; collar white; bill stout; crest ragged. Male has blue-gray band across breast. Female has second, rusty band across belly. Seldom seen away from water; dives from air for fish; has loud, rattling call.
Habitat. Rivers, lakes, ponds, and marshes.

Nesting. Eggs 5–8, white, laid on bare soil at end of tunnel, usually 3–7 ft. long, in steep riverbank. Sometimes use same burrow in successive years. Incubation about 24 days, by both sexes. Young leave nest after 33–38 days.
Food. Mainly small fish and tadpoles; also salamanders, frogs, and insects.

Green Kingfisher

Chloroceryle americana

Ever watchful, a green kingfisher perches just above the water.

Kingfishers around the world come in many colors, but all share the same general form. With their squat bodies, short legs, large heads, and stout dagger-shaped bills, they could not exactly be described as graceful. When it comes to catching fish, however, they easily live up to their name.

Green kingfishers are tropical birds that reach the northernmost limits of their range in southern Texas and Arizona. They are most likely to be found in shaded areas around quiet pools or along the slow-moving backwaters of rivers, where they perch atop boulders on the bank or even in midstream and, jerking their tails up and down, wait watchfully for their victims. Then, with a sudden, swift lunge, they deftly nab any unwary fish that ventures near the water's surface.

Yet despite their aquatic hunting skills, green kingfishers are occasionally found miles from water, securing their meals by darting from low perches to capture butterflies and other flying insects, and by pouncing on luckless grasshoppers and small lizards. If they land on the ground themselves, however, the kingfishers' prowess seems to vanish; waddling awkwardly about, they maintain a precarious balance only by jerking their tails as they go.

Recognition. 7–8 in. long. Much smaller than belted kingfisher, and lacks ragged crest. Green above and on head; collar white; bill long and stout; tail has white outer edges. Male has rusty band on breast. Female similar, but lacks breast band. Seldom seen away from water; dives for fish from the air.

Habitat. Usually rivers, streams, and ponds.
Nesting. Eggs 3–6, white, laid on bare soil at end of tunnel 2–3 ft. long in steep bank. Incubation 19–21 days, by both sexes. Young leave nest 22–26 days after hatching.
Food. Mainly small fish; sometimes insects.

Gila Woodpecker

Melanerpes uropygialis

In its native haunts—the deserts of the American Southwest—the Gila woodpecker often performs real services for its environment. Like all woodpeckers, it digs insects from woody plants for food. But when it hacks a hole into a huge saguaro cactus, it sometimes saves the giant's life.

Some of the insects the Gila feeds on are carriers of a bacterial disease that discolors the surface of the saguaro and causes internal damage. To the Gila woodpecker, the discolored skin is a sign of insect larvae waiting to be eaten. As the woodpecker excavates the burrowing worms, it cuts away the diseased cactus tissue as well. With the worms gone and the bacteria removed, the syrupy sap of the saguaro hardens around the cavity the Gila has created, and the plant heals.

Often these surgical scars become homes for other desert inhabitants. So, too, do the larger nest cavities that the Gila digs for itself. No sooner have the woodpecker pair and their young moved on than the nest is taken over by new residents, from kestrels and elf owls to flycatchers and wrens. All are happy to find a ready-made and empty shelter—as at times are homeless rats, mice, and snakes.

Besides food, the saguaro cactus also supplies the Gila woodpecker with shelter.

Recognition. 8–10 in. long. Back and wings barred black and white; head and underparts tan. Male has small red cap on top of head; female similar but lacks red cap.
Habitat. Cactus deserts, brushy woodlands and groves, and urban parks.
Nesting. Eggs 3–5, white, laid in cavity excavated 15–25 ft. above ground in saguaro cactus, or in mesquite, willow, or cottonwood along desert watercourse. Incubation about 14 days, by both sexes. Nestling period unknown. Two broods often raised per year.
Food. Insects, cactus fruit, and berries; visits feeders for suet and corn.

Black Phoebe

Sayornis nigricans

Neither a high flier nor a high percher, the black phoebe swoops from a low branch or fence post to take its prey of beetles, bees, flies, and moths on or near the ground. Agile on the wing, like all flycatchers, it is a nearly silent flier; indeed, the sharp click of its bill as it snatches an insect from the air often comes as a surprise to nearby observers.

This stealthy hunter's nest is remarkable for its strength. Often built beneath a sheltering overhang, the mud-and-grass dwelling is so securely attached to a vertical cliff or the underside of a bridge that if you try to pull it down, it will break before it becomes detached. Pairs of phoebes typically rear two and sometimes three broods in these snug homes during a single season.

The black phoebe, never far from water, commonly nests around marshes and mountain streams. But even watering tubs, irrigation trenches, and backyard garden pools are suitable for its needs. Reportedly, these phoebes have even been found living happily beside the Colorado River at the bottom of the Grand Canyon. Although a cool drink is presumably the first goal, this gentle land bird has also been known on occasion to dart to the water's surface to feed on small fish.

Like a hawk, a phoebe swoops down on insects swarming above a pool.

Recognition. 5½–7 in. long. Black on head, upperparts, and breast; belly and outer tail feathers white. Usually seen near water. Pumps tail while perched; darts after flying insects.
Habitat. Streams, canyons, and farmlands.
Nesting. Nest is a cup of mud pellets and grass firmly attached to sheltered spot on vertical cliff or wall, usually near water. Eggs 3–6, white, sometimes dotted with brown or reddish. Incubation 15–17 days, by female only. Young fed by both parents; leave nest 3 weeks after hatching. Usually 2 broods a year.
Food. Mainly insects captured in flight.

Eastern Phoebe

Sayornis phoebe

Cheerful and restless little birds, eastern phoebes have neither eye-rings nor wing bars to distinguish them. And the erectile feathers on their heads are far too short to make a respectable crest. Nonetheless, they are in no danger of being overlooked among their more colorful fellows. Returning north at the beginning of spring, sometimes before the snow has left their summer haunts, these early arrivals are unmistakable. Pumping their tails up and down as only phoebes can, the males perch among the leafless trees and announce themselves to everyone within earshot by emphatically and endlessly repeating their name: *fee-bee fee-bee fee-bee.*

Although eastern phoebes nest on the ledges of open cliffs in the wild, they also find bridges, porches, eaves, and sheds convenient housing sites. They will practically move in on favored homeowners, who may then be lucky enough to watch the construction of their moss-and-mud nests, the hatching of the eggs, the feeding of the young, and finally the fledging of the brood as they take their first awkward flights. Frequently returning to the same nest sites year after year, eastern phoebes are eagerly welcomed as harbingers of spring by those whose dwellings they have chosen as their homes.

Sheltered niches on buildings are among the phoebe's favorite homesites.

Recognition. 5–7 in. long. Dull olive-brown above, darkest on head; throat and underparts whitish; bill black. No eye-ring or wing bars. Usually seen near water. Pumps tail while perched; darts after flying insects.
Habitat. Near banks of rivers and ponds.
Nesting. Nest is a cup of moss and mud pellets attached to support on sheltered ledge over water, under bridge, or on building. Eggs 3–8, white, occasionally spotted with brown. Incubation 14–17 days, by female. Young leave nest 15–17 days after hatching. Usually 2, sometimes 3 broods a year.
Food. Insects and spiders captured in flight.

Say's Phoebe

Sayornis saya

This lively inhabitant of the arid West is rightfully named for its discoverer, Thomas Say. The entire genus of phoebes, in fact, is called *Sayornis*—"Say's bird"—in honor of the tireless investigator who first described all three North American species.

Born to a wealthy Philadelphia Quaker family in 1787, Thomas Say was the nephew of William Bartram, one of our most distinguished 18th-century naturalists. Under his uncle's influence, young Say developed the lifelong love of nature that led him, despite poor health, to trek to the Rocky Mountains on mapping expeditions in 1819 and again in 1823. Returning with specimens of mule deer, kit fox, coyote (until then unknown to science), and eight newly discovered western bird species, Say richly advanced our knowledge of North American wildlife and enhanced the already formidable scholarly reputation he had earned cataloging shells and insects.

Even in his personal affairs, Say revealed an adventurous spirit. Moving to New Harmony, Indiana, in 1825, he joined Robert Owen's experimental utopian community. Although the community failed, Say stayed on and lived there happily for the rest of his days. It seems fitting that this energetic explorer of our natural world is recalled today by the nimble and spirited bird that bears his name.

Perched on a fence, a phoebe is poised to dart after insects.

Recognition. 6–8 in. long. Gray-brown above and on breast; tail black; belly warm buff or pale rusty. Pumps tail while perched; darts after flying insects.
Habitat. Desert ponds and canyons, open brushy country, and farmlands.
Nesting. Nest is a platform of stems, grass, and moss, on ledge or in building, natural tree cavity, or old cliff swallow's nest. Eggs 3–7, white, sometimes spotted with reddish brown. Incubation 12–14 days, by female. Young leave nest 14–16 days after hatching.
Food. Insects and spiders captured in flight.

Vermilion Flycatcher

Pyrocephalus rubinus

The scientific names of birds do not always seem appropriate, but in the case of the vermilion flycatcher, *Pyrocephalus* ("firehead") could not be more apt. The males are indeed a blaze of color. In Mexico they are referred to by another vividly descriptive nickname—*sangre de toro,* or "bull's blood."

Such brilliance is rare among the generally inconspicuous flycatchers. It is also unusual among our open country birds. Typically, brightly feathered birds lurk among the foliage to avoid predators. But the vermilion flycatcher flaunts its colors, perching boldly in the open on a shrub or bare limb.

This brashness is even more pronounced during the nesting season, when the male vermilion flycatcher is a delight to watch. Stationing himself in his favored open area, often near a stream or in a dry wash, the male performs a dramatic aerial courtship ritual. Fluttering his wings energetically, with his tail fanned and cocked, he spills out a tinkling flight song as he rises from his perch, alternately climbing and hovering until he reaches a height of 50 feet or more. Then he swoops down to join his mate. It is an exuberant display, and it seems exactly right for such a boldly painted bird.

Tail fanned and pumped up slightly, the male begins his courtship song.

Recognition. 5–6 in. long. Male black above, with scarlet crown and underparts; black line through eye. Female and immature are gray-brown above; underparts whitish with fine streaks; belly tinged yellow or pink.
Habitat. Streamside woods and thickets.
Nesting. Nest is a flat, sturdy cup of twigs, stems, grass, and rootlets set deeply into fork of branch far from trunk, 6–20 ft. above ground in tree along watercourse. Eggs 2–4, white, heavily spotted with brown and lilac. Incubation about 14 days, by female. Young leave nest 14–16 days after hatching.
Food. Insects captured in flight.

Ash-throated Flycatcher

Myiarchus cinerascens

More than 30 species of flycatchers migrate northward each spring to breed in North America. And included among them are some of the most difficult of all birds to identify. Many a birder has spent hours puzzling over one small clue or another that might decisively prove an individual bird to be an ash-throated flycatcher or one of its very similar relatives, the great crested, dusky-capped, or brown-crested flycatchers.

The crucial clue in many cases turns out to be its voice. For the look-alike flycatchers themselves, sound is often a much more important tool than sight for distinguishing between friend and foe, passerby and competitor. Thus each species has a distinctive repertoire of calls and songs—sounds that the experienced birder, too, soon learns to recognize.

The sounds of ash-throated flycatchers are common in the open woodlands where they breed. These birds and their near relatives are unusual among North American flycatchers in nesting in tree cavities. Natural openings will do, as will abandoned woodpecker holes; but if need be, they will even displace smaller birds from occupied cavities. Then the nest is quickly built and lined with fur, hair, and—more often than not—a shed snake skin.

A flycatcher carries a snake skin into its tree trunk nest hole.

Recognition. 7–8½ in. long. Slender, with black bill. Gray-brown above; throat and breast gray-white; belly pale yellow; tail shows flash of rusty. Call note a rolling *ka-wheer.*
Habitat. Open woodlands, streamside thickets, and dry, brushy country.
Nesting. Nest is a loose cup of grass, rootlets, and stems at bottom of natural tree cavity less than 20 ft. above ground. Eggs 3–7, whitish, finely streaked with brown. Incubation about 15 days, by female. Young leave nest about 17 days after hatching.
Food. Mainly insects captured in flight.

Cassin's Kingbird

Tyrannus vociferans

How much is a bird's egg worth? Few people ask that question today, but during the 19th-century heyday of oology (the study of birds' eggs), eggs were avidly collected, traded, and sold. In a typical oologist's catalog of the late 1800's, prices ranged from three cents for an American robin's egg to $5 for those of rarer species, such as the trumpeter swan or great gray owl.

Today wild birds' eggs are no longer items of commerce; rather, we view them as some of nature's most exquisite works of art. Among the loveliest are those of the kingbirds. The eggs of the Cassin's kingbird, for instance, have slightly glossy, light creamy-buff shells that are decorated, mostly at the larger end, with spots of brown, gray, and lavender, like drops spilled from an artist's palette.

The color is applied to the eggs toward the end of the shell-building process by tiny pigment glands in the bird's uterus. Whether solid tones, specks, or mottles, the color presumably serves as camouflage, since only three types of birds lay plain white eggs: hole nesters, whose eggs are hidden; birds that incubate their eggs as soon as the first one is laid; and birds that cover them with down or vegetation before leaving the nest.

Typical flycatchers, kingbirds dart from a perch, snatch their prey in midair, and return.

Recognition. 8–9 in. long. Upperparts and breast dark gray; chin and throat white; belly yellow; tail black with no white edges.
Habitat. Open woodlands, groves, and streamside thickets in mountain regions.
Nesting. Nest is a bulky cup of twigs, rootlets, strips of bark, and string, placed near end of branch 8–40 ft. above ground. Eggs 2–5, white or cream-colored, spotted with brown, gray, and lavender. Incubation 12–14 days, by female. Young leave nest 14 days after hatching.
Food. Insects caught in flight; also a few berries.

Western Kingbird

Tyrannus verticalis

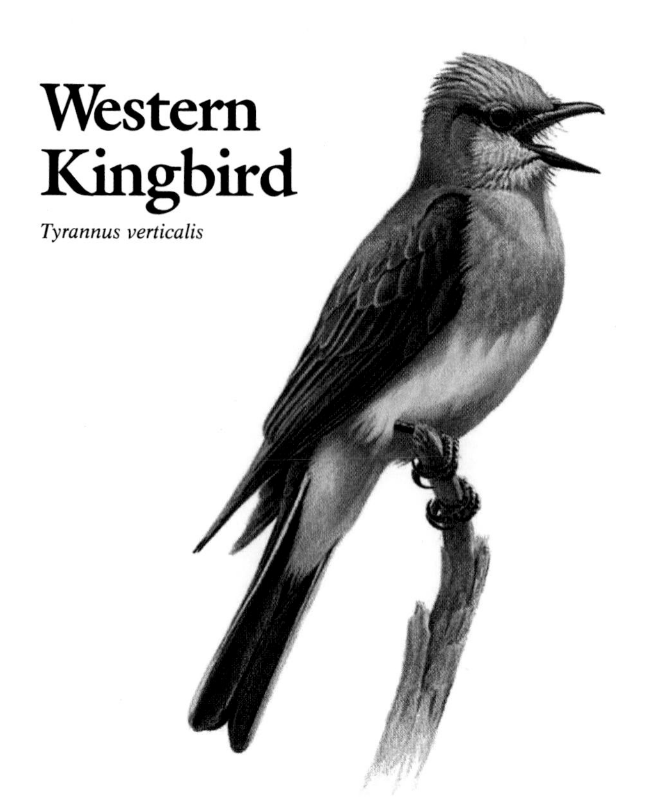

Utility-pole crossarms are among the favored abodes of the western kingbird.

Many birds are totally predictable in their choice of nest sites. No one would look for a downy woodpecker's home anywhere but in a tree cavity, nor for the domed dwelling of an ovenbird elsewhere than on the forest floor. At the other extreme are such birds as western kingbirds, notably adventurous in selecting nesting places.

Before the West was settled, western kingbirds nested almost exclusively in streamside thickets—particularly in sycamores, cottonwoods, and willows, several pairs sometimes sharing the same tree. But with human settlement came irrigation, tree planting, and the rise of towns. Kingbird nests were soon found in everything from oaks to apple trees—whatever grew in windbreak or orchard. Nor did they confine themselves to trees. Making themselves at home both in town and on the farm, western kingbirds took to building in barns, on windmills, in church steeples—even on the crossarms of utility poles.

The birds also made adjustments in their choice of building materials. Wherever located, western kingbird nests are heavily felted with animal hair. But whereas bison wool was the traditional material of choice, western kingbirds nowadays have turned to cattle hair and sheep wool to line their bulky abodes.

Recognition. 8–9 in. long. Upperparts, throat, and breast pale gray; belly yellow; tail black, with white edges.
Habitat. Open country with scattered trees and thickets or groves.
Nesting. Nest is a large, bulky cup of stems, twigs, rootlets, plant fibers, grass, animal hair, snake skins, and string, 8–40 ft. above ground in tree or on utility pole or fence post. Eggs 3–7, white, buff, or pinkish, blotched and spotted with brown. Incubation 12–14 days. Young leave nest 2 weeks after hatching.
Food. Insects caught in flight; also a few berries.

Eastern Kingbird

Tyrannus tyrannus

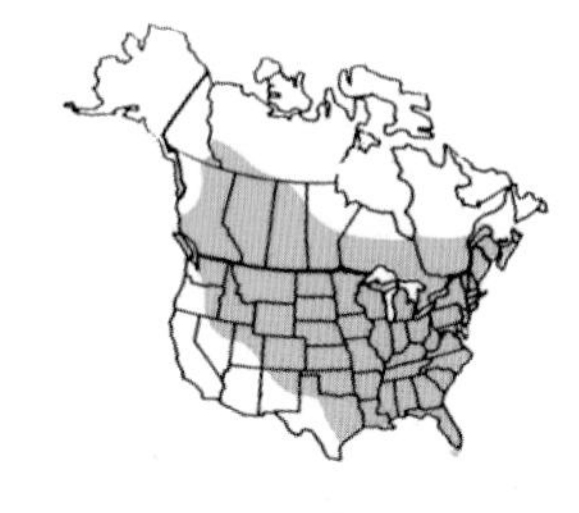

Some birds are by nature unobtrusive, slipping furtively through heavy cover like camera-shy celebrities. But not the eastern kingbird. Living up to its name, it perches with regal aplomb on fences or exposed branches, master of all it surveys. Equally appropriate is its Latin name—meaning tyrant of tyrants—with its connotations of arrogance and absolute power. For the eastern kingbird does not merely sit and survey—it calls a challenge to the whole world, with a furious tirade punctuated by harsh, explosive squeals.

And the kingbird is every bit as fierce as it seems, especially when nesting is under way. Constantly on the lookout for potential threats to eggs or young, it drives squirrels to cover, knocks jays from their perches, and takes off in hot pursuit of distant crows or hawks. Not content to chase such would-be predators, it frequently lands on the back of an intruding bird in midair and inflicts as much punishment as possible.

Ironically, the eastern kingbird undergoes a striking personality change on its South American wintering grounds. There, the bird we know as a noisy, belligerent loner wanders the countryside in quiet flocks, often suffering intimidation and pursuit by native flycatchers determined to defend their own sovereign territory.

Fiercely territorial, eastern kingbirds will attack even much larger intruders.

Recognition. 8–9 in. long. Upperparts, crown, and sides of face black; throat and underparts white, sometimes with gray tinge on breast; tail black, with white band across tip.
Habitat. Roadsides, woodland edges, orchards, and open country with scattered trees.
Nesting. Nest is a large cup of twigs, stems, and grass, up to 60 ft. above ground in tree or on stump in water. Eggs 3–5, white or pinkish, heavily spotted with brown. Incubation about 13 days, by both sexes. Young leave nest about 14 days after hatching.
Food. Insects captured in flight; also a few berries and seeds.

Gray Kingbird

Tyrannus dominicensis

Poised alertly on telephone wires along coastal Florida highways, gray kingbirds suddenly dart forward to nab butterflies, beetles, or other flying insects in midair with an adroit snap of the bill. The same display of skill impressed John James Audubon when he saw his first gray kingbirds during a trip to the Florida keys in the 1830's. They were, he noted, springing into the air "with great velocity" in hot pursuit of insects.

This lightning-quick assault on flying prey—biologists call it aerial hawking—is one of the hallmarks of kingbird behavior. But it is not their only feeding technique. They are capable of hovering long enough to pluck insects from foliage or water, and can also snatch them from leaves or blossoms while in full flight.

While people generally approve such consumption of insects, kingbirds are not always welcome, for they have also been accused of preying on domesticated honeybees. At least some studies, however, suggest that most of the bees they eat are wild species. And the birds certainly consume plenty of insect pests, from wasps to weevils. Like many other flycatchers, kingbirds also relish fruit in season, picked either from a perch or while hovering. Nor do the birds disdain an occasional small lizard—a most unexpected prey for a kingbird.

The feeding techniques of some kingbirds include hovering to pick berries.

Recognition. 8–9 in. long. Gray above with dark patch across eye; mostly white below with grayish wash across breast; tail black and notched; bill stout and black.
Habitat. Mangroves, shrubby marsh borders, and residential areas near coast.
Nesting. Nest is a large, loosely built cup of twigs, grass, and rootlets 3–12 ft. above ground in mangrove, oak, or palmetto. Eggs 3–5, pink, spotted with brown. Length of incubation and nestling periods unknown.
Food. Insects caught in flight; also a few berries.

Immature male

Scissor-tailed Flycatcher

Tyrannus forficatus

Among our most elegant birds, the scissor-tailed flycatcher is known to some as the Texas bird of paradise, although it ranges far beyond the borders of that state. The bird's most notable feature is its deeply forked tail—nine inches long on the male—which it opens and closes, scissors-style, while in flight.

In many ways the scissor-tailed is a typical flycatcher. It characteristically perches on exposed branches or fence posts and darts out to snatch passing insects in mid-flight. And like its cousins, the kingbirds, it is decidedly aggressive, boldly attacking crows and other large birds that trespass on its territory.

But the scissor-tailed differs from most other flycatchers in the extravagance of its courtship ritual. With tail trailing dramatically and gorgeous salmon-pink underwings showing to best advantage, the male performs a display flight that has been likened to an aerial ballet. Zooming up and down as if on a roller coaster, and chattering all the while, he suddenly shoots straight up, then tumbles earthward in a series of backward somersaults. The performance, repeated again and again throughout the breeding season, is a spectacle that never ceases to delight.

Zooming or tumbling, the scissor-tailed provides a thrilling and elegant display of aerial acrobatics.

Recognition. 11–15½ in. long. Pale gray, with black wings, salmon-pink sides, and very long black, forked tail. Male has longer tail than female. Often quite tame.
Habitat. Open country with scattered trees.
Nesting. Nest is a bulky cup of stems, rootlets, twigs, and other plant material 5–30 ft. above ground on horizontal branch of isolated tree or on utility pole. Eggs 4–6, cream-colored or white, spotted with brown. Incubation 12–14 days, by female. Young leave nest about 14 days after hatching.
Food. Insects captured in flight; also grasshoppers, crickets, and a few berries and seeds.

Horned Lark

Eremophila alpestris

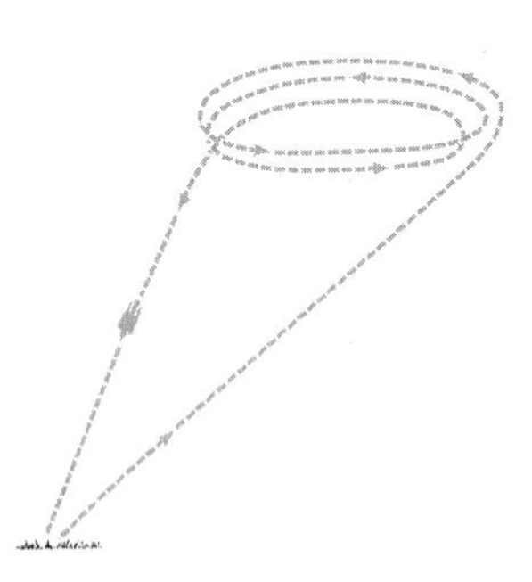

The horned lark's courtship flight is climaxed by a sudden closed-wing drop to earth.

Of the world's 75 species of true larks, only two occur in North America. The Eurasian skylark, famed for its song, was introduced on Vancouver Island in the late 1800's. But its cousin, the horned lark, is found around much of the Northern Hemisphere. In North America it ranges virtually from coast to coast, making it one of the most widely distributed of all our native birds.

Wherever it is found, the horned lark shows a preference for wide-open spaces, from Arctic tundra to prairies and airports. The destruction of forests and the plowing of croplands have helped expand its range, since mated pairs require bare ground on which to build their nests.

Arriving early at the breeding grounds, the male horned lark, like so many birds of open country, performs his courtship display on the wing. Soaring to heights of up to 800 feet, he circles for several minutes, caroling a sweet song before tucking in his wings and dropping with a sudden silent rush to the ground.

Generally inconspicuous during the summer months, the horned lark becomes much more noticeable in winter, when it gathers in flocks that may number in the thousands and heads for open terrain. In Europe the bird is so frequently seen on and around winter beaches that the British call it the shore lark.

Recognition. 6–7½ in. long. Brown, with black markings on head and black band across breast; face white or yellow; tail black, with white outer edges. Black "horns" seldom visible. Feeds on ground; sings in flight.
Habitat. Prairies, fields, dunes, and airports.
Nesting. Nest is a small hollow in ground in shelter of rock or tuft of grass, lined with feathers, grass, and hair. Eggs 2–5, white or greenish, finely spotted with brown. Incubation 11 days, by female. Young leave nest 9–12 days after hatching.
Food. Grain, seeds, insects, and spiders.

Purple Martin

Progne subis

From high overhead, streaming down like summer sunshine, the bubbling chirps and trills of purple martins descend to earth. The happy sounds are so attractive that many people in the eastern part of the country construct elaborate multichambered birdhouses especially designed to attract colonies of martins. (Some of the biggest houses have as many as 200 separate "apartments.") In the West these largest of our swallows shun such man-made structures in favor of old woodpecker holes in dead trees or, in Arizona, in saguaro cacti.

Native Americans forged a friendship with the martins long before Europeans arrived; they hung hollow gourds near their lodges in order to entice the birds to nest nearby. The voracious martins more than earned their keep by snapping up vast numbers of insects on the wing.

Sadly, these handsome birds are no longer so numerous as they once were. The principal culprits in their decline are the introduced European house sparrow and starling. Shortly after the Civil War, house sparrows began usurping martin houses. Though smaller than martins, these pesky intruders start nesting earlier in the season, and once established, are difficult to dislodge. In the West, starlings are the nemesis, monopolizing nest holes before the martins return in spring.

Purple martins are true avian apartment dwellers.

Recognition. 7–8 in. long. Wings pointed; tail notched or shallowly forked; bill very short. Male glossy purplish black. Female duller, with grayish underparts. Usually seen circling in flight, catching insects in air.
Habitat. Farmland, towns, and flat land away from trees near marshes and lakes or in deserts.

Nesting. Nest is a loose mass of grass, feathers, twigs, and other material built in woodpecker hole or, more commonly, in compartments in multi-unit martin houses. Eggs 3–8, white. Incubation about 16 days, by female. Young leave nest 26–31 days after hatching.
Food. Flying insects and airborne spiders.

Tree Swallow

Tachycineta bicolor

Sportive and cheerful, tree swallows enjoy chasing feathers and other windborne objects.

Like green-and-white beads strung between utility poles, garlands of tree swallows line the roadsides in late summer; overhead, twisting, darting masses of even more birds darken the sky. Vacationers at seaside resorts should be pleased, for without these throngs of swallows, the plague of mosquitoes and biting flies from nearby marshes might well be unbearable.

The tree swallows' annual trek to the shore—ocean, river, or lake—begins in July, soon after the young birds take to the wing. Hundreds of thousands gather over insect-rich wetlands for a time of leisure and plenty. For weeks on end, the air vibrates with motion and the swallows' high-pitched "chittering" calls. And then, on a morning when the air carries the unmistakable chill of autumn, suddenly the birds are gone.

Migrating southward, the swallows winter along our seacoasts from the Carolinas to California. But they are hardy birds. During mild winters some remain as far north as Long Island, where they subsist on the waxy fruits of bayberries. People living along the coasts, unaware that the birds were eating, long ago concluded that tree swallows went to the shrubs in order to "wax their wings." This, they explained, accounted for the swallows' shiny plumage.

Recognition. 5–6 in. long. Wings pointed; tail notched; bill very short. Male glossy greenish blue above and on sides of face; underparts pure white. Female duller, almost brownish; underparts grayish. Usually seen in buoyant flight, catching insects in air.
Habitat. Lakes, marshes, and woods near water.

Nesting. Nest is a cup of grass and feathers in woodpecker hole, another tree cavity, or birdhouse. Eggs 4–7, white. Incubation 13–16 days. Young leave nest after 16–24 days.
Food. Flying insects and airborne spiders; also bayberries in winter.

Violet-green Swallow

Tachycineta thalassina

Cloaked in iridescent green and purple, the violet-green swallow is a feathered gem. But more than merely beautiful, it is also a marvel of adaptation. Like swallows everywhere—creatures of the air that feast on insects captured on the wing—it is superbly equipped for a very special way of life.

The bills of swallows are short and wide, like a baseball glove, and they open and close with precise "snaps." Their wings—long, narrow, and pointed—are ideal for reducing the wasting drag of friction. Large hearts and flight muscles meet the demands of energetic flight. And notched or deeply forked tails facilitate fast turns in pursuit of prey. So specialized are swallows for life in the sky that, when grounded, they can barely walk. Their feet are small and weak—suitable for perching but ill-designed for getting about on the ground.

Swallows go wherever they must in search of food. In summer this may mean climbing thousands of feet to feed on insects carried aloft on warm air currents. Later in the year, when temperatures drop, they skim just above the surface of lakes and ponds, snapping up the insects that concentrate over the warm water. And, of course, in fall, the quest for prey takes them southward, to tropical regions that lie beyond the reach of winter.

Swallows sometimes ride currents of warm, rising air to lofty heights in their search for food.

Recognition. 5–5½ in. long. Wings pointed; tail notched; bill very short. Adult green above; sides of face, underparts, and sides of rump pure white. Usually seen in flight over treetops, hunting for insects.

Habitat. Mountain forests and wooded canyons; also over towns.

Nesting. Nest is a cup of grass, feathers, and stems built in woodpecker hole, cavity in cliff, or birdhouse. Eggs 4–6, white. Incubation about 14 days. Young leave nest about 10 days after hatching, but return for several more days. Pairs often nest in loose colonies.

Food. Flying insects.

Northern Rough-winged Swallow

Stelgidopteryx serripennis

If ever a creature set out to be inconspicuous, it could hardly have been more successful than this bird. Dingy brown above and whitish below, it looks like a bank swallow that has neglected to don the distinctive brown breast band. So similar are the two that John James Audubon, who discovered the rough-winged swallow in 1819, did not recognize it as a separate species.

The rough-winged swallow derives its name from the row of tiny barbules along the edge of the outer flight feather on each wing. But while a finger run along this saw-toothed margin can easily feel the roughness, ornithologists still puzzle over the function of the tiny hooks.

Perhaps they produce a distinctive whistle in flight or facilitate access to the sundry nooks and crannies that serve as nest sites. Cracked foundations, bridges, old pipes, and even drain spouts all provide suitable housing for this unobtrusive and unassuming bird. Away from man-made structures, it digs burrows in dirt banks, a talent it shares with the bank swallow. But unlike its gregarious cousin, the rough-winged swallow is a loner. Although several pairs may occupy the same bridge or crumbled foundation if there are enough nest sites, for the most part the rough-winged swallow much prefers a quiet stretch of riverbank that it can call its own.

Preening busily, a rough-winged swallow runs its barbed flight feather through its bill.

Recognition. 5–6 in. long. Wings pointed; tail notched; bill very short. Dull brown above; whitish below, with brownish tinge on throat and breast. Usually found flying near streams, giving harsh *bzzzt* call note.

Habitat. Streams, flooded gravel pits, and road cuts; often near culverts and bridges.

Nesting. Nest is a saucer of twigs, grass, and leaves at end of tunnel in bank, sometimes dug by birds themselves, or in drainpipe, cliff crevice, or similar opening. Eggs 4–8, white. Incubation 12–16 days, by female. Young leave nest 19–21 days after hatching.

Food. Flying insects.

Bank Swallow

Riparia riparia

The swallow with the distinctive brown collar is something of an anomaly. Both adults and young are cloaked in somber brown, their feathers radiating none of the iridescent colors typical of most other North American swallows. Nor do the contrary bank swallows show any interest in altering their nesting habits just for the sake of living closer to man. Unlike their relatives, they are not attracted to barns, bridges, hollow gourds, or even the most elaborate of birdhouses.

Search for them instead along steep-cut riverbanks and sandy excavations, for bank swallows are burrow-nesters. In April, colonies of "sand martins," as they are called in England, return to traditional sites and go through the ritual of moving in. Some may dig new tunnels—up to four feet long—if space can be found on the honeycombed cliff. As often as not, however, the birds simply renovate old burrows. In June their colonies are abuzz with nasal calls and the frantic comings and goings of birds tending their young. By September, however, the swallows have gone south for the winter. Their cliffs stand empty then, mute and mutilated, like the walls of an abandoned fortress that has been pockmarked by a summer-long siege.

Dwelling in tunnels, bank swallows are tempting prey for roving mammals.

Recognition. 5–5½ in. long. Wings pointed; tail notched; bill very short. Dark brown above; underparts white, with band of dark brown across breast; throat white. Looks smaller than other swallows in flight.
Habitat. Fields and marshes, especially near steep sandy stream banks and road cuts.

Nesting. Nest is a loose saucer of grass, stems, rootlets, and feathers at end of tunnel dug into vertical sandbank. Eggs 4–8, white. Incubation 14–16 days, by both sexes. Young leave nest about 21 days after hatching. Nests in colonies, sometimes of hundreds of pairs.
Food. Flying insects.

Cliff Swallow

Petrochelidon pyrrhonata

The famed swallow of Capistrano, the cliff swallow is the bird whose faithful return to the adobe walls of that California mission is cause for annual celebration. But just as noteworthy as its fidelity to nesting sites is its skill at building the nests themselves. On the rocky cliffs from which its name derives, and on the sides of barns, bridges, and other man-made structures of all sorts, these master masons mold gourd-shaped nests in colonies that may number in the hundreds.

The building material is mud, which the birds shape into beadlike bricks and carry in their bills to the nest site. There the bulging structure gradually takes form as the pieces are added carefully one to another. While the barn swallow, an equally talented builder, adds grass to its mud nest, the cliff swallow is a purist—its muddy medium is rarely tainted by foreign material.

If mud is readily available and pilferage by neighboring swallows is kept to a minimum, the nest is completed in about five days. Some grass and stray feathers are then brought in to provide a snug lining, and the birds at last settle down to the serious business of laying eggs and raising a new generation of swallows.

Dependable as the change of seasons, the cliff swallow brings joy to Capistrano.

Recognition. 5–6 in. long. Wings pointed; tail square-tipped; bill very short. Blue-black on back; forehead whitish; throat dark rusty; breast and belly white; rump buff.
Habitat. Open country, especially near barns and other buildings; also over water.
Nesting. Nest is a gourd-shaped structure of mud lined with grass and feathers, in colony under eaves of building or on cliff. Eggs 3–6, white or pinkish, dotted with brown. Incubation 12–14 days, by both parents. Young leave nest about 23 days after hatching.
Food. Flying insects and airborne spiders; occasionally juniper berries and other fruit.

Barn Swallow

Hirundo rustica

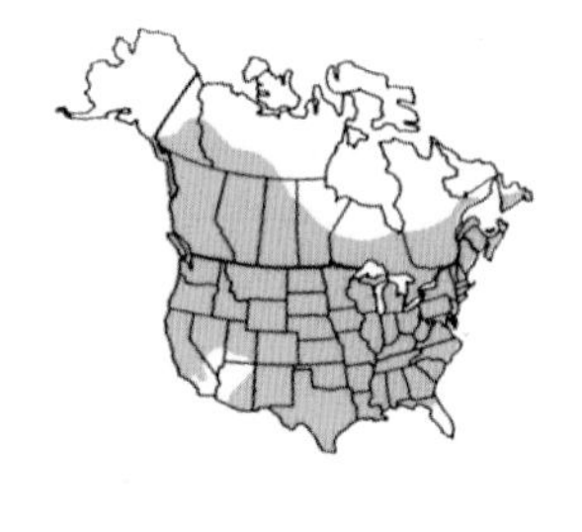

Nesting all across Eurasia as well as much of North America, the agile barn swallow is among the world's most familiar creatures. In England it is so widely known that it is called simply "the swallow."

Although it originally nested in caves and on rocky cliffs, the barn swallow was quick to perceive the advantages of man-made structures. It long ago abandoned natural nest sites in favor of docks, bridges, and of course, barns. So complete is its acceptance of civilization that today, barn swallow nests *not* built in association with some man-made structure are almost unknown.

North American colonists welcomed the swallow, in part because it reminded them of their European homelands but also because of its appetite for insects. Since swallows capture most of their food on the wing, they spend more time aloft then almost any other land bird. Alexander Wilson, America's great pioneering ornithologist, once calculated the number of miles a barn swallow might travel in a lifetime. Assuming a speed of one mile per minute for 10 hours a day and a lifespan of 10 years, Wilson came up with a total of 2,190,000 miles. The estimate undoubtedly is too high, but Wilson's admiration of the bird remains well-founded.

Insects tossed up by a tractor provide easy pickings for the experienced barn swallow.

Recognition. 5½–7 in. long. Wings pointed; tail deeply forked; bill very short. Dark blue-black above; throat dark rusty; rest of underparts buff or pale rusty. Flight fast and direct; often skims close to ground or over water.

Habitat. Open country and marshes, especially near buildings.

Nesting. Nest is a cup of mud pellets lined with grass and feathers, built under eaves of building, resting on a beam, or occasionally in niche on a cliff. Eggs 3–8, white, spotted with reddish brown and lilac. Incubation 13–17 days. Young leave nest 18–23 days after hatching.

Food. Flying insects.

Pinyon Jay

Gymnorhinus cyanocephalus

By November most birds have settled down to a winter routine of simple survival as they wait for spring. But in the large flocks of blue, almost crowlike pinyon jays that roam the pinyon-juniper woodlands of the West, something unusual is afoot. Here and there—quietly, almost unnoticed—one adult passes a small piece of food to another. Were it March or April, one might guess that courtship was getting under way.

In fact, the jays *are* courting. Whereas most American birds pair off in early spring, and nest in late spring or summer, when food is abundant, pinyon jays begin this cycle much earlier. Courtship commences in November, and nest building and egg laying take place by early February, often while snow still covers the ground.

The jays have been liberated from the seasonal schedule of other birds because of their association with the trees for which they are named. Feeding largely on the seeds of pinyon pines, they bury huge numbers of them for retrieval in winter. With this steady supply of food available, females can lay their eggs and males can feed their mates as they sit on the nest, shielding their clutch from the still-frigid winter air.

Assured of plentiful winter food, pinyon jays begin their courtship in late autumn.

Recognition. 9–12 in. long. Stocky and dull blue; bill long, pointed, and black; tail short and square-tipped. Often travels in noisy flocks; walks like crow.
Habitat. Dry forests of pinyon pine and juniper.
Nesting. Nest is a bulky cup of twigs, roots, grass, and hair, 3–20 ft. above ground in tree. Eggs 3–6, bluish or greenish. Incubation 15–17 days, by female. Young leave nest about 21 days after hatching.
Food. Pinyon nuts, seeds, berries, insects, and eggs and young of other birds.

Cactus Wren

Campylorhynchus brunneicapillus

Desert animals disappear in the midday heat. Kangaroo rats and other small mammals spend the day resting in their burrows. Lizards lie underground or in the shade. Quail shelter amid the mesquite, and tarantulas among the rocks. In the harsh, relentless sunlight the cactus wren is among the few animals that demand to be noticed. Noisy and conspicuous, it provides a very welcome sign of life in a still, still world.

In both looks and behavior, the cactus wren is an unusual wren. Husky rather than diminutive, it is distinctly larger than its North American cousins, and the boldly spotted breast and throat are unique. The peppery bird holds its tail pointed downward rather than in the cocked position of the smaller wrens, and often it hunts insects from the ground instead of perching on a plant. Foraging among the rocks and brush, it uses its beak to turn over leaves, pebbles, or any other small objects under which its prey may be taking shelter from the sun.

And, of course, the cactus wren lives among cactus—an unusual home for any bird, be it wren or not. Woven among the spines, the nest is built in a prickly situation that deters predators but not the parent birds. No one quite understands how the wrens manage to flit so freely among the treacherous spines.

The cactus wren's prickly home is soft inside but nearly impregnable to predators.

Recognition. 6–8½ in. long. Largest of our wrens. Streaked and spotted with brown and white above; crown dark rusty; eyebrow white; underparts spotted. Harsh *jug-jug-jug-jug* call often heard before bird is seen.
Habitat. Deserts with cactus and thorny shrubs; nearby gardens.

Nesting. Nest is a ball of grass and twigs with side entrance, 2–9 ft. above ground in cactus or thorny shrub. Eggs 3–7, white or pinkish, spotted with brown. Incubation 16 days, by female. Young leave nest about 21 days after hatching.
Food. Insects, berries, and seeds.

Rock Wren

Salpinctes obsoletus

Darting and bobbing over barren, sunbaked slopes, sputtering challenges to any intruder, the rock wren is apt to be heard before it is seen. But even when it is silent, hidden at home in a crevice or sheltered hole in a cliff face, its presence is unmistakably revealed.

Before constructing its nest of grasses, rootlets, and animal hair, the rock wren "paves" the bottom of its nest cavity with pebbles, and sometimes even extends a telltale walkway from the entrance for a distance of eight to ten inches. The stones, laid with the precision of a master craftsman, are as much as two inches long—an amazing size for such a small bird to carry. And the number of paving pieces is equally impressive. The floor of one nest was found to be covered with more than 1,600 bits of stone, metal, shell, and bone.

The rock wren's passion for paving is unexplained. Since many of the walkways end in a mound of pebbles at the nest's entrance, some researchers believe they are intended as barriers against predators. Others suggest that they are markers that help the bird find its home amid the visual monotony of its habitat. Or they may simply be foundations that protect the nests from dampness. Whatever the reason, it remains the rock wren's secret.

Theories abound, but no one can really explain the rock wren's curious penchant for paving.

Recognition. 4½–6 in. long. Stocky, with long, slender bill. Grayish brown above; underparts whitish with buff on flanks; tail long, with buff corners, black strip near end.
Habitat. Rocky slopes, mesas, and buttes.
Nesting. Nest is a cup of dry grass lined with hair, in crevice among rocks, on cliff, or in wall of adobe building. Nest usually has "path" of pebbles leading to entrance. Eggs 4–8, white, spotted with brown and purple. Incubation by female; incubation and nestling periods unknown.
Food. Insects, spiders, and earthworms.

Canyon Wren

Catherpes mexicanus

Though found across much of the West, canyon wrens are in large part mystery birds. Anyone traveling in western canyons or cliff areas has probably heard their song—a sweet, loud, descending series of clear whistles echoing from the rock walls. Yet the birds themselves more often than not escape detection; like many other secretive species throughout North America, canyon wrens are better known by sound than by sight.

Creeping mouselike over rocks and probing into crevices with their long bills, canyon wrens search for insects and other small prey. But the details of their diet are largely unknown. And while they nest in shallow caves or rock openings, and occasionally in abandoned buildings, other facets of their lifestyle—from incubation period to age at first flight—have yet to be discovered.

Canyon wrens are thus reminders of just how incomplete our knowledge of the natural world really is. And in a time when so much scientific research requires complex equipment, they also remind us that there still are areas—including the study of birdlife—where the most vital skill is patient observation. By unraveling even the smallest detail in a creature's life history, anyone can participate in the continuing process of scientific discovery.

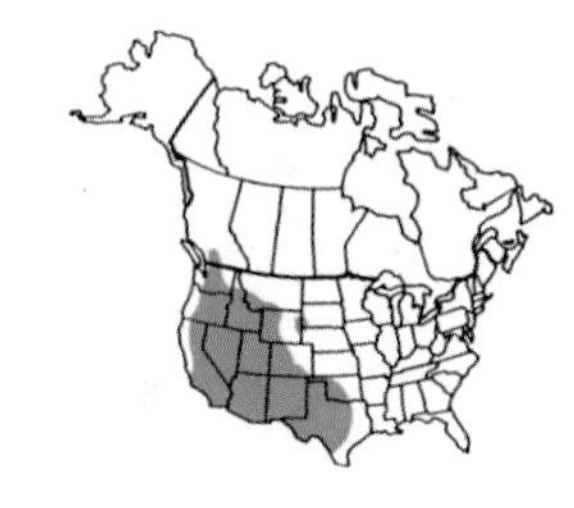

Hunched against the ground, a canyon wren uses its bill to probe rocky crevices for food.

Recognition. 5½–6 in. long. Stocky with long, slender bill. Dark rusty above and on belly; throat and breast clear white. Usually secretive, hiding among rocks and plants; loud, clear, descending song often reveals bird's presence.

Habitat. Steep, rocky, watered canyons, cliffs, buttes, and river gorges.

Nesting. Nest is an open cup of twigs, moss, stems, and leaves, on ledge or in cavity in cliff, canyon wall, or building. Eggs 4–7, white with fine, dark brown speckling. Incubation and nestling periods unknown.

Food. Insects and spiders.

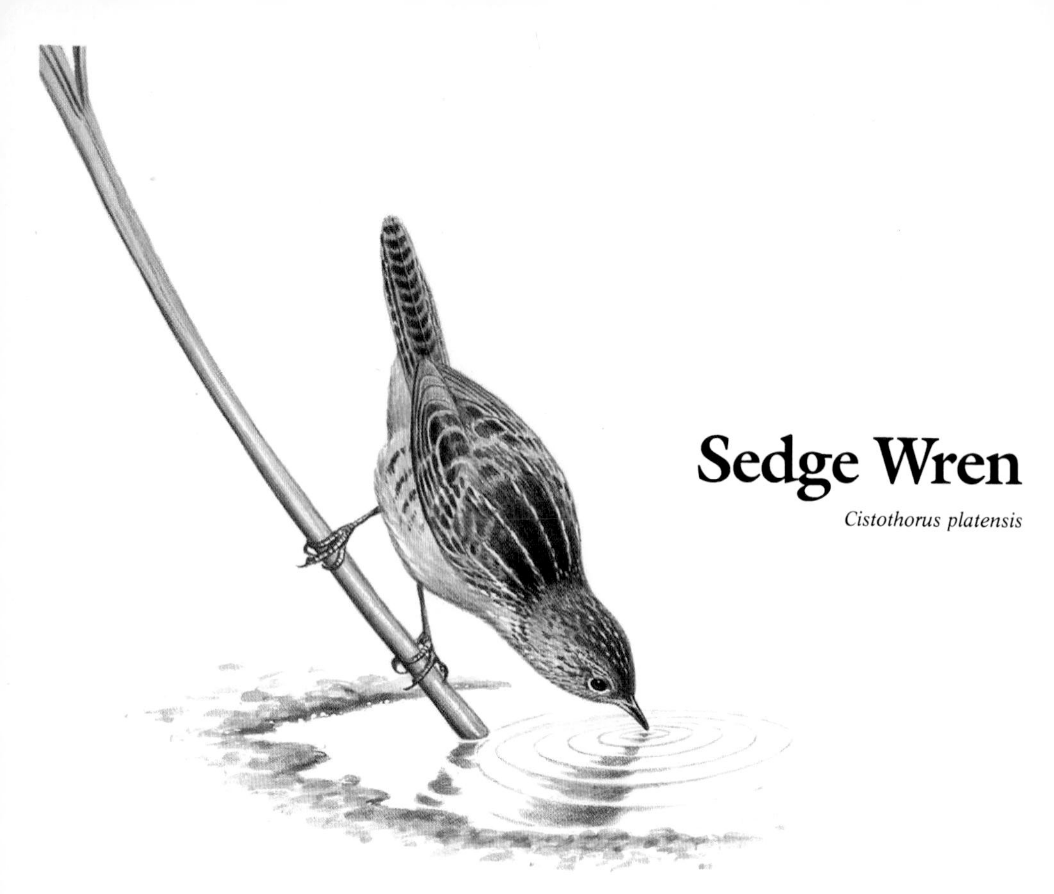

Sedge Wren

Cistothorus platensis

Birdsongs are invisible fences, fences made of music, for a bird sings to keep rivals out of the territory where it breeds. If one wren approaches the territory of another, the owner will sing out a challenge—and the interloper often sings back in response.

Wrens are a particularly musical family, and sedge wrens have a huge repertoire, of more than a hundred different types of songs. Only rarely in a territorial squabble do opposing sedge wrens duplicate one another's song. A marsh wren, in contrast, will match the song type of its rival with extreme precision. It is unlikely that marsh wrens can memorize music faster; more probably, they have had more time to learn their neighbors' songs.

In many ways, such as the males' taking more than one mate, sedge and marsh wrens are very much alike (so alike, in fact, that the sedge wren used to be called the short-billed marsh wren). But their habitats are quite different. Marsh wrens, which nest in wetlands that change little from year to year, keep returning to the same site to breed, and neighboring birds develop the same repertoire. Sedge wrens live in wet meadows that may flood or dry out from one year to the next, forcing the birds to move to a different place—one where they are far less likely to know the songs of their new neighbors.

Though few ever become homes, the dummy nests of male sedge wrens are a key to courtship.

Recognition. Tiny, 4–4½ in. long. Stocky with short bill. Buff overall; crown and back streaked; face plain, with no eyebrow. Very secretive. Often appears yellowish when flushed from grass.
Habitat. Higher parts of grassy marshes, wet meadows, and sedges; seldom found in reedy marshes favored by marsh wren.

Nesting. Nest is a ball of grass and sedges with side entrance, 1–2 ft. above water in marsh. Eggs 4–8, white. Incubation 12–14 days, by female. Young leave nest after about 2 weeks.
Food. Insects and spiders.

Marsh Wren

Cistothorus palustris

Marsh wrens embrace the breeding season with a burst of energy matched by few other species. Seek them out in late spring, and most likely you'll find them nest building, constructing not just one nest each but sometimes a dozen or even 20 or more.

With so many nests about, it might seem that marshes would be overstocked with baby birds. But in the case of this species, and of its close relative the sedge wren, most nests remain eggless, serving instead as a focus for courtship. If an interested female approaches, the male escorts her on an inspection tour of the property. Should she begin to nest, he may build a second courtship center to attract another mate—a polygamous arrangement more common in the West than in the East.

To the human eye, a marsh wren's behavior seems both zealous and jealous. Should a neighboring bird leave its nest unattended, the marsh wren is quick to sneak in and destroy the eggs, apparently not to eat them but to eliminate the competition. Fellow wrens are not the only victims; marsh wrens will even plunder the nests of such bigger birds as red-winged and yellow-headed blackbirds. Yellow-heads seem well aware of the raiding habit. Tit-for-tat, they ward off danger by chasing marsh wrens away from places where blackbirds have set up home.

Pugnacious and jealous, male marsh wrens plunder the nests of nearby birds.

Recognition. 4–5 in. long. Back dark rusty with white stripes; crown dark brown; eyebrow bold and white; underparts white, with rusty tinge on flanks. Male's rapidly trilled song is distinctive.
Habitat. Marshes of cattails, bulrushes, and tall reeds; not found in drier grassy marshes inhabited by sedge wren.

Nesting. Nest is a ball of cattail leaves and sedges, 1–3 ft. above water in marsh. Eggs 3–10, pale brown, finely dotted with dark brown. Incubation 13–16 days, by female. Young leave nest 11–16 days after hatching.
Food. Aquatic insects and small snails.

Eastern Bluebird

Sialia sialis

Well-placed birdhouses help the bluebird survive competition from other hole nesters.

Once as familiar as the robin, the eastern bluebird in many areas has become more memory than reality. Few people have ever seen even one pair of bluebirds flitting about an orchard; fewer still have witnessed the antics of a whole bluebird family splashing in a shallow bath. And we no longer hear their murmured, low and lovely song.

The bluebirds' decline has been due in part to natural causes; they have always had to contend with killing winter weather and competition from house wrens and tree swallows for nesting holes. But their major problem has been the introduction of alien house sparrows and starlings, whose overwhelming numbers have crowded the gentle bluebirds out of nearly all available nesting places.

The best hope for survival seems to be bluebird trails—long lines of bluebird houses that are being put up along rural roadsides. Carefully monitored to keep out the intrusive sparrows and starlings, the trails have brought a local resurgence in bluebird numbers wherever the birds still exist. But hundreds of additional trails must be organized and thousands of houses put up before the eastern bluebird is truly reestablished. Only then will all of us delight in the sweet-singing little bird that seems to wear the blue sky on its back and the warmth of the sun on its breast.

Recognition. 5–7 in. long. Male bright blue on back, wings, and tail; throat and most of underparts orange-rusty; belly white. Female similar, but paler, back tinged with brown; throat rusty, not gray as on western bluebird.

Habitat. Farmlands, orchards, open woodlands, and sparse trees on mountain slopes.

Nesting. Nest is a cup of grass and stems 3–20 ft. above ground in woodpecker hole, birdhouse, or other cavity. Eggs 3–7, pale blue or white. Incubation 13–16 days, by female. Young leave nest 15–20 days after hatching.

Food. Insects, spiders, and berries; visits feeders for peanut butter.

Western Bluebird

Sialia mexicana

In the Rocky Mountain states and westward, the western bluebird replaces its cousin, the eastern bluebird, with whom it shares a similar lifestyle. Both birds are residents of open country and woodland edges, and both nest in cavities, natural or man-made. Like eastern bluebirds, westerns set out to find a nest hole early in the breeding season. And once the first brood has been fledged, the parents often begin to raise a second family immediately.

During the warmer months of spring and summer, western bluebirds feed primarily on insects, which they spot from elevated perches or by hovering near the ground. Then as fall arrives, they turn their attention to gleaning a variety of fruits, including those of junipers, elderberries, and mistletoes.

Some biologists speculate that by depositing undigested mistletoe seeds on trees—which may eventually suffer death and decay of some of their limbs from the growth of the parasitic mistletoe, whose roots reach deeply into the wood—the bluebirds may create new nest holes for themselves. If so, they would unwittingly be playing a conservation role like that of the nest-box programs that have been so vital to all our bluebirds.

Parasitic mistletoe may benefit western bluebirds, which dig new nesting holes in rotted tree limbs.

Recognition. 5–7 in. long. Male bright blue on head, wings, tail, and throat; center of back and most of underparts orange-rusty; belly white. Female similar but grayer; throat gray, not rusty as on eastern bluebird.
Habitat. Open pine forests and farmlands; also brushy deserts in winter.

Nesting. Nest is a loose cup of grass and feathers 5–40 ft. above ground in woodpecker hole, other natural cavity, or birdhouse, usually in clearing. Eggs 3–8, pale blue. Incubation and nestling periods unknown; probably similar to those of eastern bluebird.
Food. Insects, spiders, and berries.

Mountain Bluebird

Sialia currucoides

Greeting the sunrise with song, mountain bluebirds fall silent once the sun has come up.

Before dawn, when the deep valleys and alpine meadows of the western mountains are still wrapped in darkness and not even the highest peak has caught its first beam of morning light, mountain bluebirds begin their daily caroling. Theirs is a gentle chorus, warbled in clear, short phrases that resemble the robin's song, their soft bluebird voices pitched just a little higher than the sweet murmur of their eastern cousins. They are, the Navajos say, the heralds of the rising sun, for they sing while the sky grows lighter and brighter. And then, soon after the moment of sunrise, they stop.

Beloved as they are, however, mountain bluebirds are in trouble. Like all bluebirds, they need sheltered places to nest—clefts and crannies, nooks and crevices—and these are being taken from them by stronger, scrappier immigrant birds. Bitter winters also decimate their numbers, and late-spring frosts kill off the insects on which they depend for food.

Thus, while the mountain bluebirds' song has always been reserved for early risers, in recent years even *they* hear less and less frequently the murmured chorus of these beleaguered sky-blue birds.

Recognition. 6–7½ in. long. Male bright blue all over. Female grayish; wings and tail blue. Has longer wings than eastern or western bluebird; often hovers in flight.
Habitat. High, open mountain country in summer; plains and prairies in winter.
Nesting. Nest is a loose cup of stems and rootlets in woodpecker hole, natural tree cavity, or cavity in earth bank or among rocks on cliff; sometimes uses birdhouse. Eggs 4–8, pale blue or white. Incubation about 14 days, by both sexes. Nestling period unknown.
Food. Insects and berries.

American Robin

Turdus migratorius

Spring never truly arrives in the northern states until the robins fly back from the south. Red-winged blackbirds may be singing in the marshes, and dandelions glowing among the meadow grasses. But it is never *spring* until a robin bounces sturdily across the lawn or sings a cheery roundelay from an apple tree in the orchard.

Taking off in enormous numbers from their wintering grounds, the northbound flocks of robins break up as they progress toward their summer homes, with smaller flocks veering off in various directions. Those flocks in turn break up into even smaller ones while, all along the way, individual birds keep dropping off as they reach the general areas where they themselves were fledged.

It is really the older males who usher in the returning season. Jaunty in jet-black caps and brick-red breasts, they head north in the first of the flocks, seeming to tow spring behind them as they fly. Before long their paler-colored mates begin to arrive, while the one-year-old birds trail along by themselves even later. But no matter what the timing or the weather, when the first robin bounces across the lawn, spring *is* here and all is right with the world.

Its head cocked, a robin seems to listen for worms but in fact hunts by sight.

Recognition. 9–11 in. long. Male has bright rusty or brick-red breast; head black with white around eye; back gray; adult bill yellow. Female similar but duller. Immature resembles female, but has black spots on breast, vague wing bars.
Habitat. Lawns, gardens, parks, open woods, forest borders, and farmland.

Nesting. Nest is a sturdy cup of mud, grass, and twigs, 3–25 ft. above ground in tree, dense shrubs, or on building. Eggs 3–6, bright blue. Incubation 12–14 days, by female. Young leave nest 14–16 days after hatching.
Food. Insects, earthworms, and berries.

Northern Mockingbird

Mimus polyglottos

Related to the catbird and the brown thrasher—notable musicians in their own right—the mockingbird has been hailed as the King of Song. Its scientific name means "mimic of many tongues"—an apt description for a creature that can imitate more than 30 birdsongs in rapid succession, along with such other sounds as squeaky hinges, barking dogs, and chirping crickets.

Warbling liquid melody, usually from a high, commanding perch, the mockingbird continues its serenade from morning to evening and through many a moonlit summer night, frequently repeating a phrase many times in rapid succession. The adaptive value of such vocal virtuosity is unclear. Perhaps, as has been suggested, it is simply a manifestation of an overwhelming urge to sing.

Like many another artist, the mockingbird has a temperamental side. Fiercely territorial, it regularly attacks crows and grackles, and even takes on cats during the breeding season. Audubon painted the mocker defending its nest against a rattlesnake; whether or not he portrayed a documented incident, he did dramatize the bird's aggressive nature. For, like the famed minstrel of Greek mythology, Orpheus, the mockingbird proves itself to be at once both gently musical and boldly fearless.

Even cats do not intimidate an angry mockingbird during its nesting season.

Recognition. 9–11 in. long. Slender and long-tailed. Gray above, paler below; 2 wing bars on each wing; large white patches under wings visible in flight; tail black, with white borders. Runs on ground or skulks in bushes; often swings tail from side to side and quickly opens wings, flashing its white feathers.

Habitat. Lawns, gardens, open woods, farmland, brushy deserts, and streamside thickets.
Nesting. Nest is cup of twigs, other material, usually 4–12 ft. high in shrub or tree. Eggs 3–6, blue or green, with brown spots. Incubation 12 days, by female. Young leave nest after 10–12 days.
Food. Insects, spiders, fruit.

Sage Thrasher

Oreoscoptes montanus

In the classic western novels of Zane Grey, men who needed to get out of town, whether heroes or villains, would ride out "into the sage," rolling deserts of gray-green plants that seemed to go on forever. The clumps of sagebrush stood low enough to enable a man on horseback to see for miles, but were tall enough so that someone on the ground could hide. There a cowboy could rest, scan the horizon, and sniff the air, "fragrant with the breath of sage." And if he listened for it, he could hear the song of the sage thrasher.

Smaller than other thrashers, the sage thrasher spends its time hunting quietly for insects on the ground like a robin. Approached too closely, it melts like an outlaw into the wilderness of sage. But if left undisturbed, at dawn and dusk it climbs to the top of a clump of sage and sends its sweet, warbling song echoing across the land.

Since winter brings harsh conditions to the northern parts of sagebrush country, most sage thrashers migrate south for the season. The majority move into the juniper-covered mesas, canyons, and foothills of the Southwest. But occasionally there are mavericks: like rodeo cowboys traveling east for a performance, roving sage thrashers sometimes wander all the way to the Atlantic Coast.

Like an army bugler, the sage thrasher trills reveille at dawn and taps at day's end.

Recognition. 8–9 in. long. Shaped like a robin. Grayish brown above; whitish with black streaks below; tail feathers dark, with white tips; eyes yellow; 2 white wing bars visible in fall. Usually seen perched on top of shrub or sagebrush.
Habitat. Brushy deserts and sagebrush plains.
Nesting. Nest is a solid cup of twigs, stems, and sage leaves, 2–3 ft. above ground in shrub; sometimes on ground. Eggs 4–7, blue or greenish blue, coarsely spotted with brown. Incubation by both sexes; incubation and nestling periods unknown.
Food. Insects, spiders, and berries.

Bendire's Thrasher

Toxostoma bendirei

On July 28, 1872, Lieut. Charles E. Bendire of the U.S. Army was exploring the desert near Fort Lowell, Arizona. A former divinity student from Germany, Bendire was also an avid student of birds. So when he came across an unfamiliar female thrasher, he shot the bird and sent it to the Smithsonian Institution in Washington, D.C.

There the specimen was examined by Elliott Coues, one of the foremost ornithologists of the time. Coues, too, was puzzled; he showed the bird to a colleague, who concluded that it was a female curve-billed thrasher. But Coues had his doubts. Suspecting that the bird might be unknown to science, he asked Bendire for more details. Yes, Bendire replied, it was indeed something different, with distinctive eggs and habits. A second specimen, this one a male, was enough to convince Coues. He described the thrasher as a new species and named it in honor of Charles Bendire.

Bendire later became an honorary curator at the Smithsonian and earned a reputation as an ornithologist in his own right. When he died in 1897, he was buried at Arlington National Cemetery. But he is best memorialized today by the desert bird that bears his name—the bird that many say has the most beautiful song of all the thrashers.

Adult Bendire's

Young curve-billed

Its short beak makes a Bendire's easy to mistake for a young curve-billed thrasher.

Recognition. 9–11 in. long. Slender, with long tail and short bill. Grayish brown above; underparts paler, with faint streaks; tips of outer tail feathers white; eyes yellow.
Habitat. Brushy deserts, valleys, dry farmlands, and nearby gardens.
Nesting. Nest is a cup of twigs lined with grass, stems, and rootlets, 2–12 ft. above ground in cactus, thorny desert shrub, or small tree. Eggs 3–4, pale greenish or bluish, spotted with pale brown and purple. Incubation and nestling periods unknown.
Food. Caterpillars, beetles, and other insects.

Curve-billed Thrasher

Toxostoma curvirostre

The desert is a rough place. Just as its plants are armed with spines and thorns, many of its animals, from horned lizards to scorpions, have a sharp-edged appearance and a prickly nature. Even some of the songbirds seem tough. The curve-billed thrasher, for instance, carries a heavy truncheon of a bill, while its glaring orange eyes give it an expression that is fierce, even slightly crazed.

And the impression of toughness and craziness seems confirmed when the thrasher suddenly flies headlong into the bristling heart of a cholla cactus—among the spiniest of all the desert's plants. But the bird is undaunted by the fearsome armament: it builds its nest and raises its young right in the center of the cactus.

The curve-billed thrasher is not punctureproof of course. Agility is the key to its success: it is practiced at slipping in among the cholla branches without hitting the spines, and once inside, it is safe from most predators, whether they are hawks swooping in from above or snakes and rodents creeping up from below. Occasionally a young thrasher climbing from its nest is impaled on the spines. But for the most part, the cholla's formidable array of spines furnishes a haven of security for the bird that has learned to live with such hazards.

Perched to avoid the sharp cactus spines, a curve-billed thrasher pours out its song.

Recognition. 10–12 in. long. Slender, with long tail and long, sickle-shaped bill. Pale grayish brown; underparts vaguely spotted; tips of outer tail feathers white; eyes yellow or orange. Often secretive, but forages on ground in gardens.
Habitat. Deserts with brush or cactus, thickets, dry farmlands, and nearby gardens.

Nesting. Nest is a loosely made cup of thorny twigs, 3–12 ft. above ground in cactus or small tree. Eggs 2–4, bluish green, speckled with brown. Incubation 13 days, by both sexes. Young leave nest 14–18 days after hatching.
Food. Insects, seeds, berries.

California Thrasher

Toxostoma redivivum

Experts at foraging, California thrashers sweep the ground with their bills in search of food.

In brushy California thickets, the movements of a phantom creature often are betrayed by the rustling of dry leaves on the ground. But even a close approach seldom produces a clear view—just a glimpse of a long-tailed, dusty-brown bird scurrying away. More often than not, the skulking noisemaker is the California thrasher, hunting for food and making a getaway in its usual fashion.

A bird that spends most of its life on the ground, the thrasher forages amid dense brush and normally escapes danger by running instead of flying. Its scimitar-shaped bill is a perfect tool for finding food. Using it as a rake, the bird brushes leaf litter aside to scare insects out of hiding; using it as a pick, it pounds the soil to dig out grubs.

The thrasher builds its home in a bush where it can reach the nest by hopping up through tangled branches. When the chicks leave the nest, they too travel on foot. Unlike most songbirds, young California thrashers usually spend several days walking, running, and clambering up branches before they even begin to practice flying. And they never become very good at it. Leaving mastery of the air to other birds, the California thrasher is content with a more down-to-earth way of life.

Recognition. 11–13 in. long. Slender and long-tailed, with long, black, sickle-shaped bill. Dark brown above with blackish mustache; paler below; throat whitish. Usually seen on ground. Common, but shy and often overlooked.
Habitat. Brushy hillsides, thickets, and neighboring gardens.

Nesting. Nest is a cup of twigs and rootlets, built near ground in low shrub or tree. Eggs 2–4, pale blue or green with light brown spots. Incubation 14 days, by both sexes. Young leave nest 12–14 days after hatching.
Food. Insects, spiders, seeds, and fruit.

American Pipit

Anthus rubescens

Beyond the tree line in the Far North lies the Arctic tundra, a bleak, often marshy plain where soil only inches below the surface remains permanently frozen. Farther south another kind of tundra—alpine tundra—occurs in isolated patches on mountaintops. Although lacking permafrost, it is every bit as forbidding as the Arctic variety. And one of the masters of this demanding habitat in summer is the American pipit: whether hundreds of miles beyond the Arctic Circle or on a few acres atop a mountain, it nests on tundra, and nothing else.

The American pipit is superbly attuned to a world where summer is brief and all living things must reproduce quickly before winter returns. Ready to begin nesting as soon as they arrive on the breeding grounds, males often set up territories while snow still lingers all about. Nests are sunk into the soil, oriented toward the sun and away from the wind, and the females sometimes simply refurbish old nests instead of building new ones. With nesting completed by early September, chill winds and plunging temperatures send the birds southward or down the slopes to milder climates. Few of us ever see pipits on the tundra, but when we do see them, we see birds finely adapted to one of the world's harshest environments.

American pipits walk instead of hopping, perhaps to avoid being blown about by tundra winds.

Recognition. 5–6½ in. long. Sparrowlike but with slender bill; tail has white outer edges; legs dark. Fall birds grayish, streaked brown above; underparts whitish, heavily streaked. Spring birds plain grayish above; underparts buff. Seldom perches; walks on ground, bobs or swings tail.
Habitat. In spring and fall migration and in winter, bare fields, dunes, rocky shores, plowed cropland; nests on Arctic and alpine tundra.
Nesting. Nest is a cup of grass and hair sunk in ground near tussock. Eggs 3–7, whitish with brown blotches. Incubation 14 days, by female. Young leave nest about 14 days after hatching.
Food. Insects, spiders, and small snails.

Sprague's Pipit

Anthus spragueii

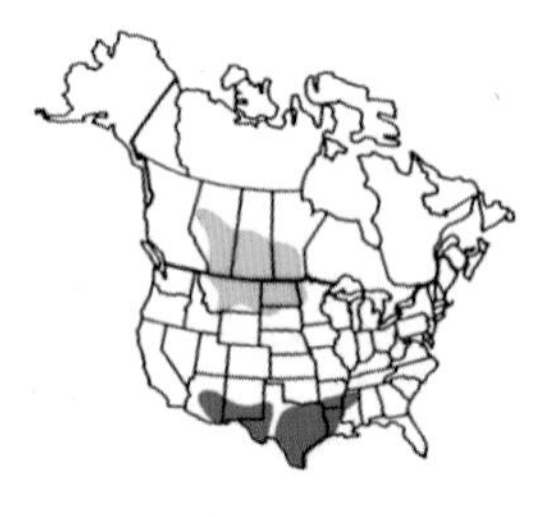

Birders who frequent the prairie in June know that most of the action takes place near ground level, in the carpet of fragrant grasses and colorful wildflowers. Overhead there seems to be nothing but the sun. Listen carefully, however, and you may hear a song ringing down from somewhere in the sky—a soft, chiming sound that spirals downward while its source remains unseen.

Finally, against the glare, it may be possible to pick out the singer, a tiny speck hundreds of feet in the air. With wings and tail fully spread, the pipit circles as it pours out a cascade of song. Then the music stops, and the bird plummets like a rock until, at the last moment, it spreads its wings and swoops to a graceful landing.

Although we may enjoy the spectacle, the Sprague's pipit does not sing for our benefit. In the nesting season, birdsong has purposes of its own. Staking his claim to a piece of land, the male bird sings to warn rival males to keep out, and to beckon a mate to join him. Singing is thus a kind of advertising. To get the message out, the singer usually chooses a high perch—the top of a tree, perhaps, or a utility pole—making himself as conspicuous as possible. But because the Sprague's pipit lives on the open prairie, it must adopt a different ploy. And so it takes to the sky. Singing high above his chosen territory, the courting pipit guarantees that his message will not go unheeded.

Sprague's pipits sing in full flight, claiming a territory with song.

Recognition. 5–6½ in. long. Sparrowlike but with slender bill; tail has white outer edges; legs yellowish or pale pinkish. Grayish brown, with heavy streaks above; underparts whitish, with fine streaks on breast; eyes black and conspicuous. Walks on ground but does not bob tail.
Habitat. Plowed fields and shortgrass prairie.

Nesting. Nest is a cup of grass sunk in ground. Eggs 3–6, whitish, blotched with purple and brown. Incubation period unknown. Young leave nest 10–11 days after hatching.
Food. Insects and seeds.

Phainopepla

Phainopepla nitens

The desert mistletoe is a common sight in Arizona, growing as a parasite on the branches of mesquite, palo verde, and other trees and shrubs. And where the mistletoe is found, so too is the phainopepla, for its red berries are one of the bird's staple foods.

The relationship between bird and berry is a symbiotic one. Just as the bird depends on the plant for food, the mistletoe depends on the bird to carry its seeds from one tree to another. For mistletoe seed must lodge on a live branch in order to send roots into the bark and begin its growth. The red color of the fruit attracts the phainopepla, and the pulp provides a nutritious reward.

Other birds eat mistletoe fruits, but the phainopepla has a special, thin-walled gizzard that removes the pulp without injuring the seed as it passes through the digestive system. Moreover, the seed has an adhesive coating that makes it stick wherever it lands. Most fall to the ground, but enough of them adhere to the branches of new host trees to ensure another generation of the species.

Thus, the best way to find phainopeplas in the Arizona desert is to look for clusters of mistletoe. If the time of year is right—between November and April—the mistletoe will be covered with bright red berries, and these elegantly crested birds will be on hand to eat them.

The distinctive white wingtips of the male phainopepla reveal themselves in flight.

Recognition. 6½–7½ in. long. Slender, with crest and long tail. Male glossy black; wings have white patches visible in flight. Female gray with pale wing patches.
Habitat. Open woodlands or brushy deserts.
Nesting. Nest, built by male, is a cup of plant fibers and twigs held together with spider webs; 4–50 ft. above ground in tree. Eggs 2–4, whitish, spotted and scrawled with brown and black. Incubation 14–16 days, by both sexes, but mainly by male. Young leave nest about 19 days after hatching.
Food. Berries and insects.

Northern Shrike

Lanius excubitor

Feeding mainly on insects in summer, the shrike depends on mice and birds for winter food.

Though classified as songbirds—most of which pose no threat to anything larger than an insect—northern shrikes are fearsome predators indeed, with a diet that includes mice, snakes, frogs, and other birds. They knock their avian victims out of the air with sharp blows from their bills, then bite their necks to sever the vertebrae.

Shrikes dispatch land animals in much the same way, for they have a formidable crushing weapon—a toothlike structure on the upper mandible with a matching notch on the lower bill, similar to the bills of falcons. When dining on insects, they catch their prey in midair and swallow it in a single gulp. Insects make up about two-thirds of the shrikes' summer diet, while the birds are nesting in the Far North; but in winter, of necessity, they eat mostly mice and birds.

Shrikes are also called butcher birds—and with good reason. Before eating their vertebrate prey—and often large insects as well—they characteristically hang the animal, head up, on a thorn or barbed-wire fence, or wedge it into the fork of a branch. There they either down their catch immediately or allow it to hang, as in an open-air pantry, for a day or two—sometimes even for a week or more—before returning for a meal.

Recognition. 9–11 in. long. Large-headed, with dark hooked bill. Pale gray above, white below, with faint bars on breast; wings black with white patches; tail black with white outer edges. Black mask through eyes does not extend across bill. Loggerhead shrike is slightly smaller, darker, with solid black bill and no barring on breast.

Habitat. Open country with scattered trees.
Nesting. Nest is a bulky cup of twigs, 5–20 ft. above ground in tree or shrub. Eggs 2–9, whitish or greenish, spotted with brown. Incubation 15 days, by female. Young leave nest about 20 days after hatching.
Food. Chiefly insects, mice, and birds.

Loggerhead Shrike

Lanius ludovicianus

At first glance the story seems an all too familiar one: the population of loggerhead shrikes is diminishing, especially in the northeastern part of their range, largely because of the growing number of people in the area and the decline in open land. As it turns out, however, this particular story may have an ironic twist.

Except during the nesting season loggerheads are loners who never inhabit an area in large numbers—and it might not just be other shrikes they're snubbing. Unlike many birds driven from their habitats by the side effects of human settlement, loggerheads may find people themselves the undesirable element. Settlers' livestock seem an acceptable replacement for herds of grazing buffalo (which carried insects in their hair and stirred others out of the ground), and barbed wire fences provide numerous prongs on which shrikes can hang their prey. But the humans who accompany these changes are a different matter.

One Canadian study showed that each breeding pair of shrikes needs 25 acres of territory for nesting. Within that area there must be hunting perches with a clear view of open country, well-thorned bushes on which to impale prey, and an absolute minimum of human interference. In effect, it appears that even when other conditions are just right, a loggerhead will balk at making its home too close to any people who are doing likewise.

The distinctive tooth on its bill underscores the shrike's fearsome reputation as a hunter.

Recognition. 7–9½ in. long. Plump and large-headed, with black, hooked bill. Gray above, white below; wings black with white patches; tail black with white outer edges. Black mask extends from sides of head through eyes and across bill. Northern shrike is larger and paler, with longer bill and mask that does not meet over bill.

Habitat. Open country, thickets, and deserts.
Nesting. Nest is a cup of twigs lined with animal hair and plant fiber, usually 3–20 ft. above ground in bush or tree. Eggs 4–7, whitish or buff with brown spots. Incubation 10–12 days. Young leave nest about 20 days after hatching.
Food. Insects, mice, and small birds.

European Starling

Sturnus vulgaris

Brazen and belligerent, starlings often rob other hole-nesting birds of their homes.

"I'll have a starling [that] shall be taught to speak nothing but 'Mortimer.' " If William Shakespeare had known the fateful consequences of that line (from *Henry IV*) he might have stayed his pen. Some 300 years later, in 1890, a Shakespearean zealot named Eugene Scheifflin turned loose 60 starlings imported from England in New York City's Central Park. The tens of millions of starlings that plague North America today are descended from these birds and 40 more released the following year—part of a larger scheme by Scheifflin to transplant all the birds mentioned in Shakespeare to the New World.

These unwelcome immigrants consume the fruit and grain of farmers, befoul the buildings and cars of city dwellers, and usurp the homes of bluebirds, purple martins, and other popular hole-nesters. Once, in 1960, an airliner was taking off from Logan Airport in Boston when one of its engines inhaled a tightly packed flock of starlings, causing a crash that killed all 62 people aboard.

Their knack for destruction makes it easy to forget that starlings devour insect pests with a voracity unmatched by any native species. There is also no denying that, in their iridescent spring finery, they are hardly unpleasant to behold, and their skills as vocal mimics are worthy of the stage. Small wonder, then, that so keen an observer of life as William Shakespeare took notice!

Recognition. 7–8½ in. long. Stocky; bill long and pointed; tail short and square-tipped. Glossy black with yellow bill in summer. Speckled with white, bill blackish, in winter. Immature dusky brown. Waddles about on lawns, often seen probing ground with bill in search of food.
Habitat. Farms, cities, towns, and gardens.

Nesting. Nest is a loose cup of grass and twigs in woodpecker hole, natural cavity, or birdhouse. Eggs 2–9, white, pale blue, or greenish. Incubation 12 days, by both sexes. Young leave nest 21 days after hatching.
Food. Insects, spiders, worms, fruit, and seeds.

Lucy's Warbler

Vermivora luciae

In the Sonoran Desert of the Southwest, creatures mold their lives along the rim of existence. A harsh yellow sun bakes the land by day; cold starlight glimmers through the chilly night. Desert creatures whose lives are bonded to daylight confine their activity to dawn and dusk. Caution is ever a mind-set in the desert; reticence, a virtue. These are the standards that the Lucy's warbler lives by.

This small and active bird flits through the mesquite thickets that flank valleys and washes. The call is sharp and dry, the song elusive. Its plumage is as pale as a Sonoran sunrise; the crown and rump are washed with a blush of sandstone—soft desert colors, nothing to catch and hold a searching eye. It may be hard to believe that this furtive gray shadow is related to the colorful warblers of the forest. But it is, and where they could not survive, the Lucy's warbler thrives.

The bird was discovered in 1861 by Dr. J. G. Cooper at Fort Mojave in Arizona. Desert warbler or mesquite warbler would have been equally appropriate names for the bird. Instead, Cooper chose to honor Spencer F. Baird, an eminent zoologist at the Smithsonian Institution and a key figure in arranging that government surveys of the Southwest would include scientific studies of the region's wildlife and natural history. Mingling respect and affection, Cooper named the slight, shy, discreet warbler for Baird's 13-year-old daughter, Lucy.

The crisp desert air resonates with the song of a Lucy's warbler greeting the Sonoran sunrise.

Recognition. 4–4½ in. long. Small and thin-billed. Gray above; underparts whitish; rump rusty. Immature similar, but tinged with rusty. Flicks tail while foraging among leaves.
Habitat. Brushy deserts amid mesquite and cottonwoods, and thickets along streambeds.
Nesting. Nest is a compact cup of bark and stems lined with hair and fur, up to 15 ft. above ground in tree cavity such as a woodpecker hole, crevice in bark, or old verdin's nest. Eggs 3–7, white, finely dotted with brown. Incubation and nestling periods unknown.
Food. Insects.

Prairie Warbler

Dendroica discolor

Despite its name, the prairie warbler is most at home in trees and bushes, feeding on insects.

What's in a name? The common nighthawk is not a hawk; the black-headed gull's head is brown; and the screech-owl's quavering whistle is far from a screech. The prairie warbler, too, labors under a misnomer. Its habitat is brushland, not grassland, and its breeding range lies east of the tree-poor prairies.

Little is confusing about this bird's identity, however, for it boasts a veritable checklist of field marks—those trademarks of nature that help both to differentiate species and to group them by genus and family. The markings it shares with other warblers give it a family resemblance to its cousins. But as with any cousin, the prairie warbler has those family traits put together in a combination that is distinctively its own.

Like the blue-winged warbler, the male prairie warbler has a thick black eyeline. Its curled mustache stripe is only slightly less prominent than a Kentucky warbler's. Bold, black spots, recalling those of the Kirtland's warbler, march in ranks down the bird's yellow sides. And the twin wing bars that appear on warblers of almost every kind are prominent here too. Although the prairie warbler is a less famous tail bobber than the Kirtland's warbler or the palm warbler, it is just as much a master of the sprightly bob that makes so many warblers such a joy to watch.

Recognition. 4½–5 in. long. Upperparts olive-green with faint rusty stripes on back, black streaks on face and sides; underparts yellow. Female and immature similar but duller. Usually feeds above ground; pumps tail.
Habitat. Scrubby woods and overgrown pastures; mangroves in Florida.

Nesting. Nest is a cup of plant down, bark, and grass, 2–15 feet above ground in bush, small tree, or mangrove. Eggs 3–5, white or faintly greenish, spotted with brown. Incubation 12–14 days, by female. Young leave nest 8–10 days after hatching.
Food. Insects and spiders.

Palm Warbler

Dendroica palmarum

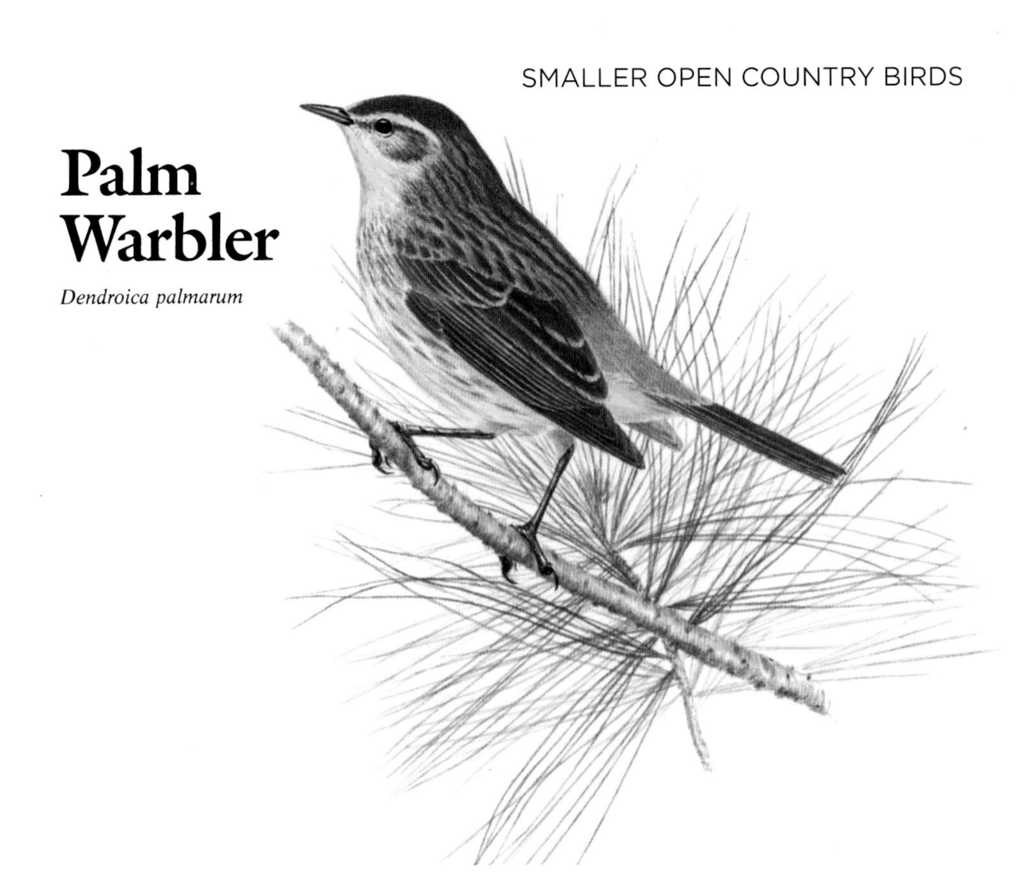

They can't wait for spring, these ground-loving members of the warbler tribe. Long before the leaves come out, as much as a month before the first waves of other warblers flood forests and meadows, palm warblers are inching their way north, testing the limits of spring. Even the hardy yellow-rumped warblers, which winter much farther north, are a poor second in the race to summer grounds.

These impatient heralds of spring could well have been called wagtail warblers. Walking much of the time with the brisk gate of wagtails or larks—in contrast to the distinctive hop of most warblers—they flit their tails up and down with the persistent monotony of a metronome. Bog warbler would have made a fine name too, for the birds' long journey northward will usually end at the brushy edge of a spruce bog in Canada.

As it turned out, however, the palm warblers' somewhat misleading name was drawn from their southern wintering ground, where a specimen was first discovered on a Caribbean island in 1789. Ironically, palm warblers care little for palm trees, unless the various scrub palmettos are considered trees. In reality they show greater fondness for lawns, parks, pastures, and golf courses—all especially plentiful in Florida, where palm warblers are among the most common wintering birds.

Golf courses and other close-cropped lawns are favorite winter habitats for the palm warbler.

Recognition. 4–5½ in. long. Upperparts brown; crown chestnut; eyebrow and throat yellow; underparts grayish or, in some birds, yellow, streaked with chestnut. Immature much duller. Walks on ground; pumps tail frequently.
Habitat. Weedy fields and low bushes; nests in northern spruce-fir bogs.

Nesting. Nest is a cup of bark, grass, and stems, in hummock of moss or 1–2 ft. above ground in low bush. Eggs 4–5, white with spots and blotches of brown. Incubation about 12 days. Young leave nest about 12 days after hatching.
Food. Insects and berries.

Common Yellowthroat

Geothlypis trichas

If nature had set out to design a feathered mouse, she could hardly have done better than the yellowthroat—or, as the tiny warbler is still called in many places, the Maryland yellowthroat. The bright-eyed bird with the bandit's mask is a skulker, a fugitive figure creeping from shadow to leaf. In wet tangles of Spanish moss or the brushy edges of marshes, it may be maddeningly difficult to see, frustrating even the most patient observers.

But where patience fails, guile sometimes succeeds. There is a trick to seeing yellowthroats (and many other birds as well). Soft squeaking noises made by kissing the back of a thumb will usually draw the bird into view. Some may fairly charge from hiding with a raspy "growl," but most are simply curious to see where the strange noise is coming from.

Despite this elusiveness, yellowthroats are so common throughout their southeastern range that their song is among the easiest to learn. But no two yellowthroats seem to sing it quite the same way—and in the 19th century no two bird students seemed to hear it the same way, either. One observer transcribed the song as *Whitit'iti! Whitit'iti!* Another writer suggested several versions, including *Follow me, follow me* and *I beseech you, I beseech you*—as well as the less poetic but more plausible *Wichity, wichity,* which is the way most people hear the song today.

Among the most numerous of its family, the common yellowthroat favors wild, marshy terrain.

Recognition. 4–5½ in. long. Male has bold black mask, bright yellow throat, no wing bars or streaks. Female and immature similar, with throat bright yellow, but lack mask. Skulks in thick vegetation.

Habitat. Brushy swamps, wet thickets, tangles of weeds or berry bushes, and marshes.

Nesting. Nest is a bulky cup of grass, stems, and leaves, in tuft of grass or reeds or 1–3 ft. above ground in dense shrub. Eggs 3–6, white or cream-colored, speckled with brown, gray, and blackish. Incubation 12 days, by female. Young leave nest 8 days after hatching.

Food. Insects and spiders.

Yellow-breasted Chat

Icteria virens

Audubon thought this bird should be classed with the manakins. Others have suggested vireos, honeycreepers, and tanagers. But on at least one point all would agree: if the ranks of birds can claim a jester, the chat owns clear title. Not only does it make a mockery of ornithology's thoughtful ranking system, but this "raucous polyglot" hardly seems able to take itself seriously.

Its bill is bulbous, its proportions portly, and its bespectacled expression falls somewhere between guileful and comic. Any lingering pretense of dignity disappears as soon as the bird opens its mouth. The chat's repertoire is an astonishing string of chuckles and gurgles, wheezes, warbles, squeals, and squawks. When it tires of its own nonsense, the chat may mimic the rattle of a kingfisher, the whistle of a yellowlegs, the cawing of a crow, or even the bleat of a car horn.

But this feathered buffoon's premier routine is, unquestionably, its disappearing-into-the-hedge trick. With a ventriloquist's voice, the chat seems to hurl abuse from all directions at once. Approach too close and the noise stops, only to begin anew, with a razzing chortle, when the observer moves off in another direction. A quick glance may find the bird perched atop a bush, but most observers will be lucky to catch a fleeting glimpse of tail feathers.

This largest woodland warbler is an ungainly flier who sometimes mimics the songs of other birds.

Recognition. 6½–7½ in. long. Large for a warbler. Olive-green above; face blackish with white spectacles; underparts bright yellow; bill thick and dark. Skulks in thickets.
Habitat. Brushy thickets and weedy tangles.
Nesting. Nest is a cup of grass, leaves, and strips of bark, 2–8 ft. above ground in dense bush or tangle of vines. Eggs 3–6, white, speckled and dotted with brown and lilac. Incubation about 12 days. Young leave nest 8–11 days after hatching. Sometimes nests in loose colonies of several pairs.
Food. Insects and berries.

Pyrrhuloxia

Cardinalis sinuatus

The brightly colored fruit of a Christmas cactus is a favorite winter treat for the pyrrhuloxia.

At first they look like a pair of strangely drab parrots desperately misplaced from their tropical homes. Then suddenly the male is alert. The feathers on his head flash upward to form a flame-colored crest as he calls shrilly, batting his tail to the beat of each cry. Now it is clear that these are pyrrhuloxias, the cardinals of the West.

The name *pyrrhuloxia* comes from two Greek words—*pyrrhos* (flame colored) and *loxos* (slanting)—which refer to the reddish highlights in the bird's feathers and to its curved, parrotlike beak. For birders in the Southwest, where the pyrrhuloxia's range overlaps that of the northern cardinal, this beak is an important field mark in distinguishing the two species' females, which otherwise are nearly identical in their buff plumage.

And it is not only in appearance that pyrrhuloxias resemble cardinals. Flitting over the mesquite in early spring after the winter flocks have disbanded, brightening the landscape with his flashing feathers, a male pyrrhuloxia may nibble on a Christmas cactus, then dart away to perch near his mate and delicately place the bright red fruit in her beak. This courtship feeding, as such gallant behavior is called, is also typical of northern cardinals, both species showing deep loyalty to their mates throughout the year.

Recognition. 7½–8 in. long. Crested, with stout, yellow, parrotlike bill. Male mainly gray with red on crest, face, throat, breast, wings, and tail. Female similar, but more grayish buff.
Habitat. Thorny desert brush, streamside thickets, and nearby gardens.
Nesting. Nest is a small, neat cup of twigs, bark, and grass, 5–15 ft. above ground in mesquite or thorny desert shrub. Eggs 2–5, whitish or greenish with brown spots. Incubation 14 days, by female. Young leave nest 10 days after hatching.
Food. Fruits, seeds, and insects.

Blue Grosbeak

Guiraca caerulea

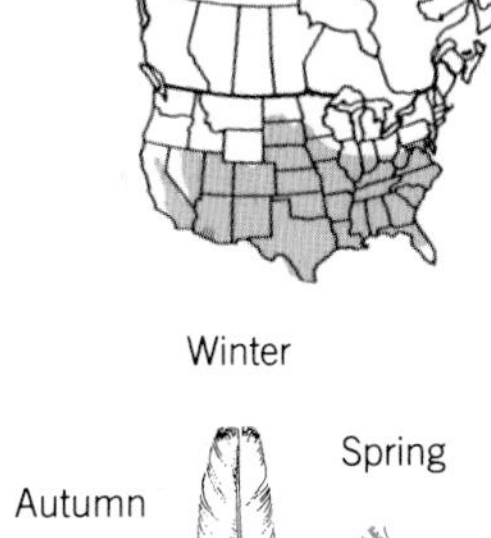

Birds get their colors in a variety of ways. Brown, yellow, and red hues come from pigments in the bird's feathers. But blue is special: there is no pigment for that, and when a single feather of a blue bird is lit from behind or pulverized, it does not appear blue at all. The color we see actually arises from the way light is filtered by thin layers on the feather's surface.

There are also different ways in which blue birds get to be blue. Like most birds, male blue grosbeaks and indigo buntings molt in the fall, losing their vibrant spring plumage and growing the brown winter feathers that resemble those of their females. Indigo buntings will molt again in spring, regrowing their summer garb. But blue grosbeaks do not molt anew. Instead, their trick of changing color is "feather wear." The gradual abrasion of the brownish tips of their body feathers during the winter reveals the feathers' blue bases by spring. Starlings also change color through feather wear, and even the complex beauty of snow buntings and bobolinks is hidden in fall, waiting to be revealed by months of winter's buffeting.

Winter wear and tear on feather tips leave the blue grosbeak with his splendid springtime garb.

Recognition. 6–7 in. long. Stout-billed. Breeding male dark blue with 2 rusty wing bars. Female and fall male brown with 2 rusty wing bars. Forms flocks in late summer and fall before migration.
Habitat. Brushy thickets and weedy tangles along roads and streams; edges of woodlands; visits croplands after nesting season.

Nesting. Nest is a cup of grass, rootlets, and stems, 2–14 ft. above ground in dense tree, shrub, or tangle of vines. Eggs 2–5, very light blue, unmarked. Incubation 11 days, by female. Young leave nest 13 days after hatching.
Food. Insects, spiders, grain, and fruits.

Indigo Bunting

Passerina cyanea

Indigo buntings often raise the young of cowbirds who have left their eggs in the buntings' nests.

Like an irrepressible church-choir tenor, the indigo bunting is joyful but not especially melodic. The observation that "he sings just as well as he can, and I believe just as loud as he can" was all the praise one well-mannered naturalist could muster. Yet what this songster lacks in quality he makes up in persistence, singing well into the late summer, when most birds have long since fallen silent.

To the casual listener, one indigo sounds like another. In fact, their songs vary greatly in melody and sequence—so much that these fiercely territorial birds know each other by song patterns, and react more strongly to strangers than to neighbors with whom they share a fragile peace.

Nonetheless, biologists have found these signature songs repeated by several birds within a single colony. It appears that some young males copy the songs of more established neighbors, and that these impersonators have better luck holding territory and gaining a mate than their less imitative peers. Mistaken for his song's originator, a talented copycat can avoid battles with other male indigo buntings in the area, since the elder he imitates has already done his fighting for him.

Recognition. 4½–5½ in. long. Bill conical and sparrowlike. Breeding male bright blue, darkest on head. Female and fall male mainly buff; breast faintly streaked with brown.
Habitat. Brushy pastures and woodland edges.
Nesting. Nest is a well-made cup of grass, leaves, bark strips, and paper, 5–15 ft. above ground in bush, small tree, or clump of weeds. Eggs 3–6, bluish, unmarked. Incubation about 13 days, by female. Young leave nest 8–10 days after hatching.
Food. Insects, seeds, and berries.

Lazuli Bunting

Passerina amoena

Whether it is spotted along the humid Pacific Coast or up on a high, dry slope in the Rocky Mountains, a lazuli bunting is likely to be busy foraging in one of those rich, weedy areas wildflower guides call waste places. Low plants such as Spanish lettuce and alfilaria present no feeding problem, but the tall, delicate grasses are another matter. A lazuli will flutter to the head of an oat or canary grass stalk, grasp it in one foot, and ride it to the ground, there to consume the seeds at leisure.

The conical beak typical of lazulis and other finches is often taken to indicate a strictly vegetarian diet. In most cases, though, that simply is not true. Nearly half of the lazuli's summer diet is thought to be made up of insects—grasshoppers, caterpillars, beetles, bees, and ants. Most of these are caught on or near the ground, but occasionally buntings pluck tidbits from beneath leaves and even chase their prey through the air like flycatchers.

Insects are also the food the parents give to their chicks, and young grasshoppers come with a special benefit. As they mature through the nestling period, the grasshoppers provide progressively larger meals to satisfy progressively larger baby bunting appetites.

Clever feeders, lazuli buntings alight atop grasses and ride the edible seed heads to the ground.

Recognition. 4½–5½ in. long. Bill conical and sparrowlike. Breeding male has light blue head and upperparts, with 2 white wing bars; breast orange-rusty; belly white. Female and fall male grayish brown with 2 pale wing bars; breast buff.
Habitat. Brushy hillsides, thickets, open woodlands, chaparral, and sagebrush.

Nesting. Nest is a cup of grass, 1½–10 ft. above ground in thick shrub, clump of ferns, or tangle of vines. Eggs 3–5, pale blue, unmarked. Incubation 12 days, by female. Young leave nest 10–15 days after hatching.
Food. Insects and seeds.

Painted Bunting

Passerina ciris

Brightly plumed painted buntings were once treasured as pets. Today their capture is illegal.

Spanish speakers call it *mariposa,* the butterfly, while in the South it is traditionally known as nonpareil for its peerless beauty. According to an American Indian legend, when the Great Spirit was giving all the birds their colors, he ran short of dye, so he gave the very last one—the painted bunting—a coat of many colors made from dabs of whatever was left.

Most of us still tend to think that all color is made up of pigments. Red, for example, is a pigment that reflects the light waves we perceive as red, while absorbing every other wavelength. Not all hues in bird plumage, however, come from pigmentation. Blue and iridescent colors are actually a kind of optical illusion produced by the microscopic structure of the feathers. The painted bunting's striking blue head—or the rich, deep hue of the blue grosbeak's body—results from the way light is scattered by particles in the feathers and then reflected. (Iridescent colors, on the other hand, such as the blue of hummingbirds and the sheen of some blackbirds, are conjured up in a different way. Like the rainbows on a soap bubble, they are produced by light waves playing on a thin, transparent film.) In a sense, the Indian legend was on the right track: the Great Spirit had run out of pigment—so he simply invented a whole new method of creating colors.

Recognition. 5–5½ in. long. Male has blue head, green back, and red rump and underparts. Female mainly green; underparts tinged with yellow. Usually shy and difficult to see.
Habitat. Woodland and streamside thickets, roadsides, and gardens.
Nesting. Nest is a cup of grass, stems, and leaves, normally 3–6 ft. above ground in tangle of vines, dense bush, or tree; in hanging Spanish moss may be some 25 ft. above ground. Eggs 3–5, pale blue spotted with brown. Incubation about 12 days, by female. Young leave nest about 2 weeks after hatching.
Food. Seeds, insects, and spiders.

Dickcissel

Spiza americana

Maps of breeding and wintering ranges depend on the assumption that birds stay put—or at least that they hold to consistent patterns. But some species don't play by the rules. Dickcissels, for example: in colonial times, they seem to have been restricted to grasslands from South Dakota and Indiana down to Texas and Louisiana. Then, early in the 19th century, they spilled over the Appalachians, establishing a satellite breeding population in New England and the mid-Atlantic states. By the late 19th century, however, this population had mysteriously disappeared, while at the same time the species was *expanding* its midcontinent range to the north, west, and east. Where the dickcissels may turn up next is anyone's guess.

These wanderers are equally adventuresome in migration. Most simply fly south through Central America to wintering grounds in Venezuela and then return. But a few invariably turn up in the Southeast every spring, and in the Northeast every fall. And some of them just keep going. About 100 miles off the coast of Nova Scotia lies Sable Island, famous as the last landfall for lost vagrants heading into the open Atlantic. A recent study disclosed that the most common vagrant to reach that remote island was, not too surprisingly, the wayward dickcissel.

Notorious for their wanderings, dickcissels have been found 100 miles offshore in the Atlantic.

Recognition. 6–7 in. long. Sparrowlike. Breeding male streaked brown above; breast yellow with black bib; eyebrow yellow; shoulders rusty with chestnut wing patch. Female and winter birds lack black bib; have yellow on face and sometimes on breast. Female's wing patch less prominent.
Habitat. Croplands and pastures.

Nesting. Nest is a cup of grass and stems, usually concealed on ground, but may be 2–14 ft. above ground in bush or tree. Eggs 3–5, pale blue, unmarked. Incubation about 13 days, by female. Young leave nest about 9 days after hatching. Usually 2 broods a year.
Food. Seeds, grain, and insects.

Green-tailed Towhee

Pipilo chlorurus

A coyote lopes across a gentle mountain slope, then pauses to look and listen. Suddenly, what seems like a fluffy chipmunk darts from the base of a bush. The coyote lunges to give chase, but only briefly: the creature has already vanished into the underbrush.

More than one green-tailed towhee has been mistaken for a chipmunk, not only by coyotes but by humans as well. These higher-elevation finches are masters of the "rodent run," a distracting bit of behavior that serves them well during nesting. Green-tails nest on or near the ground in sagebrush country, and when an intruder draws near they don't simply sit tight, hoping to go unnoticed. Instead, they literally spring into action. Dropping from their nests with wings closed, they hit the ground on the run, their tails cocked high, looking for all the world like a streaking chipmunk or other small mammal. And that is precisely their aim: a predator convinced it has just flushed a meal is unlikely to pause to search for a nearby bird's nest.

The smallest and most migratory of the towhees, green-tails leave their nesting range entirely as fall sets in, most bound for Mexico. But a few regularly wander off toward the East, and many a winter feeder as far away as New England has been brightened by the presence of one of these handsome strangers.

Resembling a fleeing chipmunk, a green-tailed towhee darts from its nest to distract a predator.

Recognition. 6–7 in. long. Back and tail greenish, especially in strong light; crown reddish brown; throat white with black streaks at sides. Underparts and sides of face gray. Shy; escapes by darting behind tree trunk or into brush.
Habitat. Dry, brushy slopes, open pine forests, and sagebrush plains.

Nesting. Nest is a deep cup of grass, bark, and twigs, built on ground or up to 2½ ft. above ground in dense shrub or cactus. Eggs 2–5, white, densely spotted with brown. Incubation and nestling periods unknown.
Food. Seeds, berries, and insects.

Canyon Towhee

Canyon Towhee *Pipilo fuscus*
California Towhee *Pipilo crissalis*

With their muted brown plumage and sedentary ways—pairs mate for life and remain year-round in the same territory—canyon towhees might appear to be rather bland, even boring. But first impressions can be misleading, and an observer with a bit of luck may see a very different side of these sedate birds: the "squeal duet."

As towhees move about their territory searching for seeds and insects, the male and female occasionally are separated. Soon one flies to the top of a shrub and is joined by its mate. Then the ceremony begins. The two face each other, posturing and bobbing their heads in an explosion of loud squeals that reaffirm their lifelong bond. Once they have finished their display, the devoted mates resume foraging—but they'll probably repeat the duet several more times before the day is done.

Such overt affection seems only fitting for birds with the towhees' prolific nesting habits. From the interior Southwest, where the peaceful canyon towhees are found, to the West Coast with its more aggressive California towhees (until 1989 the two were classified as one species, the brown towhee), pairs nest two or three times a season, driving away the young from the previous clutch when a new one is laid. Members of the season's last clutch may stay with their parents longer, the whole group moving about as a family until late fall or early winter.

Aroused by his own reflection, a towhee will attack a window in defense of his domain.

Recognition. 8–10 in. long. Mainly brown. California towhee dull gray-brown with buff under tail. Canyon towhee paler with rusty crown; often a dark spot on breast.
Habitat. Brushy and weedy hillsides and canyons, open woodlands, gardens, and lawns.
Nesting. Nest is a cup of twigs, grass, and stems on ground or up to 35 ft. above ground in dense shrub or tree. Eggs 2–6, pale blue-green, spotted with brown and black. Incubation 11 days, by female. Young leave nest 8 days after hatching. Sometimes 3 broods a year.
Food. Seeds, grain, and insects.

Cassin's Sparrow

Aimophila cassinii

A Cassin's sparrow ends his flight song with head held high, tail cocked, and talons flared.

They are, without question, among the most undistinguished-looking sparrows in all of North America. But when the courting season arrives, these desert stoics make up in effort for what they lack in sartorial style. Residents of the dry, grassy plains, where scattered shrubs, cacti, or yucca supply the males with singing perches, Cassin's sparrows remain undisturbed by the intense heat of spring and early summer that drives most visitors to cover. Nonetheless, they are not imperturbable. When challenged by a neighbor, a male defending his territory soars from a perch to begin his skylarking flight song. Reaching a peak of some 15 or 20 feet, he utters a long, sweet trill and floats downward with head upheld, tail fanned, and feet reaching threateningly for the ground. Most intruders are duly impressed. Not content with daytime singing at this highly charged time of year, the males can be heard at almost any time of night.

When nesting is complete and the young are fledged, however, Cassin's sparrows seem to vanish. They do move south from the northern parts of their breeding range; but even where they remain they can be exceedingly difficult to find. A dull blur flashing across an open space or a mouse-like rustle amidst the tall grass may be all there is to see until the skylarking begins anew in spring.

Recognition. 5–6 in. long. A grayish, long-tailed sparrow. Gray-brown and streaked above; underparts plain buff; eyebrow buff. Best distinguished by fluttery song flight over nesting area.
Habitat. Dry, shortgrass western plains with scattered bushes.
Nesting. Nest is a cup of stems and grass hidden on ground in tuft of grass or at base of shrub, or very close to ground in tangle of cactus or brush. Eggs 3–5, white, unmarked. Incubation and nestling periods unknown.
Food. Insects, seeds, and flower buds.

Rufous-crowned Sparrow

Aimophila ruficeps

Naming birds can be a risky business, especially if the creature in question hasn't actually been seen first-hand. When the naturalist John Cassin of Philadelphia first described this sparrow in 1852, he named it the western swamp sparrow. A more ill-suited name could hardly have been coined for a bird that lives on dry, rocky hillsides. But Cassin had not seen the bird in its habitat—only specimens received from a California collector, whose brief notes suggested that the bird lived near water.

Bird names, both English and Latin, are meant to impart information. But sometimes, through faulty observation or slight changes in the names themselves, they can confuse more than they clarify. Philadelphia vireos and Tennessee warblers, for example, only migrate briefly through the areas whose names they bear. We have a summer tanager but no winter one; it was originally called the summer red bird to distinguish it, not from any winter bird, but from the red bird that resides year-round, the cardinal.

Some names express the namer's perplexity. Wrentits are neither wrens nor tits (as chickadees are called in Old World parlance), but apparently the namer felt they were somehow both. Other namers seem to favor words that most people don't know. *Rufous,* for example, is merely an obscure synonym for "reddish."

Slow to take flight, the rufous-crowned sparrow would rather feed—and flee—on foot.

Recognition. 5–6 in. long. Brown above with rusty crown; face with light eyebrow, white eye-ring, white line below eye, and black whisker mark; underparts plain gray.
Habitat. Rocky hillsides and open, often rocky, woodlands with little undergrowth.
Nesting. Nest is a cup of twigs, strips of bark, and grass, sunk into ground under edge of rock, in tuft of grass, or at base of sapling, or 1–3 ft. above ground in small tree or sagebrush. Eggs 2–5, pale bluish, unmarked. Incubation and nestling periods unknown.
Food. Insects and seeds.

American Tree Sparrow

Spizella arborea

Why the word "tree" was added to the name of this ground-loving bird remains a mystery. It spends the winter in open and brushy fields across the United States and the summer on the southern edge of the Canadian tundra—where it is more likely to nest atop a mossy hummock or in a tuft of grass than in the scruffy conifers or willows that punctuate these vast plains. Even during its winters to the south the tree sparrow shows no special interest in trees and can more often be found flitting about brushy thickets and hedges.

But the name "sparrow," which is used for small, brownish, seed-eating birds, fits without a wrinkle: beyond a doubt the tree sparrow is among our most prodigious seed-eaters. Weighing scarcely an ounce, it devours an average of one-quarter of an ounce of weed seeds every day. Experts have estimated that tree sparrows wintering in Iowa consume more than 875 *tons* of weed seed each year. And this hardy creature is one of the most common winter birds in the central and northeastern states. Even when winds are at their bitterest and temperatures hover near zero, flocks of tree sparrows dot the frozen fields and prairie, tittering musically as they browse among the heads of weeds, grasses, and sedges that poke above the snow.

American tree sparrows scavenge for feathers and lemming fur, both prized nest liners.

Recognition. 5½–6½ in. long. Upperparts brown and streaked, with 2 white wing bars; crown chestnut; pale gray below with black spot in center of breast; bill dark above, yellow below.
Habitat. Brushy fields, marshes, and roadsides; nests in Arctic thickets.
Nesting. Nest is a cup of grass, strips of bark, stems, and rootlets, on ground or 1–5 ft. above ground level in stunted willow or other shrub. Eggs 3–6, pale blue with brown spots. Incubation about 13 days, by female. Young leave nest about 10 days after hatching.
Food. Seeds, insects, and spiders.

Chipping Sparrow

Spizella passerina

Almost all New World sparrows line their nests with hair if they possibly can, but the chipping sparrow uses so much that it is referred to as the hair bird in many parts of the country. During the era when horses were a part of the American scene, chipping sparrows used mostly horsehair, often building their nests with the long hairs from the manes or tails and then lining them with shorter hairs. They still make nests this way in horse country, where even today one of these dainty, chestnut-capped birds can sometimes be seen tugging at a swishing tail, then flying off with a long, dark strand of hair trailing from its bill. But chippys now mostly build their nests out of thin rootlets, then line the little cups with precious hairs from horses, dogs, cattle, deer, rabbits, raccoons, or even humans—whatever donor they can find.

The same holds true for other hair-using sparrows. Horsehair is generally the first choice; dog hair, the second. But all the sparrows make do. Fox sparrows use caribou hair, golden-crowned sparrows use moose hair, tree sparrows up in the tundra use lemming fur, and the little black-throated sparrows sometimes use porcupine hairs (but not the quills). Where there is no hair available, sparrows substitute feathers or fine rootlets and grasses, which also make a warm, soft, resilient pillow on which to cradle the birds' eggs and tender, featherless hatchlings.

Pesky and tireless, chipping sparrows nip at horses' tails for hairs to be woven into nests.

Recognition. 4½–5½ in. long. Streaked with brown above; crown chestnut; eyebrow white; black line through eye; underparts plain gray. Immature has crown streaked with brown.

Habitat. Farms, yards, and pine woods.

Nesting. Nest is a neat cup of grass, stems, and rootlets, usually lined with hair, 1–40 ft. above ground in bush or dense tree. Eggs 3–5, pale blue-green, marked with brown, lilac, and black. Incubation 11–14 days, by female. Young leave nest after 8–12 days. Usually 2 broods a year.

Food. Seeds and insects.

Clay-colored Sparrow

Spizella pallida

At first glance, it seems that when nature's treasures were being distributed among North America's birds, sparrows in general (and the clay-colored in particular) must have waited near the end of the line. Next to the party-dress finery of the wood warblers, sparrows seem to be outfitted like chain-gang workers, clothed mostly in a blend of uniform gray and commonplace brown. As songsters, too, they are far behind the music hall voices of orioles, or the exquisite, liquid calls of thrushes. The clay-colored sparrow typifies the generally undistinguished musicianship of the sparrow family, announcing himself with a sonorous, insectlike call of three or four buzzes.

For all these seeming drawbacks, however, there is nothing hand-me-down about this bird. Perched atop a bush on a summer morning, this little sparrow has an unmistakably gentrified air. Its soft gray and crisp brown feathers are subtly woven like a fine English tweed, well cut and expertly fitted. The younger birds wear a warm, buff-brown sweater with a gray collar before growing into the tweedy look, and both parents tend to their welfare in the meantime. The male sometimes feeds his mate while she nests, and the female is an adept impersonator, luring predators away from her hatchlings by feigning an injury that promises an easy meal—but hardly ever delivers it.

Like other mother birds, a clay-colored sparrow will feign injury to lure intruders from her nest.

Recognition. 4½–5 in. long. Small and boldly streaked with brown above and on crown; ear patch brown, narrowly bordered with black above and below; underparts plain grayish-buff. Immature like immature chipping sparrow but with brown rump.
Habitat. Open, brushy country.

Nesting. Nest is a cup of grass and twigs, on ground or up to 5 ft. above ground in tuft of weeds or dense shrub. Eggs 3–5, pale blue, marked with brown. Incubation 11–14 days, by both sexes. Young leave nest 7–9 days after hatching. Sometimes 2 broods a year.
Food. Seeds and insects.

Brewer's Sparrow

Spizella breweri

The Great Basin lies like a sagebrush sea between the Rocky Mountains and the Sierra Nevada–Cascades. It is a strikingly spare land, where the skies are bleached clean of everything except air and clouds. This is the home of the Brewer's sparrow, a patently unadorned bird in tune with its environment, with its dress of pale underparts, thinly streaked back, and an overall color once described as "desperate gray." Though the sparrow is found in various grassy and woodland habitats, it seems most at home on sagebrush flats. Wherever observers search for it, their main clue is its single white eye-ring.

Plain-looking as it may be, this little bird carries a distinguished ornithological name. Thomas Mayo Brewer was a 19th-century Bostonian who pursued multiple careers as a physician, newspaperman, and book publisher, and remained an avid birder through it all—one held in particularly high regard by John James Audubon.

The Brewer's sparrow bears other distinctions as well. It is hardy enough to survive a three-week dry spell on nothing but seeds, making up for this enforced abstinence with joyous and frequent bathing when the rains finally arrive. Joyousness takes another form, too: at dawn, dusk, and in midday heat the trills of the Brewer's sparrow add a tiny grace note of beauty to the endless roll and march of the sage.

Brewer's sparrows can survive for weeks without water, drawing needed moisture from seeds.

Recognition. 4½–5 in. long. Small and finely streaked with brown above and on head; white eye-ring; lacks strong eyebrow, ear patch, or chestnut crown; underparts plain grayish-buff. Immature grayer than immature clay-colored; lacks head pattern.
Habitat. Sagebrush plains and weedy fields.

Nesting. Nest is a tightly made cup of grass, stems, and rootlets, on ground or up to 4 ft. above ground in sagebrush or dense cactus. Eggs 3–5, bluish green, with scattered brown, black, and lilac spots. Incubation 13 days. Nestling period unknown.
Food. Seeds and insects.

Field Sparrow

Spizella pusilla

Judged solely by its appearance, this buff-colored bird would not fare particularly well in a contest with its more brightly decorated peers. But colorless garb does not mean colorless habits, and field sparrows definitely know how to grab attention.

Incessant battles over territory begin in the spring and persist through the entire nesting season, males chasing one another frantically through the bushes and often ending up on the ground, grappling with bills and claws. When eggs or young are in the nest, this fierce activity can become even more ferocious as the male constantly chirps sharp challenges, sounds alarms, gives chase, or does battle in defense of his realm. But, in spite of this diligence, the field sparrow's ground-level nest is often destroyed, and the male and his mate must begin again. With characteristic persistence, some pairs have rebuilt as many as seven times before successfully rearing a brood.

Yet through it all they have music. Although the female does not sing, the male is tireless. His clear, silvery jingle rings loud and clear over the weedy fields and pastures from spring into mid- or even late summer, when nesting is over and most birds have fallen silent. Then, as if feeling relieved of all other duties, he sings repeatedly in his own individual style, a series of notes on one pitch but accelerating in speed—still not especially eye-catching, but a vibrant delight to the ear.

Song may stake the claim, but field sparrows must often fight to defend their territory.

Recognition. 5–6 in. long. Warm brown and streaked above, with 2 white wing bars; crown rusty; bill pink; face gray with white eye-ring; underparts pale grayish.
Habitat. Weedy and brushy fields and pastures.
Nesting. Nest is a cup of grass and leaves concealed on ground or up to 10 ft. above ground in bush or dense tree. Eggs 3–6, pale green or bluish, densely covered with fine spots of reddish brown. Incubation 11–17 days, by female. Young leave nest about 8 days after hatching. Up to 3 broods a year.
Food. Insects, spiders, and seeds.

Black-chinned Sparrow

Spizella atrogularis

Junco or sparrow? It's not easy to tell as these shy gray-heads flit by, and their skittish habits don't give an observer much time to decide. They are indeed sparrows, but black-chins have the fluffy gray crown and body feathers reminiscent of juncos. This makes the distinctively sparrowish brown of their wings and tails seem almost out of place, as if pasted to the birds' bodies. Favoring chaparral slopes with a dense growth of bushes, black-chins avoid detection by taking low, short flights near the ground or through small openings in the brush. Quick to start, they can be gone before an observer is sure what they were.

During the courtship and breeding season, however, this diffidence seems to vanish. Living in loose colonies, the normally wary males make themselves conspicuous, moving from one tall perch to the next every few minutes as they spill forth their spectacular song, a trilling *sweet-sweet-sweet-weet-trrrrrr* that echoes loudly through the valley. In the open like this it is easy to see their black chins and masks—which are much smaller or absent altogether on their mates, a means of distinguishing between the sexes that is unusual among our sparrows. Incubating females also become bolder. No longer easy to flush, they seek protection in concealment rather than the ploy of distracting an intruder's attention away from the nest.

Agile on the wing, black-chinned sparrows can maneuver through narrow passages in the brush.

Recognition. 4½–5 in. long. Adult male mainly gray; back streaked brown; bill pink; face and chin black. Female and immature similar, but lack black on face and throat.
Habitat. Brushy slopes, desert flats, and sagebrush plains.
Nesting. Nest is a tightly made cup of grass and stems, up to 4 ft. above ground in low dense shrub. Eggs 2–4, pale blue, sometimes with scattered spots of black or dark brown. Incubation 13 days. Nestling period unknown.
Food. Insects and seeds.

Vesper Sparrow

Pooecetes gramineus

This aptly named singer greets the approach of dusk with a lovely serenade.

As the spring afternoon fades into sunset and deepening shadows creep across hayfields and pastures, the plaintive song of the vesper sparrow rises sweetly on the evening air. Vespers sing freely at most any time of day, but as their name suggests, their warm, fluid cadences seem to belong especially to the quiet time of evening. They sing mostly from a perch or while foraging—but sometimes, unpredictably, one will suddenly flutter some 50 feet into the air and give voice to a wild and ecstatic melody.

These startling, dramatic flight songs are thought by some scientists to be inborn relics of the species' primitive past, erupting in rare and spontaneous performances that afford a glimpse of the whole sparrow clan's ancient inheritance. For all musical sparrows sound distinctively *sparrowy*. Whatever kind is encountered—whether the vesper, the song sparrow, the lark sparrow, or any other—though it revels in variations on its own species' unique air, its song is a rippling, tinkling, summery melody.

That is not to say that all sparrows are musical. Some buzz, some squeak, and others rattle one unchanging note. Yet most of them, talented or not, sing without restraint through the spring and into the summer, and many sing occasionally at night. But only the vesper sparrow serenades us regularly at twilight under the evening star.

Recognition. 5–6½ in. long. Mainly pale grayish brown with brown streaks; outer tail feathers white; eye-ring white; shoulders chestnut; underparts finely streaked, without central spot; bill partly pinkish. Sings sweetly in evening.
Habitat. Prairies, fields, and meadows.
Nesting. Nest is a cup of grass, stems, and rootlets sunk into ground in concealed spot. Eggs 3–6, pale blue or green, spotted with brown. Incubation 11–13 days, usually by female only. Young leave nest 7–12 days after hatching. Usually 2 broods a year.
Food. Insects, seeds, and grain.

Lark Sparrow

Chondestes grammacus

A charming oddball with a beautiful voice and handsome plumage, the male lark sparrow behaves as though convinced that he's not a sparrow at all. During the mating season he struts about like a proud turkey cock—colorful head held high, wings drooped and dragging in the dust, tail cocked stiffly over his back. Yet while the turkey puts on this display only in his own fiercely defended mating area, the lark sparrow feels no such constraint. Although he generally takes but one mate during the season, he is likely to break into his ostentatious dance whenever an interesting female passes by. Swaggering forward and back, trilling broken bits of his song, he's a dandy with an eye for all the ladies.

Another of his markedly unsparrowlike traits is the habit of singing now and again while hovering in the air—like the lark for which he is named. In contrast to the vesper sparrrow, however, whose ecstatic song in flight differs from normal earthbound melody, the lark sparrow sings his ordinary song from the sky. No matter where he appears—in open badlands or on sagebrush plains, in upland pastures or on scraggly hillsides, in farmlands or in orchards—his sprightly, offbeat manners and joyous songs stir up and enliven all the empty spaces.

Courtship competition among male lark sparrows can erupt into an aerial melee.

Recognition. 5½–6½ in. long. Streaked with brown above; crown stripes and ear patch rusty; eyebrow, eye-ring, and whisker mark white, with black whisker mark below white one; underparts white with black breast spot.

Habitat. Farmland and dry open country with scattered trees and bushes.

Nesting. Nest is a cup of grass, twigs, and stems, often on ground, or 3–30 ft. above ground in bush or small tree. Eggs 3–6, whitish with spots and scrawls of black and dark brown. Incubation 12 days, by female. Young leave nest about 10 days after hatching.

Food. Seeds and insects.

Black-throated Sparrow

Amphispiza bilineata

Thriving where few others can, black-throated sparrows must still confront predators.

They are a hardy and adaptive breed, the sparrows of North America. For every natural niche, there seems to be a sparrow custom-tailored to fill it. Many are specialists, suited for particular environments—some so harsh they test the limits of survival. Snow buntings eke out an existence in the Arctic, where summer and winter meet with scarcely a pause for spring. Seaside sparrows live in a marginal world between the tides. Hardiest of all is the desert's black-throated sparrow, thriving in a land where temperatures soar and rainfall is usually just a rumor.

These "desert sparrows" seem immune to the rigors of their terrain. The rockier the hillside, the sparser and sharper the vegetation, the better they seem to like it. In 1932, Joseph Grinnell observed a bird in Death Valley "less than fifty yards from the very edge of the lowest part of the sink." In such environs, black-throated sparrows survive by extracting all the moisture they need from the insects, plants, and seeds they eat, a talent that enables them to occupy territory the competition shuns.

In the first gray light of morning, when the desert stirs to life, male black-throated sparrows take to the tops of cacti and creosote bushes and send abroad their loud, clear song. To birders awed by the unrelenting dryness of the desert, the song sounds like a challenge hurled back at an inhospitable world, one that declares, "I live—I, a sparrow, here on the edge of existence, doing just fine."

Recognition. 4½–5½ in. long. Gray and unstreaked above; crown and ear patch gray; eyebrow and whisker stripe white; throat black; underparts white. Immature similar, but lacks black throat.

Habitat. Desert hillsides with scattered low brush or cactus.

Nesting. Nest is a loose cup of grass, twigs, stems, and plant fibers, well-hidden 6–18 in. above ground in dense shrub. Eggs 3–4, white or pale blue. Incubation and nestling periods unknown.

Food. Seeds and insects.

Sage Sparrow

Amphispiza belli

From a single tangled bit of sagebrush, indistinguishable from the millions around it, a creature drops to the hard-packed ground and skitters away. A bird? A mouse? The elusive wraith—whatever it is—runs a short distance, inviting pursuit, then disappears behind a curtain of branches. Accept its challenge, and it then speeds off in a new direction, holding its head low like a sprinter leaving the blocks, its tail cocked high in a seeming gesture of disdain. Pursuit is futile, for this imp, the sage sparrow, is a past master at the game of hide-and-seek. The best an observer can hope for is a view of a gray-headed bird perched atop some distant bush; many will have to settle for a whispered *psst* and a twitching tail that drops quickly back into the sage.

Back where the chase began, somewhere between 6 and 18 inches above the ground, may be a nest holding three or four eggs. Much of the material woven into the thickish, cup-shaped nest is typical of sparrows' nests—dry twigs, assorted sticks, withered grass, and weed stalks. But while many species of sparrow cushion their nests with nothing more than grass or shredded bark, sage sparrows have a weakness for cozy comfort. Their nests are typically lined with feathers, rabbit fur—and, quite frequently, tufts of wool, an available extra wherever sheep and sagebrush are found together.

Mouselike in demeanor, a sage sparrow may find itself jousting with a real mouse for food.

Recognition. 5–6 in. long. Mainly grayish, streaked with dusky above and with 2 pale wing bars; face with thin white eye-ring and dark area between eye and bill; throat white with blackish whisker mark; underparts whitish with dark breast spot and streaked flanks.
Habitat. Brushy desert and sagebrush plains.

Nesting. Nest is a cup of twigs, grass, and strips of bark, up to 4 ft. above ground in shrub, sometimes on ground. Eggs 3–5, pale blue, spotted and scrawled with brown and blackish. Incubation 13 days. Nestling period unknown. Apparently 2 broods a year.
Food. Insects and seeds.

Lark Bunting

Calamospiza melanocorys

Packed side by side at a popular watering trough, lark buntings adjust to each other's needs.

Late in the prairie farmer's day, but still before vesper sparrows begin their evening song, the lark buntings come down to drink. As windmills spin above, the thirsty birds line the metal troughs below, streaky females and sooty black males shoulder to shoulder with other sparrows, longspurs—and the cattle that graze where great buffalo herds once roamed the land. Life on the Nebraska prairie was nomadic then: grass followed the spring, and buffalo followed the grass. Today the buffalo is gone, but the lark bunting continues to live the life of a nomad.

During good years lark buntings may be abundant in a particular region, then the next year they may be all but absent. Indeed, the arrival of lark buntings has come to be regarded as a sign of good fortune among farmers. "If they arrive in great numbers and every fence and weed stalk is hanging full of them," said one, " . . . we know that they know a good season is ahead, and there will be plenty of everything for both us and them."

Beginning in late July, lark buntings move south in large, tumbling flocks. But with the still uncertain warmth of another spring, farmers will look again for the returning birds, who sing both trills and single notes as they fly upward with a new day's sunshine on their wings.

Recognition. 5½–7 in. long. Breeding male black with bold white patches on wings; bill whitish. Female and winter male brownish, coarsely streaked with black; pale wing patches; bill whitish. Forms large flocks in winter.

Habitat. Dry grasslands, sagebrush plains, and desert scrub.

Nesting. Nest is a loose cup of grass, stems, and rootlets, hidden on ground in grass. Eggs 3–7, pale blue-green, sometimes spotted with brown. Incubation 12 days, probably by female only. Nestling period unknown.

Food. Insects, seeds, and grain.

Savannah Sparrow

Passerculus sandwichensis

Millions of them live out a part of their lives on salt meadows—on those quiet, sunlit spaces that lie within earshot of the sound of ocean surf along shorelines from Labrador to Florida. But these small birds also live just as contentedly on inland pasturelands and prairies across the continent and up to the Arctic. And like other species that cover exceptionally large ranges, the Savannah sparrows include a number of varieties, or subspecies, that have developed different plumages or proportions.

One of these, the Ipswich sparrow, breeds only on a 20-mile-long strip of land on Sable Island, Nova Scotia, then winters along portions of the Atlantic seaboard, including the dunes of Ipswich, Massachusetts, where it was first discovered. Many biologists believe that when a small group of nesting birds is isolated long enough from a main population, inbreeding produces such definite changes in appearance (as with the larger and paler Ipswich) that mating with the main group ceases entirely.

That, however, is the biologists' business. Most observers are content simply to enjoy the sight of the small Savannahs bustling about in locations as diverse as the sea grass behind people-packed New York beaches and the remote shores of Sandwich Bay in the Aleutian Islands, where the first of these birds were collected by the scientists who would later identify them as a new species.

Coast to coast and season to season, spiders are a mainstay of the Savannah's diet.

Recognition. 4–6 in. long. Finely streaked with brown above and below; eyebrow usually yellow; tail long and slightly notched. Pacific Coast birds darker; eastern Ipswich form larger and paler, with whitish eyebrow.
Habitat. Prairies and fields near water, freshwater and salt marshes, and grassy dunes.

Nesting. Nest is a cup of grass, plant stems, and moss concealed in vegetation on ground. Eggs 3–6, whitish or bluish, speckled and blotched with brown. Incubation 12 days, by both sexes. Young leave nest 14 days after hatching. Sometimes 2 broods a year.
Food. Seeds, insects, and spiders.

Grasshopper Sparrow

Ammodramus savannarum

In any fair-minded compilation of attractive birdsongs and calls, the grasshopper sparrow's efforts deserve to be about as far down on the list as the peacock's, which is very far down indeed. Some birders, in fact, do not even consider this sparrow's grasshopperlike buzzing to be any kind of song at all. But the female of the species clearly has a better opinion of her prospective mate's musicianship as he pursues her in flight, even while emitting sounds that are too high-pitched for most human ears to hear.

These are sparrows of open field and meadow, well known to the early homesteaders who cleared woodlands to make room for farms and grazing areas. Now as then, dried flower stalks serve as favorite vantage points for these birds, which feed on both insects and seeds in places where grasses grow in clumps with good foraging space in between. Although grasshopper sparrows prefer meadows and fields that have not recently been harvested, they pay a costly price for such space whenever their tiny, ground-level nests are ripped apart by mowing or cultivation.

More recently, the grasshopper sparrow population has been cut back by urban development and the reduction of pasturelands. In earlier days mother birds often succeeded in distracting predators from their broods by feigning injury, then making short flights away from their nest and young. But today's bulldozers and giant backhoes are not tricked by such a ploy.

In courtship pursuit, the grasshopper sparrow buzzes at a pitch that humans cannot hear.

Recognition. 4–5 in. long. Brown and streaked above, with distinctive flat-headed look; under-parts plain buff; dark eye conspicuous on plain buff face; bill larger than that of other sparrows.
Habitat. Meadows, prairies, and grain fields.
Nesting. Nest is a cup of grass lined with soft roots or hair and sunk into ground at base of tuft of grass or other vegetation; often arched on top to hide eggs. Eggs 3–6, whitish, speckled with brown. Incubation about 12 days, by female. Young leave nest 9 days after hatching. Often nests in loose colonies.
Food. Insects, spiders, grain, and seeds.

Sharp-tailed Sparrow

also known as

Saltmarsh Sharp-tailed Sparrow

Ammodramus caudacutus

The lush green carpet of salt hay may look devoid of life. But if some prowling hawk starts a ruckus among the resident red-winged blackbirds, just watch! Suddenly a sharply pointed face, highlighted by a bright orange triangle, pokes theatrically through the grass; soon half a dozen more sharp-tails appear as spectators. The necks stretch for elevation and finally, unable to stand the strain and excitement, one member of the audience takes off. With head held high and sharply pointed tail held low, it manages a fluttering rise of about a foot and a distance of perhaps 20 feet before diving back into the marsh.

Sharp-tailed sparrows are, to put it mildly, not enthusiastic fliers once settled into a territory. They rarely fly when they can run, and they run a great deal. As a result, although they may be quite common in their salt hay or sedgy fortresses, they are not easily seen. Nor are they often heard, for their unmusical voices carry little farther than their stunted flights and are easily lost in the cacophony of sounds emerging from coastal and freshwater marshes.

But this reticence above the marsh is forgotten amid its grasses, where the sharp-tailed sparrows are avid, tireless carnivores. Nearly all of their summer diet consists of insects, spiders, and small snails—a far cry from the seeds on which most sparrows are content to dine.

Awkward fliers, sharp-tailed sparrows will flutter only short distances before landing again.

Recognition. 4½–5½ in. long. Brown and streaked above; eyebrow and broad whisker mark bright buff; underparts whitish with fine streaks on breast on most coastal birds, rich buff and unstreaked on inland birds.
Habitat. Salt marshes along coast; grassy marshes and wet meadows inland.

Nesting. Nest is a tidy cup of grass and sometimes seaweed, hidden in clump of grass or sedges, or under fallen stalks from previous year. Eggs 3–7, greenish, speckled with brown. Incubation 11 days, by female. Young leave nest 10 days after hatching.
Food. Insects, spiders, sand fleas, and seeds.

Seaside Sparrow

Ammodramus maritimus

Seaside sparrows thrive in a narrow habitat amid marshes and pools at the ocean's edge.

With a range extending more than 2,500 miles along the Atlantic and Gulf coasts of the United States, and an ample choice of diet to go with it, seaside sparrows would seem perfectly situated to exist healthily and happily into the distant future. They are creatures of the salt marsh: their relatively large feet provide support on the soft muck, and their eggs rest inconspicuously in nests that are often woven into live marsh grass.

But seaside sparrows also illustrate the special vulnerability of many coastal species. The fact is that their habitat, although thousands of miles in length, is no more than a few hundred feet wide in many places. One such area in east-central Florida was home to a local population known as the dusky seaside sparrow, which is now believed to be extinct. As recently as 1968, 1,800 birds were thought to exist, but construction projects, marsh drainage, fires, and mosquito-control programs all took their toll, leaving only a handful of males by the early 1980's.

The demise of one local population of a bird rarely attracts widespread public attention. And yet each of these groups is a genetic reservoir of differing tolerances—to hot or cold, wet or dry, high or low elevations, and other factors—not exactly duplicated by other populations. And any of these traits could prove vital in allowing species to adapt to changes in climate, habitat, or other conditions they may face over the long term.

Recognition. 5–6 in. long. Dark olive-gray above, with short yellow stripe above and in front of eye; throat and short whisker mark white; underparts whitish streaked with dusky. Birds in some areas paler in color.
Habitat. Salt marshes.
Nesting. Nest is a cup of marsh grass and rushes, in grass tussock 3–9 in. above mud. Eggs 3–6, white or pale green, spotted with brown. Incubation 12 days, by female. Young leave nest about 10 days after hatching. Sometimes 2 broods a year.
Food. Small crabs, snails, insects, spiders, and seeds.

Song Sparrow

Melospiza melodia

Unchallenged virtuoso of the sparrow clan, the male song sparrow will sing as many as 20 different melodies on a morning in May. And by the time the day is over he may have improvised nearly 1,000 variations on his basic themes.

Thus the courtship-minded bird claims his territory and serenades his mate. But he does not stop there: a musician to the core, he continues to sing through much of the heat of summer and right on into the mellowing days of autumn. If he remains in the North, he even carols freely when winter's snowflakes drift gently all around. While the male song sparrow is the family's star performer, the female may also join in the chorus in the weeks just before nesting begins, though with a song that is shorter and softer than her mate's.

Song sparrows begin to hone their vocal skills even as they learn to fly. The young bird at first adds short calls to its random warblings, until by early autumn it is singing songs that more or less resemble the adult's. By spring these rough-hewn melodies have been dropped, and the budding virtuoso is singing the musical phrases of his primary song in their proper sequence. Before long he develops his own repertoire of variations on this theme—and emerges on his own as an artiste.

Feathers puffed, wings raised, a song sparrow adopts a tough stance to defend his territory.

Recognition. 5–7 in. long. Large and long-tailed. Upperparts brown and streaked; underparts whitish with dark streaks and spot in center of breast; tail usually tinged with rusty. Birds in some areas darker, others paler, more grayish. Pumps tail in flight.
Habitat. Thickets, roadsides, and gardens.

Nesting. Nest is a cup of grass and stems, on ground early in year, up to 30 ft. above ground later in season. Eggs 3–6, pale green with brown speckles and spots. Incubation 12–15 days, by female. Young leave nest about 10 days after hatching. Up to 3 broods a year.
Food. Insects, seeds, grain, and berries.

Lincoln's Sparrow

Melospiza lincolnii

On June 4, 1833, Thomas Lincoln—no relation to Abraham—boarded the schooner *Ripley* in Eastport, Maine. Lincoln was 21 years old, and for him it was the beginning of a summer adventure in the company of America's foremost naturalist, John James Audubon. Together they sailed up the coasts of Nova Scotia and Cape Breton to the southern shore of Quebec. It was a trip of trials—of damp weather, poor supplies, and seasickness. But it was also a trip of excitement, a chance to visit vast colonies of seabirds and observe new plants in a little-known region.

Near the mouth of the Natashquan River in Quebec, Audubon's party went ashore to hunt. Audubon heard an unfamiliar birdsong, and Lincoln finally chased the bird down and collected it. It was indeed unfamiliar—a new species, the only one for the expedition. In Lincoln's honor, Audubon named it "Tom's finch."

Known today as Lincoln's sparrow, the bird summers on the brushy bogs of the Far North or the wet meadows of western mountains at elevations up to 11,000 feet. It is not uncommon, but it is shy and often skulks in deep cover, even while migrating from its winter range in the southern United States and Central America. Indeed, it may be that Audubon had walked by many a quiet Tom's finch during his wanderings long before the expedition to Quebec.

To find its favorite insects and seeds, a Lincoln's sparrow scratches the soil vigorously.

Recognition. 5–6 in. long. Olive-gray and streaked above; crown stripes rusty; eyebrow and sides of neck clear gray; narrow eye-ring white; underparts whitish, with buff band across breast marked with fine streaks. Usually secretive.
Habitat. Northern bogs and wet meadows; thickets in winter.

Nesting. Nest is a cup of dry grass, on ground or in bush or tussock. Eggs 3–6, pale green, spotted with brown. Incubation about 14 days, by female. Young leave nest about 20 days after hatching. Sometimes 2 broods a year.
Food. Insects, grain, and seeds.

Swamp Sparrow

Melospiza georgiana

A dark, secretive, and timid sprite, the swamp sparrow hides among the reeds of marshes and the boggy edges of lakes or slow-moving streams. Spending most of its time walking about on the spongy earth, it is seldom seen unless it is singing from an alder bush or the very top of a swaying reed. But if, while singing, it becomes aware that it is being watched, it flits away with nervous haste, drops into a patch of sedges—and vanishes.

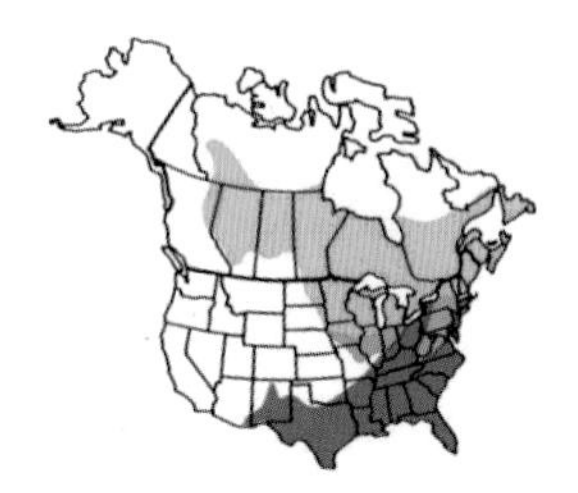

Peculiarly water-loving for a sparrow, it often wades about, feeding on floating insects and seeds, and usually builds its grassy nest close to the water. The nest may be as high as six feet up in an alder bush, but it is usually situated low in a tussock of sedges or among the cattails only inches above the ripples. Whatever its location, the nest is a well-hidden little cup with a special feature—its entrance is not on top but on the side.

Hopping through the branches to the top of a bush, this accomplished singer has the purposeful air of a virtuoso mounting the stage. Its common song is a tinkle of rapidly repeated notes much like that of the chipping sparrow—but sweeter. Its second song, heard more rarely, is slower and richer, a rising repetition of *peet-peet peet-peet.* But for real virtuosity little can compare with its ability to sing two notes simultaneously—at different tempos!

Constructed of coarse marsh grasses, the swamp sparrow's nest has a side entrance.

Recognition. 4½–5½ in. long. Reddish brown and streaked above; crown rusty; face and breast gray; throat white; flanks tinged with buff. Immature similar, but with blackish stripes visible on crown.

Habitat. Cattail and sedge marshes, wet meadows, and bogs.

Nesting. Nest is a cup of marsh grass, woven into clump of cattails or sedges, up to 6 ft. above water. Eggs 3–6, pale green, spotted and blotched with brown. Incubation 12–15 days, by female. Young leave nest after about 13 days.

Food. Insects and seeds of marsh plants.

Golden-crowned Sparrow

Zonotrichia atricapilla

Many a Pacific Coast bird watcher has long suspected that golden-crowned sparrows—among other species—return year after year to the same small wintering ground. But just how strong is this "winter site tenacity"? To try to find out, biologists trapped 480 golden-crowns at banding stations across central California, moved the birds to different stations, and released them. Many of the adults that were moved around early in the winter returned to the sites where they had been trapped, but first-winter sparrows did not. And few birds of any age that were moved late in the winter returned, which suggests a gradual weakening of site tenacity as the season progressed.

The next winter, not one adult from this latter group returned to its new station. Neither did any of the young birds moved after mid-January, though some that were moved earlier did. It seems, then, that golden-crowns "imprint" or fix on a territory in the *middle* of their first winter. This gives them time to wander and explore, to search out an agreeable home for winters to come.

As one curious birder learned, golden-crowns are faithful to more than a wintering area. He trapped a number of them, painted them with distinguishing marks, and recorded their activity on a large feeder tray. Sure enough, his photographs showed that each bird had its own spot and used it exclusively, even when no other birds were around.

As well as seeds and tender seedlings, the golden-crowned sparrow is fond of fall flowers.

Recognition. 6–7 in. long. Large, slender, and long-tailed. Upperparts brown and streaked; head has broad black eyebrows, bright yellow patch on crown; underparts plain gray.
Habitat. Dense thickets in winter; alpine meadows and forest clearings in summer.
Nesting. Nest is a cup of stems, grass, dried fern fronds, and leaves, on ground at base of tree or sunk in earth bank. Eggs 3–5, whitish or very pale blue, finely dotted and coarsely spotted with brown. Incubation and nestling periods unknown.
Food. Buds, flowers, seeds, and some insects.

White-crowned Sparrow

Zonotrichia leucophrys

Self-promotion is not a trait often ascribed to sparrows, which are generally so shy and secretive that they are hard to tell apart in the field. The Ipswich and Savannah sparrows, for instance, can scarcely be distinguished from one another, while sharp-tailed sparrows are hard to see at all, since they scurry through the grass like mice. The Baird's sparrow can be identified by its song—but it doesn't sing in winter or during migration. The Henslow's sparrow has a song so short that the bird is scarcely noticed even when it does sing. And the tundra-nesting tree sparrow visits enough feeders during migrations to get itself recognized, but it also refrains from singing in its southern wintering grounds.

The great exception to these fugitives from the limelight is the white-crowned sparrow. Although it too nests in remote alpine regions and far up in the Canadian sub-Arctic, the white-crowned is among the most familiar of all sparrows. With the feathers of its domed crown spread into a low crest and its trim, slender body held regally upright, it is also among the loveliest of the sparrow legions. Yet appearance is not its only claim to our attention. Visiting winter feeders in flocks, these big, venturesome sparrows brighten the late winter with their songs and continue caroling all the way north, their beauty and virtuosity gathering renown.

In its northern habitat, the white-crowned sparrow often has to scratch for seeds in the snow.

Recognition. 5½–7 in. long. Large and slender. Back brown and streaked; crown boldly striped with black and white; bill pink; face and underparts plain gray. Crown stripes on immature birds brown and buff.

Habitat. Brushy thickets and woodland edges, lawns and gardens, and northern spruce forests.

Nesting. Nest is a cup of grass, rootlets, and twigs, on ground or up to 35 ft. above ground in tree. Eggs 3–5, pale bluish, spotted with rusty. Incubation 11–16 days, by female. Young leave nest 10 days after hatching.

Food. Mainly insects, seeds, and grain.

Harris' Sparrow

Zonotrichia querula

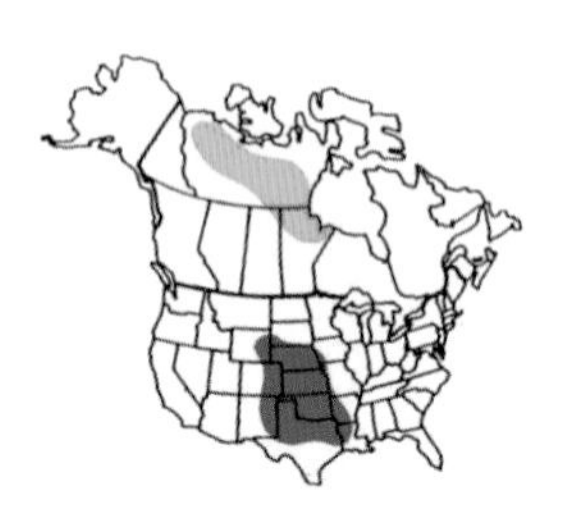

By the turn of the 20th century, the life histories of most North American birds had been well documented. One of the notable exceptions was the Harris' sparrow: as late as the 1930's no one had ever seen the eggs of this elusive bird. As fate would have it, the attempt by two birding teams to discover them became not only a professional challenge but a matter of national pride.

In 1931 an American, George Miksch Sutton, put together a crew of four veteran birders and headed north in search of the mysterious eggs. They arrived on the western shore of Hudson Bay on May 24, and found themselves knee-deep in snow. The first Harris' sparrow appeared three days later—but was followed shortly by something far less welcome: four Canadian ornithologists, arriving with the announcement that they too were there to discover the unknown eggs. The race was on.

For 15 days the two teams combed the rugged woods of stunted spruce. Then, as Sutton was slogging through a bog just before 9 A.M. on June 16, he suddenly flushed an incubating sparrow. Searching in the mat of sphagnum and Labrador tea, he found a nest containing four brown-blotched, whitish-blue-green eggs. Taking the adults, the nest, and the eggs as evidence, he ecstatically "hippity-hopped across the bogs" to announce the find to his companions, the Canadian competitors, and the world.

The retiring Harris' sparrow builds its nest in the remote bogs of northern Canada.

Recognition. 7–7½ in. long. Large and long-tailed. Back streaked with brown; crown, face, and throat black; sides of face buff; bill pink; underparts white. Immature lacks black on crown and face.

Habitat. Thickets, woodland edges, and brushy areas; nests in northern spruce forests.

Nesting. Nest is a well-insulated cup of moss, grass, and stems, hidden in moss at base of small tree or shrub. Eggs 3–5, white or greenish white, heavily spotted with brown. Incubation about 13 days. Nestling period unknown. Very few nests ever found.

Food. Seeds, grain, insects, and berries.

McCown's Longspur

Calcarius mccownii

Throughout the 19th century, the U.S. military played a key role in studying the natural history of the American West. Whether exploring, mapping railroad routes, or simply holding the fort, numerous army officers found time for some research on the side, describing or collecting animal and plant specimens for experts back east to examine.

The discovery of McCown's longspur, for one, could be described as a tale of two captains. In June 1805 the Lewis and Clark expedition was in present-day Montana, and Capt. Meriwether Lewis was exploring one fork of a river when his party noted numbers of a "small bird which in action resembles the lark," with "a dark brown color with some white feathers in the tail." The male, Lewis noted, "rises into the air about 60 feet and supporting itself in the air with a brisk motion of the wings sings very sweetly" for about a minute before making its descent.

To a modern ornithologist this sounds very much like McCown's longspur. But no one kept a specimen, and so what might have been Lewis' longspur became something else nearly 50 years later, when Capt. John Porter McCown fired into a flock of horned larks on a Texas prairie and found among the victims a bird he hadn't even seen: a small brown bird with some white in its tail, and no name.

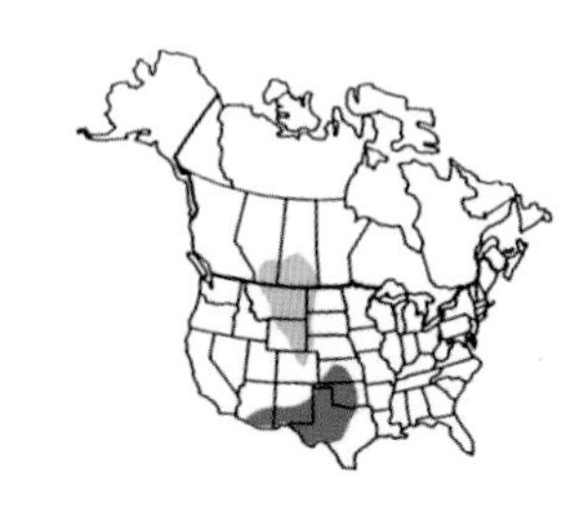

To stake its claim, a McCown's descends like a parachutist—singing all the while.

Recognition. 5½–6 in. long. Streaked brown above; tail white with black inverted T. Breeding male gray on face and underparts; crown and breast patch black; throat white; shoulder rusty. Female and winter birds streaked brown with rusty shoulder patch. Travels in flocks in winter.
Habitat. Shortgrass plains and prairies.

Nesting. Nest is a cup of grass sunk in ground. Eggs 3–6, buff, olive, or pinkish, spotted and scrawled with brown and lilac. Incubation 12 days, by female. Young leave nest about 12 days after hatching.
Food. Weed seeds and insects.

Lapland Longspur

Calcarius lapponicus

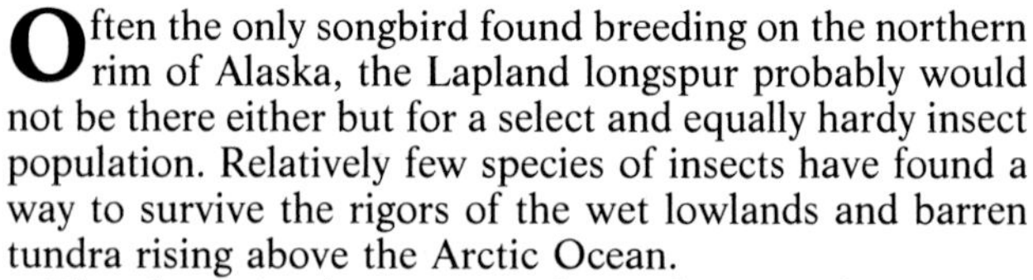

Often the only songbird found breeding on the northern rim of Alaska, the Lapland longspur probably would not be there either but for a select and equally hardy insect population. Relatively few species of insects have found a way to survive the rigors of the wet lowlands and barren tundra rising above the Arctic Ocean.

The short Arctic summer demands an early start to nesting, and female longspurs are already laying in early June, only days after snowmelt bares the first patches of tundra. Soon the abundant nests have four or five mouths to be fed, but with the exception of tiny midges and a few spiders there is apparently nothing to feed them. The parents, however, know better. Just below ground among the mosses and lichens lie maturing crane fly pupae and sluglike larvae, with which the parents stuff their young nestlings' gullets through the long Arctic day.

Then, just at fledging time in early July, the tundra explodes with insect life. Few tables in nature are more abundantly set or more accessible. Mosquitoes, midges, and newly emerged crane flies—slow-flying, long-legged creatures—dance in the millions low over the vegetation prior to mating, and provide easy pickings for the young, still-clumsy birds. The feast is short-lived, ending abruptly within 20 days, but that brief period of plenty carries the new generation from fledging to independence.

A Lapland longspur feels perfectly at home in her nest on a bare patch of tundra.

Recognition. 5½–6½ in. long. Streaked brown with white underparts, white outer tail feathers. Breeding male has black face and throat; white stripe behind eye; chestnut nape. Female and winter birds have buff face, dark ear patch.
Habitat. Bare fields, lakeshores, and beaches; nests on Arctic tundra.

Nesting. Nest is a cup of grass and moss, sunk in ground near tundra shrub. Eggs 3–7, greenish or buff, spotted with brown. Incubation about 13 days, by female. Young leave nest 10–12 days after hatching.
Food. Insects and seeds of weeds and grasses.

Chestnut-collared Longspur

Calcarius ornatus

Ground nests are so vulnerable to both predators and foul weather that some ground-nesting species have over time eliminated a nestling stage in which infants are cared for in the nest. The offspring of these species hatch sighted, alert, and ready to walk soon after their down has dried. No songbirds, however, have managed to do away with this stage. Their young hatch blind and helpless, making a dangerous nestling period inevitable. The result, demanded by the pressures of survival, is an astounding pace of chick development in ground-nesting songbirds.

Biologists studying the nest life of chestnut-collared longspurs in Manitoba and Saskatchewan found a typical incubation period of 12 days followed by a nestling period of only 10. A newly hatched chick—weighing about 1/19 ounce—can barely open its mouth, yet on the second day it holds its head erect to gape for food. By its fifth day it can see, and feathers are erupting. Three days later the primary wing feathers are showing, and by day 10—having increased its hatching weight tenfold—it is likely to scramble from the nest. Two days later the fledgling is making low but sustained flights over the prairie grasses, and after two more weeks of parental feeding it achieves independence, well before it is a month old.

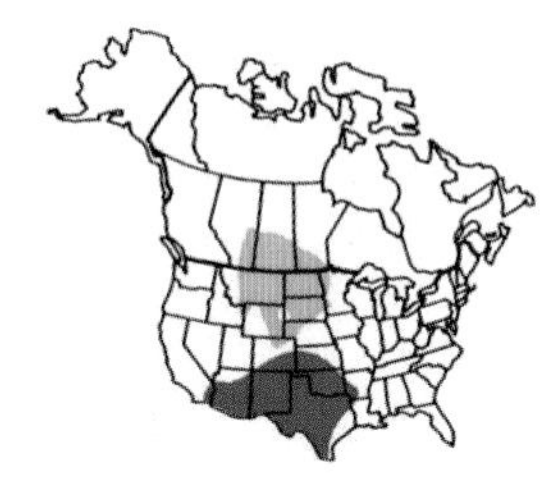

The convenience of a ground nest is paid for by its constant vulnerability to predators.

Recognition. 5½–6½ in. long. Brown and streaked above; tail white with triangular black patch. Breeding male has black and white pattern on head; nape chestnut; breast black. Female and winter birds dull and streaked; nape with trace of chestnut; face plain buff. Travels in flocks in winter.

Habitat. Tallgrass prairies and plains.
Nesting. Nest is a cup of dried grass on ground surface or sunk into soil. Eggs 3–6, whitish or pale blue, spotted with brown and blackish. Incubation about 12 days, by female. Young leave nest about 10 days after hatching.
Food. Seeds and insects.

Snow Bunting

Plectrophenax nivalis

Breeding male

Winter female

Camouflage is a bunting's best defense, its speckled summer colors blending into the tundra.

Bitter winds strip branches of their last leaves and push row after row of ragged, black-bottomed clouds across a November sky crowded with snowflakes and retreating geese. It seems another squall is approaching, a tumbling wave of flakes. Then suddenly a call note is heard overhead . . . *Peur! Peur! . . . PeurEEE,* followed by wild tittering. These are not snow *flakes,* but snow birds! The snow buntings have arrived—and so, with this hardiest of songbirds, has winter.

No songbirds are found farther north than these long-winged Arctic breeders, which even nest on Ellesmere Island and northern Greenland. Though snow buntings usually venture no farther south than northern California and Virginia, their migration is as arduous as that of warblers and thrushes who winter in Central and South America. Many miles lie between the Arctic and the mid-band of the United States.

Snow buntings are aptly named, for snow is most certainly their element. Their summer nests are at the very edge of the icy Arctic frontier, and they winter no farther south than the limits of the snow. Fluffy drifts are for them friendly insulation in which to sleep snugly through the cold of winter nights. So long as winds or weak Arctic sunshine expose patches of bare earth and the seeds of hardy plants, the snow bunting is content.

Recognition. 5½–7 in. long. Mainly white. Breeding male has black on back, wingtips, and center of tail. Female and winter birds tinged with brown on crown, face, and back.
Habitat. Barren areas, fields, beaches, and dunes; nests on Arctic tundra.
Nesting. Nest is a cup of grass, moss, and lichen concealed in a crevice among rocks. Eggs 3–9, bluish or white, marked with brown and lilac spots and speckles. Incubation 10–15 days, by female. Young leave nest 10–17 days after hatching. Often 2 broods a year.
Food. Seeds, insects, spiders, and sand fleas.

Bobolink

Dolichonyx oryzivorus

Up from Brazil and Argentina come the bobolinks each spring, traveling in small flocks of smartly clad males and sparrowy drab females. They funnel up through Florida and on toward the North, feasting on dandelion seeds and filling the air of fields and pastures with their bubbling conversations. These small flocks drop off on nesting grounds through the northern United States and southern Canada. They rear one brood of young and spend the rest of the summer bobbing about and going through a molt that leaves every bobolink wearing sparrow drab. Heading back south in the fall, they look like different birds—so much so that for years people thought they were.

They travel differently, too, forming mammoth flocks that feed on harvested grainfields and fallow land. For decades they swooped down on South Carolina rice plantations and grew as fat as little butterballs. And every fall they were killed by the tens of thousands to be sold and eaten as butterbirds. So many were slaughtered that they have never recovered their vast numbers. But most of the rice fields are gone now, the birds are protected by laws, and most people know the autumn butterbirds are the springtime bobolinks that fill the air with tinkling music.

Bobolinks are more welcome, for good reason, when they gobble up weed seeds instead of grain.

Recognition. 5½–7½ in. long. Bill short and sparrowlike. Breeding male mainly black; nape yellow; rump and shoulders white. Female, immature, and winter male buff-yellow; back and flanks streaked; crown striped.
Habitat. Meadows, grainfields, and marshes.
Nesting. Nest is a thin-walled cup of grass hidden on ground or in dense tuft of grass or clover. Eggs 4–7, pale gray to tan, blotched with brown and lilac. Incubation about 13 days, by female. Young leave nest 10–14 days after hatching.
Food. Insects, seeds, and grain.

Eastern Meadowlark

Eastern Meadowlark *Sturnella magna*
Western Meadowlark *Sturnella neglecta*

Eastern Meadowlark

Western Meadowlark

So look-alike are our two species of meadowlarks, the eastern and the western, that if they were perched side by side only an expert would know they were not the same species. In flight the similarities persist: both birds show patches of white in their tails and keep their short, stiff wings tipped downward as they flap or glide.

The two species not only look alike and fly alike—they live alike. They are birds of the open grasslands, nesting in shallow depressions at the roots of grassy clumps and weaving the tops of the blades together to make a dome above the nest. Their eggs are alike, their nestlings are alike, and the males of both species tend to be polygynous, sometimes maintaining several nests on the same territory.

But when these two birds open their bills and sing, even the rankest amateur knows they are not the same species. Their songs are so dissimilar and so opposite in tone that the nearly complete similarities in other respects seem almost a trick. The voice of the eastern bird is clear and high-pitched. His songs are usually from three to five notes long, and they have a lonely, wistful, poignant sound. The western bird usually sings five to seven notes; his voice, too, is clear, but it is also rich and fluting and gurgling, and his songs sound confident and self-assured. Listen to their songs. Their lovely voices will reveal their true identities.

Recognition. 8–10½ in. long. Stocky, streaked above; outer tail feathers white; breast yellow with black V. Western paler than eastern, but best distinguished by song: eastern makes series of clear whistles; western produces a jumble of flutelike notes.
Habitat. Grasslands, pastures, and marshes.

Nesting. Nest is a domed cup of grass and stems, hidden on ground in grass or weeds. Eggs 3–7, white or pinkish, spotted with brown and lilac. Incubation 13–14 days, by female. Young leave nest about 12 days after hatching. Usually 2 broods a year.
Food. Insects, spiders, grain, and seeds.

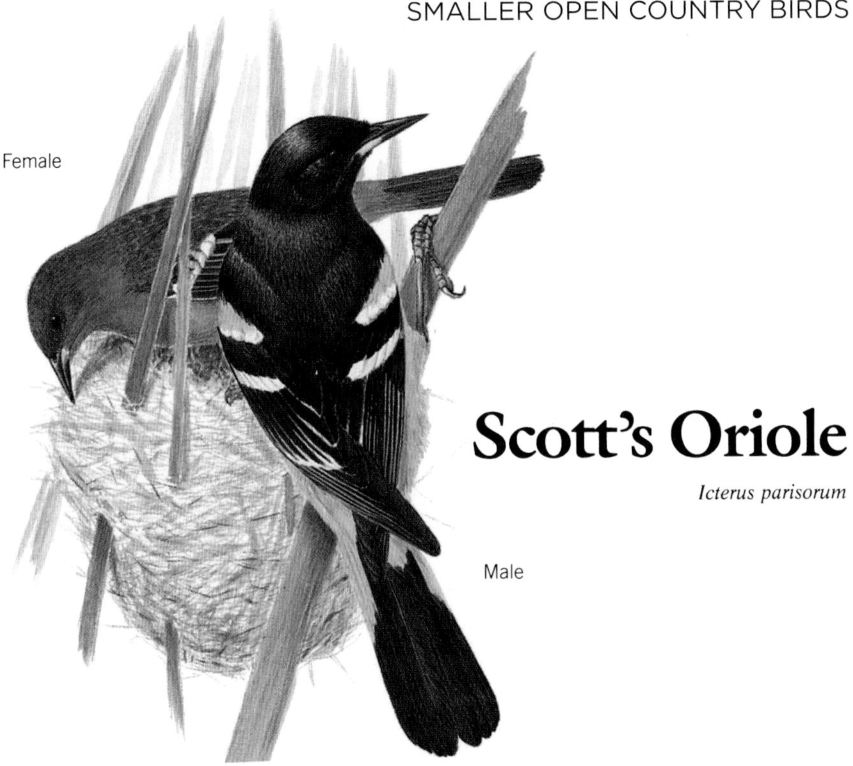

Scott's Oriole

Icterus parisorum

Dressed in eye-catching lemon-yellow and black, a half dozen male Scott's orioles, accompanied by their green-yellow mates, flit across a narrow canyon, their bright colors gleaming against the deep blue of the sky and the dark rock of the canyon walls. The whole flock slips into the foliage of a cottonwood tree—and vanishes. All their brilliant colors are swallowed up in the multitude of gray-green leaves, which seem not so much to cover the birds as to absorb them. Perhaps their blacks become shadows, their greens melt into the foliage, and their yellows become glints of sunlight among the branches. (The same vanishing magic seems to work also, to a lesser extent, on the fiery oranges and glowing reds of other birds. They gleam in flight but vanish into foliage, their colors toned down by the shadowing green around them.)

Scott's orioles feed in these treetops, picking insects from twigs, stems, and leaves, never fluttering among the limbs but creeping from one level to another. Their slow, deliberate movements are not visible to uninitiated eyes. So nature has these orioles truly protected—except for one thing. Every male bird is singing, ringing out songs so loud and clear that any interested ear in the canyon knows exactly where these birds are feeding. Even the females sing softly near their nests, no more concerned than the males with such things as color absorption and invisibility.

Music pours forth from a quartet of Scott's orioles perched atop a blooming ocotillo shrub.

Recognition. 6½–8 in. long. Male yellow (not orange), with black back, head, throat, wings, and most of tail. Female grayish-green, streaked above; underparts greenish-yellow.
Habitat. Dry woods, desert scrub, and gardens.
Nesting. Nest is a deep, well-woven cup of grass attached to leaves of yucca or outer branches of trees. Eggs 2–4, pale blue, marked with brown and lilac. Incubation about 14 days, by female. Young leave nest about 14 days after hatching. Usually 2 broods a year.
Food. Insects, berries, and nectar from flowers.

Red-winged Blackbird

Red-winged Blackbird *Agelaius phoeniceus*
Tricolored Blackbird *Agelaius tricolor*

Red-winged Blackbird

Tricolored Blackbird

Pale March sunlight spreads across a world bleached free of color. In a ravaged marsh a glossy blackbird climbs along a cattail stalk and pauses, waiting for some unseen signal. Abruptly it flexes its wings, flashing bright red epaulets in winter's face. *Tonk-a-leee,* the bird sings. *Tonk-a-leee!* The challenge hangs in the air and is taken up by a second red-shouldered sentry, then another. Across northern states, the red-winged blackbirds have returned, signaling the beginning of a new season.

These widespread and familiar birds winter in southern states, sometimes gathering in large flocks along with grackles and cowbirds. But at the first hint of spring, mature males head north to stake out their claims. The females arrive several weeks later, touching off frenzied courtship displays. Marshes fairly explode with bursts of red as rival males try to prove that *they* are the biggest, flashiest red-wings around, sometimes settling matters by reckless, high-speed chases. Then, as soon as domestic lines have been drawn, the footloose first-year males appear and strain the tranquillity. But older males are dogged defenders of territory, and females are occupied with nesting duties. Unable to win territory or brides, young males gather in bachelor flocks and wait until next year.

Recognition. 7–9½ in. long. Male black with red shoulders; female heavily streaked with dark brown. Border of shoulder patch buff on male red-winged, white on male tricolored; female tricolored much darker than female red-winged.
Habitat. Marshes and grasslands.
Nesting. Nest is a cup of grass, lashed to reeds or in small bush. Eggs 3–5, pale blue marked with zigzag lines of brown or blackish. Incubation about 12 days, by female. Young leave nest 10–13 days after hatching. Tricolored nests in huge colonies.
Food. Seeds, grain, insects, and spiders.

Yellow-headed Blackbird

Xanthocephalus xanthocephalus

In a Wyoming marsh a burly blackbird hitches himself to a bulrush stalk, his golden hood glistening in the sun. Slowly, deliberately, he fans his glossy tail. With the eyes of his intended upon him, he opens his wings in a dramatic gesture of appeal. Then bowing low, till his golden crown is nearly flush with his tail, he opens his mouth and emits about the nastiest sound ever heard from a bird's mouth.

Even the most generous critics concede that as a vocalist, the yellow-headed blackbird is unmusical—and a few might agree with W. L. Dawson, who described the song as "a wail of despairing agony which would do credit to a dying catamount." It seems that no one who has written about this bird of marshes and reedy lakes can resist commenting on the disparity between its spectacular courtship display and its horrific, croaking voice.

Yellow-heads nest in colonies, some quite large. They favor deeper water than the smaller red-winged blackbirds, and where the two birds occur together, the red-wings are shunted to the shallower marshland edge. Deep water is a deterrent to such predators as raccoons and skunks, while colony life is a defense against marauding hawks and crows. Any harrier that approaches a colony too close must soon contend with an angry black and yellow cloud of birds. It is a wise hawk that retreats before it is put to rout.

Bowing in his courtly manner, a yellow-headed blackbird seeks to impress his lady.

Recognition. 8–10 in. long. Adult male mainly black; head and upper breast bright yellow; small patch on wings white. Female browner, with only face and breast yellow.
Habitat. Freshwater marshes; fields and stockyards in winter.
Nesting. Nest is a deep cup of stems and grass, woven into standing reeds 1–6 ft. above water. Eggs 3–5, pale bluish, speckled and blotched with brown and reddish-brown. Incubation about 13 days, by female. Young leave nest after 9–12 days. Usually nests in colonies.
Food. Insects, small snails, seeds, and grain.

Rusty Blackbird

Euphagus carolinus

An oddball among blackbirds, the rusty flies like a thrush, has a bill like a thrush, and feeds in shallow water like a sandpiper. Wading about in water to the tops of its thin black legs, it captures aquatic insects and their larvae, digs out small crustaceans, and fishes for tiny tadpoles and fish. And as if to underline its fondness for watery places, its common call note is a low-pitched, hoarse *chuck* so like the croak of a wood frog in spring that it is difficult to know which creature is making the sound.

Rusty blackbirds travel in great noisy flocks with other blackbirds during the fall, winter, and early spring, but in nesting season they are usually solitary, building their bulky nests in wilderness areas of Canada, Alaska, and the northeastern fringes of the United States. When their young are fledged, the summer molt is over, and autumn has come, they return to spend winter over most of the eastern United States and in scattered areas of the West.

But they look like different birds entirely. Gone is the male's coat of black-glossed-with-green and the female's slate gray. They are all, including this year's young, newly garbed in feathers of rusty brown that show off their pale yellow eyes to perfection. But long before spring, those rust-colored fringes will have worn away, and the rusty blackbirds will go back to their northland homes with not a hint of their autumn attire.

Wading in shallow waters, a rusty blackbird dines on small fish as well as aquatic insects.

Recognition. 8½–10 in. long. Bill slender; eye yellow. Breeding male glossy black; fall male tinged with rusty on head, breast, and back. Female blackish, or else dull gray like Brewer's, but with yellow eye.

Habitat. Wooded swamps, river groves, and flooded fields; nests in northern bogs.

Nesting. Nest is a cup of twigs and grass 2–20 ft. above water in shrub or spruce tree. Eggs 4–5, pale blue, blotched with brown. Incubation about 14 days, by female. Young leave nest 14 days after hatching.

Food. insects, snails, fish, tadpoles, seeds, grain, and berries.

Brewer's Blackbird

Euphagus cyanocephalus

Of all blackbirds, the Brewer's are probably the least daunted by human beings. Walking with that familiar forward-jerking head-bob common to birds afoot, they feed as nonchalantly on small-town lawns and in city parks as they do in open grasslands or mountain meadows. They have even been known to settle into a farmer's barnyard, intimidate the resident poultry, and fatten up on the furnished grain.

All through the breeding season Brewer's blackbirds keep to their own kind in small colonies. But as soon as the young can fly well, neighboring colonies join together and pick up flocks of red-wings, cowbirds, starlings, and grackles until the vast, feeding mob may number tens of thousands of shimmering dark-colored birds. Landing, they blacken the fields; taking off, they rise like thick smoke. Flying en masse, they will rise and fall, bend and turn, speed and slow with graceful synchronization, flowing like an airborne river above the contours of the land.

But for all its smoothness, this river does not flow silently. Each bird in the huge flock constantly gives the call of its own species, until the multitude produces an unbelievable creaking, squeaking, rasping, sputtering cacophony—a startling call of the wild that inspires a human listener to awe.

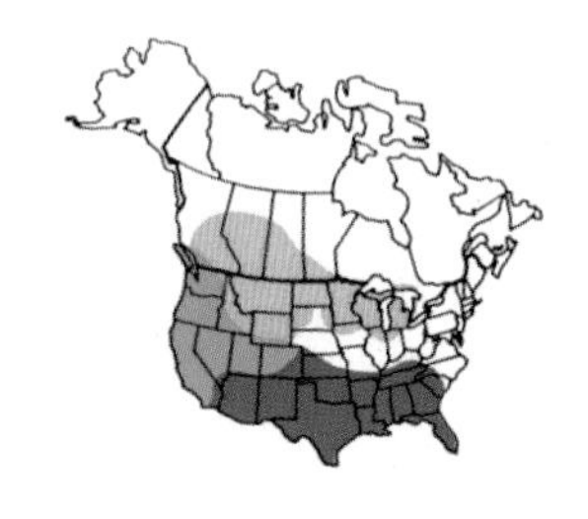

Driving off a farmer's poultry flock, Brewer's blackbirds can snatch the feed for themselves.

Recognition. 7½–9½ in. long. Bill slender. Breeding male glossy black; head tinged with purple; eye whitish. Female dull gray with dark eye. Head jerks forward when bird walks.
Habitat. Open country with scattered trees, farmyards, parks, and lawns.
Nesting. Nest is a cup of twigs, pine needles, and grass, up to 150 ft. above ground in conifer. Eggs 3–7, gray or greenish, spotted with brown and gray. Incubation 12–14 days, by female. Young leave nest about 14 days after hatching. Nests in colonies.
Food. Insects, grain, and seeds.

Great-tailed Grackle

Great-tailed Grackle *Quiscalus mexicanus*
Boat-tailed Grackle *Quiscalus major*

Great-tailed Grackle

Boat-tailed Grackle

It is a mystery that these two great grackles have remained distinct species, with no known interbreeding. In appearance they are almost identical, as are their habits. Both live in open country close to water, where they feed on small fish and other aquatic creatures, and both prey on the eggs and young of other birds. The males of both species are polygynous and seek their multiple mates by bobbing and bowing, spreading their wings and tails, uttering series of short grating notes, and singing courtship songs—one of the few aspects in which the birds differ. The songs of the great-tailed grackle are filled with loud, piercing tones followed by high, falsetto squealing, while those of the boat-tailed are lower-toned, more rolling, rattling, and throaty. Yet even where the ranges of the two species overlap in Louisiana and Texas, though they often nest together in large colonies, the birds never interbreed.

This is curious indeed, considering the interbreeding of birds with such different appearances as the lazuli and indigo buntings, or the rose-breasted and black-headed grosbeaks. But in these hybridizing pairs the voices are very much alike—which may be just the point so far as grackles are concerned. It seems that for many female birds, it's not the clothes that make the man, but his voice.

Recognition. 12–17 in. long. Male great-tailed glossy black with long, keeled tail; eyes white or yellow. Female brown, paler on breast. Boat-tailed similar, but male less glossy; eyes usually pale; female has buff breast, brown eyes.
Habitat. Great-tailed in streamside woodlands, groves, and towns. Boat-tailed in coastal salt marshes; also inland lakes and streams in Florida.
Nesting. Nest is a cup of twigs and grass 2–50 ft. above ground. Eggs 3–4, pale blue, marked with brown, purple, or black. Incubation about 14 days, by female. Young leave nest about 21 days after hatching.
Food. Mainly insects, small birds, and grain.

Common Grackle

Quiscalus quiscula

Head up, his glossy black feathers glinting in the sunlight, a male common grackle strides elegantly across the new spring grass and pauses before three females. Lifting his head still higher, he drops his wings, ruffles his shoulder feathers, opens his long black bill—and sings a song that has all the melody of a creaking garden gate. Although none of the females is interested at the moment, none is upset either, for his vocal performance *is* singing—it is the spring mating song of all common grackles, no matter how raucous it may sound to human ears.

Common grackles are classified as songbirds not because of the beauty of their songs but because they have all the vocal equipment a songbird needs. Birds have no vocal cords, and apparently neither their bills nor their tongues enter into the shaping of their songs. Rather, all the sounds and all the inflections come from a resonating voice box down at the bottom of the windpipes, just where the bronchial tubes branch off to the lungs. By varying the pressure of air from the lungs as it passes through this voice box, a bird changes both the loudness and the pitch of his notes. For this reason some ornithologists say that birds cannot make consonant sounds and sing only vowels. True or not, the common grackle makes all of his vowels sound like rusty hinges.

Unabashed by his squeaky song, a male common grackle puffs up proudly to impress a female.

Recognition. 10–12½ in. long. Large, with stout bill, keeled tail, and yellow eyes, but smaller than great-tailed. Male black, highly glossed with purple, greenish, or bronze. Female less glossy, shorter-tailed, than male.
Habitat. Groves, towns, and farmland.
Nesting. Nest is a sturdy cup of twigs, stems, grass, and sometimes mud, up to 45 ft. above ground in tree. Eggs 4–7, bluish or pinkish, scrawled and blotched with brown. Incubation about 14 days, by female. Young leave nest 18–20 days after hatching.
Food. Insects, seeds, grain, salamanders, eggs and young of other birds, and even small fish.

Brown-headed Cowbird

Molothrus ater

It's an unlikely-looking villain, this small blackbird with a brown hood like a monk's. The female, somberly clad in gray, seems even more innocuous. Yet the cowbird, abetted unwittingly by man, is believed responsible for reducing the population of many eastern woodland birds by half.

Cowbirds are vagabonds that once followed bison and cattle through the plains, but as forest barriers fell to the settler's ax, they extended their ranges to both coasts. Since the life of a rover has no room for home and family, female cowbirds deposit their eggs in the nests of other birds, then leave the care of cowbird young to the unlucky foster parents. Usually, the unsuspecting host is a smaller species, such as a warbler or vireo, and the larger, early hatching cowbird out-competes its nestmates for food and space. When fledged, young cowbirds join roving cowbird flocks. How they recognize their own kind is a mystery.

If cutting forests opened the door for this prairie tramp, suburbanization has rolled out the red carpet. Because the cowbird needs open spaces for feeding and access to woodland tracts for breeding, housing developments, highways, and the like have opened the way for its expansion. In some eastern woodlands, entire species of birds have all but vanished because of this parasite. But the cowbird cannot succeed alone; it is only an opportunist, taking advantage of changes man has wrought in the environment.

Domestic rivalry can be dangerous when a baby cowbird crowds the nest of a red-eyed vireo.

Recognition. 6–8 in. long. Small, with stout, dark, sparrowlike bill, short tail. Male glossy black; head lustrous brown. Female uniform grayish-brown. Often forages on ground with other blackbirds and starlings.
Habitat. Woodland edges, thickets, roadsides, and towns.

Nesting. Eggs 10–12 per female, white speckled with brown, each laid in nest of another species, with warblers, vireos, finches, and small flycatchers preferred. Incubation by host about 12 days. Young leave nest about 10 days after hatching.
Food. Grain, seeds, berries, and insects.

Rosy Finch *Leucosticte arctoa*

also known as

Gray-crowned Rosy-Finch

Leucosticte tephrocotis

Gray-crowned male

From high on a Colorado mountainside, a single call note sounds. It hangs momentarily in the cold, thin air at 12,000 feet, then plummets into the valley below. Following close behind is a rippling flock of dark-bodied finches. Pale wings flash in the sunlight as they land at the edge of a snowfield and begin to feed. There is no mistaking the birds for anything but the hardy rosy finch, which thrives in the tundra of high elevations.

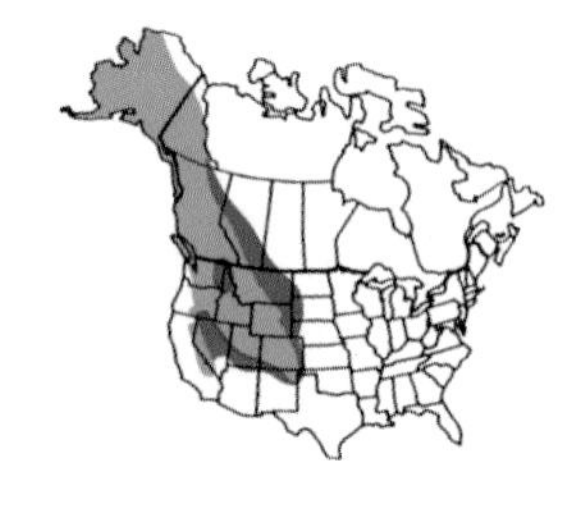

Not long ago an observer in the Rockies might have had difficulty identifying the feeding birds. Before 1983 the American Ornithologists' Union recognized three species of rosy finch in North America: the wide-ranging gray-crowned rosy finch, and the more localized brown-capped and black rosy finches. Despite differences in range and plumage, the three birds are now regarded as subspecies of a single one: the rosy finch.

Taxonomy is a tidy science but hardly interpretation-free. The line between a species and subspecies is often a fine one, with classifications made on the basis of supporting evidence. Because of new evidence, and a recent fundamental shift in the way scientists classify birds and other creatures, there is reason to believe that the three rosy finches will once again be split into three separate North American species. This will make identifications more difficult—but more interesting too.

Like many in its family, the rosy finch has a bill well adapted for cracking seeds.

Recognition. 5½–6½ in. long. Male mainly brown; sides, belly, shoulders, and rump tinged with pink. Female duller. Most birds have gray nape, black forehead. Southern Rockies' birds have head all brown; those of northern Rockies mostly black tinged with pink.
Habitat. Rocky alpine tundra; plains in winter.

Nesting. Nest is a bulky cup of grass, feathers, and moss, hidden in crevice among rocks or in cliff above timberline. Eggs 2–6, white. Incubation 12–14 days, by female. Young leave nest about 20 days after hatching.
Food. Mainly seeds; a few insects.

House Finch

Carpodacus mexicanus

Many wild house finches were once sold into captivity by unscrupulous bird traffickers.

Residents of western states have long been familiar with this small, friendly linnet. In urban, suburban, and farm areas alike, the raspberry-headed male and brown, streaky female may be the most common birds around.

Back in 1941, people living in the vicinity of New York City began making the bird's acquaintance too. That year Dr. Edward Fleisher of Brooklyn, New York, was surprised to find 20 house finches for sale in a pet shop under the name "Hollywood finches." House finches, like all native North American species, are protected by the Migratory Bird Treaty, which forbids their capture and sale. An alerted National Audubon Society made inquiries of 20 local pet shops and found they *all* carried the bird. Shipments of house finches from California were subsequently banned, but not before thousands of birds had been trafficked to eastern markets. Amid the furor, at least one proprietor evidently released his stock. On April 11, 1941, a brightly colored male house finch was sighted at Jones Beach, on Long Island, and in March 1942 a group of seven birds was found in nearby Babylon.

Now, many generations later, the progeny of those transplanted birds are abundant over much of the East and Southeast and continue to spread, pioneering a path back across the continent. At last estimate, the stay-at-home western house finches and their eastern cousins were separated by less than 100 miles.

Recognition. 5–5½ in. long. Male streaked and sparrowlike; eyebrow, forehead, and breast rose-red; flanks finely streaked. Female brown streaked overall, with no red.
Habitat. Towns, suburbs, farmlands, deserts, and thickets.
Nesting. Nest is a cup of grass, stems, and leaves 5–7 ft. above ground in woodpecker hole, other cavity in tree or building, or in dense foliage. Eggs 2–6, bluish with fine speckling. Incubation 12–14 days, by female. Young leave nest 11–19 days after hatching.
Food. Seeds, fruit, buds, and bread crumbs.

Common Redpoll

Carduelis flammea

It is February in northern Minnesota, the season of deep cold. Skies are blue, and the air is sharp and brittle, filling the lungs like broken glass. Nothing moves in the icy stillness. Then suddenly, from out of the north, comes a lisping, raspy storm of notes that sweeps ahead of a tumbling flock of birds. The pale swarm bears down upon a birch and festoons itself over the branches. Like a brigade of animated ornaments the red-capped birds dangle from branches and rip into catkins, sending husks raining to earth. Cold? What cold? They seem positively oblivious, delighted to have the frozen world all to themselves.

Long after the last of autumn's migrants have fled south to avoid winter, common redpolls sweep out of Canada and into the northern United States. Some years these plump, finely featured finches may appear in frenetic hordes, inundating woodlands, fields, and backyard bird feeders. Other years, they may absent themselves entirely. Active and nervous, the flocks seem driven by a common will. If one takes wing, all take wing. Where one feeds, all feed. If disturbed, they take off amid a flurry of call notes but, as often as not, will return to the same spot and begin feeding again. Actually quite tame, they often seem unaware of humans. In March, when winter begins to loosen its grip on the world, the wandering flocks depart. They head north to nest where spruce forests thin and become tundra, and winter is never far away.

When a common redpoll feasts on feeder seed, the approach of a human is of little concern.

Recognition. 4½–5½ in. long. Small, pale, and streaked, with conical yellow bill; forehead red; chin black. Male has pink breast. Female similar, but lacks pink on breast.
Habitat. Weedy fields and thickets; nests in Arctic scrub and tundra.
Nesting. Nest is a cup of twigs and grass in shrub, 3–6 ft. above ground. Eggs 3–7, pale blue, finely speckled with brown, gray, and lilac. Incubation 11–15 days, by female. Young leave nest 12–14 days after hatching.
Food. Seeds, especially of birches, alders, and willows; also insects in summer.

Lesser Goldfinch

Carduelis psaltria

"Handsome is as handsome does" could be the motto of the female lesser goldfinch when it comes to selecting her mate, for on the basis of appearance alone, she has a bewildering array of choices. In mature adult males, the backs vary from olive green to solid black, and the underparts from very pale to very bright yellow, while the young males—in their first breeding season—still wear their light gray-green colors, similar to those of the female herself. Still, the lady readily considers them all.

It has long been theorized that male birds in general developed such striking characteristics as bright plumage, brilliant combs and wattles, plumy heads, and colorful pouches to attract females. Yet in some species—the lesser goldfinch and the American redstart, for instance—less colorful young males seem to mate as successfully as older ones in brilliant plumage. In those few species that interbreed, the females look very much alike, while the plumage of the males varies; but the male songs and their courtship behavior are nearly identical.

Can it be that some females do not even consider the colorful male finery? Perhaps they are captivated instead by correct songs and proper courtship behavior. Thus the young lesser goldfinch male, which goes courting in a pale coat, probably wins his mate by singing his sweet, plaintive song especially to her and by feeding her such favored tidbits as thistle seeds.

A male lesser goldfinch proffers his mate a tidbit as proof of his devotion.

Recognition. 4–4½ in. long. Small; underparts yellow; wings black with white patch. Male has black crown and back (back greenish in Southwest and California). Female greenish on head and back. Flight bounding; often flies in flocks.
Habitat. Thickets, brushy country, open woodlands, and gardens.

Nesting. Nest is a tidy cup of silky plant fibers, fine grass, and plant down on tree limb, 2–30 ft. above ground. Eggs 3–6, pale blue or greenish, unmarked. Incubation about 12 days, by female. Nestling period unknown.
Food. Mainly seeds.

Lawrence's Goldfinch

Carduelis lawrencei

As all birds do, Lawrence's goldfinches live where the climate and the terrain suit them and where they can find enough food to sustain them. But because of fairly specialized needs, their lives are more restricted than those of other finches. During the nesting season, for example, their principal source of food is the seeds of the fiddleneck plant, a hairy-leaved annual that grows in dry, open areas. And so their nesting is limited to the inland reaches of California, southwestern Arizona, and Baja California, where these plants can be found. After a moist winter, fiddlenecks grow abundantly, spreading vast fields of yellow or orange-yellow flowers across the land. Even while the budding tops are still unfurling, tiny brown seeds are already ripening to produce a profusion of food for nesting goldfinches. After a dry winter, however, fiddlenecks are sparse, and the small birds' territory narrows even further.

Finding sufficient water presents another problem in their semiarid habitat. They gravitate to the banks of streams or creeks—any trickle they can find—and when these dry up they may have to take up housekeeping near an overflowing tank, a leaking garden faucet, or even a birdbath to supply their needs. But Lawrence's goldfinches are gypsies at heart; if they don't find food and water in one valley, they will look for them in the next.

An avid bather, the Lawrence's goldfinch will head for water wherever it happens to be.

Recognition. 4–4½ in. long. Small, mostly gray, with 2 yellow wing bars. Male has black face and chin; yellow breast. Female lacks black on face, has less yellow on breast. Flight bounding, like that of other goldfinches.
Habitat. Dry open woods and thickets.
Nesting. Nest a neat cup of grass, lichens, and stems 3–40 ft. above ground in shrub or tree. Eggs 3–6, bluish and unmarked. Incubation by female, period unknown. Young leave nest about 13 days after hatching. Often nests in loose colonies.
Food. Mainly seeds; some insects.

American Goldfinch

Carduelis tristis

From coast to coast and from northern Mexico into southern Canada, flocks of American goldfinches add a special liveliness to wide-open country with their bright, yellow and black coloring (paler in the West), their roller coaster flights over invisible hills and valleys of the air, and their sweet twitterings of *Just-look-at-me! Just-look-at-me!*

In the West these goldfinches may nest in May or June, but in the East they usually don't start families until August or even September, when a good supply of wild seeds, especially thistle seeds, can be counted on. For though most seedeaters feed their nestlings insects, these goldfinches nourish their youngsters on seeds that have been shelled and partially predigested. The parents fill their crops with seeds—and perhaps pick up a few tiny caterpillars and tender aphids for added protein and flavor. After a while, they regurgitate them into the open mouths of their hungry babies, not as a liquid mixture but particle by particle—most unusual food for birds so young. Carrying such a mouthful, a parent can feed every nestling, instead of the usual one or two, with each return to the nest.

And what an elegantly constructed nest those birdlings occupy! Its walls are so thick and so tightly woven that in a teeming downpour, which often occurs in late summer, its cup will retain and fill with rainwater unless a parent bird is there to shield nest and nestlings alike under its feathered, yellow and black umbrella.

An American goldfinch shields her tightly woven nest to keep it from filling with rainwater.

Recognition. 4–5 in. long. Small. Breeding male mainly bright yellow, with black forehead, wings, and tail. Female and winter birds duller and grayer; lack black cap, but have black wings and tail. Flight bounding; often travels in flocks.
Habitat. Fields, groves, thickets, and farmlands.
Nesting. Nest is a thick-walled cup of plant fibers and plant down, 1–60 ft. above ground in shrub or tree. Eggs 4–6, bluish white and unmarked. Incubation 12–14 days, by female. Young leave nest 11–17 days after hatching.
Food. Mainly seeds; a few insects and berries.

House Sparrow

Passer domesticus

Preferring to nest close to man—in fact part of its Latin name, *domesticus,* means "belonging to a household"—the house sparrow is an undeniably friendly bird. But when Brooklynite Nicolas Pike first imported a group of them from England in the early 1850's, no one dreamed that this aggressive newcomer would soon drive out bluebirds, tree swallows, and other gentle, cavity-nesting birds, or prove so prolific that its very name would become synonymous with "nuisance."

The first small flock flourished and multiplied. Soon they had spread into the surrounding towns and countryside, and before long were radiating north, south, east, and west—not only by reproducing abundantly, but even by riding the rails. For while pecking at grain on the floors of freight cars, they were sometimes shut in and not released until the cars were reopened down the line.

Because they were known in Europe for eating the caterpillars of the snow-white linden moth, a pest to city shade trees, other American cities imported them too—St. Louis in 1875, and towns in Texas, Utah, California, and some New England states. In less than 50 years, this boisterous bird extended its range from New York across the United States and well into Canada. Today these brown and gray chirpers, which squabble fiercely in competition for mates and gather in noisy congregations, are a familiar sight to city, suburban, and country dwellers alike.

During its early days in America, the imported house sparrow at times traveled by rail.

Recognition. 5–6 in. long. Streaked brown above. Male has gray crown, whitish cheek, black chin. Bill and breast black in summer; bill yellow, breast gray in winter. Female has brown crown and plain breast.
Habitat. Cities, suburbs, and farms.
Nesting. Nest built of grass and string, in bird house or hole in tree or building. Eggs 3–7, white or bluish marked with brown. Incubation 11–14 days, by female. Young leave nest after 15–17 days. At least 2 broods a year.
Food. Insects, spiders, berries, seeds, and grain; city birds eat bread crumbs.

Smaller Woodland

Woodlands come in many varieties, from sunlit aspen groves, to misty cedar forests, to the mixed hardwoods of the Appalachians—and each contains its own world of diversity. A shady forest floor may be home to ovenbirds and other ground-nesters, while cardinals and wrens live in the undergrowth at its borders; woodpeckers scour tree trunks for insects, while thrashers and orioles turn the leafy treetops into a canopy of song.

Birds

Blue-throated Hummingbird

257 Wrentit
258 Long-billed Thrasher
259 Brown Thrasher
260 Gray Catbird
261 Cedar Waxwing
262 White-eyed Vireo
263 Bell's Vireo
264 Solitary Vireo
265 Yellow-throated Vireo
266 Hutton's Vireo
267 Warbling Vireo
268 Red-eyed Vireo
269 Blue-winged Warbler
270 Golden-winged Warbler
271 Tennessee Warbler
272 Orange-crowned Warbler
273 Nashville Warbler
274 Virginia's Warbler
275 Northern Parula
276 Yellow Warbler
277 Chestnut-sided Warbler
278 Magnolia Warbler
279 Cape May Warbler
280 Black-throated Blue Warbler
281 Yellow-rumped Warbler
282 Black-throated Gray Warbler
283 Townsend's Warbler
284 Hermit Warbler
285 Black-throated Green Warbler
286 Blackburnian Warbler
287 Yellow-throated Warbler
288 Grace's Warbler
289 Pine Warbler
290 Kirtland's Warbler
291 Bay-breasted Warbler
292 Blackpoll Warbler
293 Cerulean Warbler
294 Black-and-white Warbler
295 American Redstart
296 Prothonotary Warbler
297 Worm-eating Warbler
298 Louisiana Waterthrush
299 Northern Waterthrush
300 Ovenbird
301 Kentucky Warbler
302 Mourning Warbler
303 Hooded Warbler
304 Wilson's Warbler
305 Canada Warbler
306 Red-faced Warbler
307 Painted Redstart
308 Hepatic Tanager
309 Summer Tanager
310 Scarlet Tanager
311 Western Tanager
312 Northern Cardinal
313 Rose-breasted Grosbeak
314 Black-headed Grosbeak
315 Evening Grosbeak
316 Rufous-sided Towhee
317 Bachman's Sparrow
318 Fox Sparrow
319 White-throated Sparrow
320 Dark-eyed Junco
321 Yellow-eyed Junco
322 Orchard Oriole
323 Hooded Oriole
324 Northern Oriole
325 Pine Siskin
326 Purple Finch
327 Pine Grosbeak
328 White-winged Crossbill
329 Red Crossbill

Black-billed Cuckoo

Coccyzus erythropthalmus

Occasionally the black-billed cuckoo chooses the home of a chipping sparrow for her egg.

Both black-billed and yellow-billed cuckoos are perfectly capable of building nests and rearing their own young. But occasionally one of them lays an egg in another bird's nest, often with dire consequences for the host's offspring.

Baby cuckoos, born naked and coal-black, must be kept warm by adults until they are about six days old. Then suddenly they sprout bristlelike feather-tubes all over their bodies and immediately start pulling these transparent casings off the enclosed plumage until they stand snugly clothed in their own first feathery coats. Within the next few days the rambunctious nestlings will jostle their way out of the nest and begin clambering with awkward grace about the tree or the bushy tangle in which their home is hidden. If one falls to the ground, as happens now and then, it is strong enough to pull itself back to safety. No wonder, then, that when a cuckoo egg is laid in another bird's nest, this robust activity at so early an age results in the unceremonious ousting of the host bird's rightful heirs.

Generally, however, American cuckoos do not foist their eggs on other, unsuspecting birds, and no one knows what triggers such behavior when they do. Nonetheless, these wayward parents seem to have a shrewd eye for real estate. Although the cuckoos' own nests are only thin, see-through, twig platforms, the nests in which they illicitly deposit their eggs are firmly and securely built.

Recognition. 11–12½ in. long. Slender and long-tailed. Brown above; underparts white; tail feathers with small white tips visible from below and in flight. Bill black. Wings show no rusty. Usually shy.

Habitat. Woodlands, margins of forests, orchards, and thickets.

Nesting. Nest is a flimsy saucer of twigs and grasses, 2–10 ft. above ground in tree or dense thicket. Eggs 2–5, blue-green. Incubation about 14 days, by both sexes. Young leave nest about 2 weeks after hatching.

Food. Mainly caterpillars; also other insects and some berries.

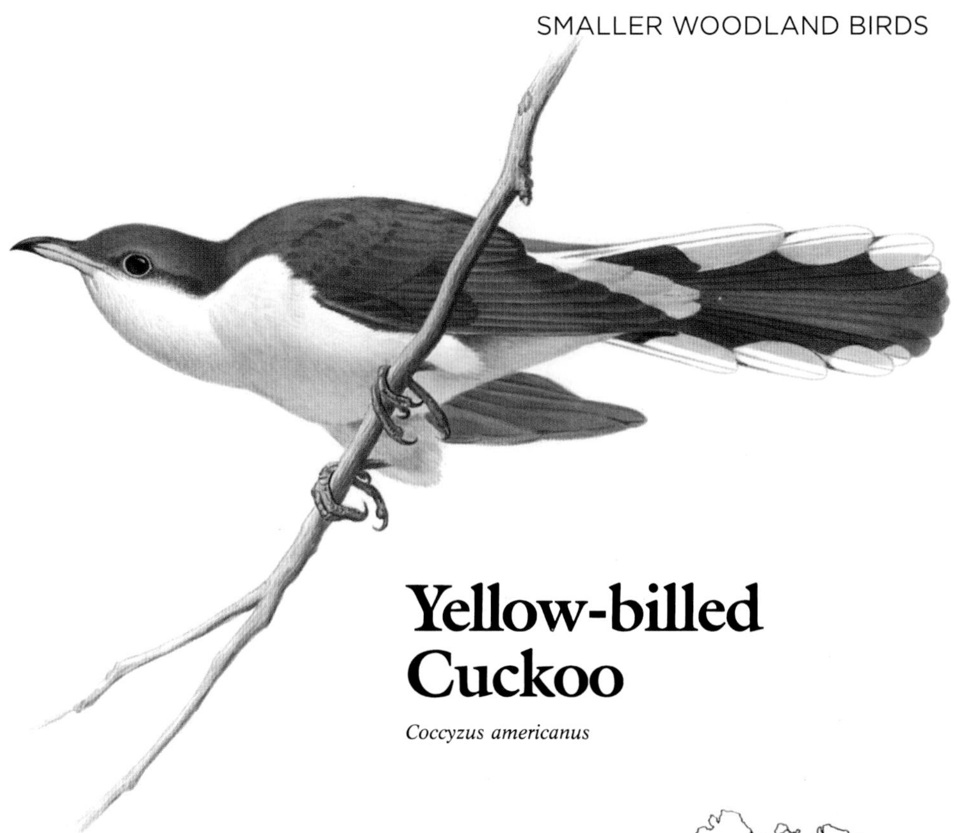

Yellow-billed Cuckoo

Coccyzus americanus

American cuckoos can't actually sing, but the unusual sounds they make so often precede summer storms that for generations the birds have been called "rain crows" by country folk who have probably never even glimpsed them. For these slender, secretive birds are seldom seen by anyone. Slipping furtively among trees and thickets, they fly briefly across clearings with swift and purposeful grace, heading straight from one sheltering tangle to another. Once hidden again by the foliage, they search for insects, particularly hairy caterpillars, climbing about with stealthy, almost snakelike movements that only the most fortunate observers have espied.

But throughout the countryside nearly everyone has heard these birds. From May to October, when they are resident in much of North America, their strange songs are apt to erupt from woods and brushy roadside tangles at almost any time of night or day. *Kuk-kuk-kuk-kuk* calls the black-billed cuckoo, repeating that syllable from six to several hundred times with hardly a change in pitch or rhythm. His yellow-billed cousin's song may be shorter and more varied—descending to a guttural *cowk-cowk* at the end—but its effect is no less eerie as it rings facelessly in hollow tones from the dense cover of the trees.

For the yellow-billed cuckoo, one of life's supreme pleasures is to gorge on hairy caterpillars.

Recognition. 11–13 in. long. Slender and long-tailed. Brown above; underparts white; tail feathers with large white tips visible from below and in flight. Bill partly yellow. Wings show rusty in flight. Usually shy.
Habitat. Woodlands and thickets; also orchards and brushy farmlands.

Nesting. Nest is a flimsy saucer of twigs, 2–12 ft. above ground on horizontal branch in dense tree or thicket. Eggs 1–5, blue-green. Incubation about 14 days, by both sexes. Young leave nest about 2 weeks after hatching.
Food. Mainly caterpillars and other insects; also berries, frogs, and small lizards.

Chuck-will's-widow

Caprimulgus carolinensis

The largest of North American birds in the nightjar family—a group noted for its loud, distinctive calls uttered between dusk and dawn—the chuck-will's-widow has an enormous mouth. As it glides mothlike through the night air, it scoops up beetles and other insects, and sometimes even small birds, between its gaping jaws.

The ancients thought that nightjars sucked goats' milk at night, hence the Latin name, which means "milker, or sucker, of goats." Undoubtedly, there were plenty of good insect meals to be found around the animals.

Like the whip-poor-will and others of this family, the chuck-will's-widow is also known for its marvelous protective coloring, which makes it nearly invisible to predators. Humans who seek out this songster of the night may see it along roadways in the evening, where it perches lengthwise on low tree limbs and its eyes gleam from the reflected light of an automobile or the moon.

The male of the species goes courting with a routine that includes a confident strut toward the female, wings adroop and tail outspread, plus a few calls and puffs thrown in for good measure. Later on, he may assume some family responsibility by helping to raise the first brood, if his mate lays a second clutch of eggs.

The chuck-will's-widow could be called bigmouth, so obvious is its gaping orifice.

Recognition. 11–12 in. long. Reddish brown finely marked with black. Throat and chin brown. Male has white on outer tail feathers. Nocturnal; usually rests on ground during day. Best distinguished by call, a mellow *chuck-will's-wid-ow.*

Habitat. Pine woods and streamside forests.

Nesting. Nest is a simple depression among leaves on forest floor. Eggs 2, glossy, pinkish or buff with brown and lilac spots. Incubation about 20 days, by female only. Young leave nest about 17 days after hatching.

Food. Beetles, moths, and other large insects caught in flight.

Whip-poor-will

Caprimulgus vociferus

Whatever else naturalists say about the whip-poor-will, all agree it is a bird that never tires of hearing its own voice. One night, for example, the great outdoorsman John Burroughs counted 1,088 consecutive renditions by one bird of the haunting *whip-poor-will* call that gives this rarely glimpsed creature its name. Whip-poor-wills use this call from May through early August to help locate a present or prospective mate or a rival.

This bird is a master of camouflage, too. So perfectly do its feathers blend with the ground cover where it nests that the female does not even bother with other concealment—confident that she can incubate her eggs without ever being discovered. The bird's eyes, usually little more than slits as it sits on the forest floor or on a bough during daylight hours, open round and large at night as it takes to the air. Moving rapidly, and usually less than 25 feet above the ground, it flies and darts with mouth extended wide, and traps flying insects with the speed and dexterity of a bat.

In the whip-poor-will's lifestyle nature works another of its astonishing miracles, conforming the rhythm of the bird's reproductive cycle to the waxing and waning of the moon. Arriving when the moon is at its brightest, newly hatched young are easily satisfied by their parents, who hunt most effectively in full moonlight.

A whip-poor-will enjoys a good dust bath, which reduces excess oil and moisture in its feathers.

Recognition. 9–10 in. long. Grayish-brown finely marked with black. Throat and chin blackish. Male has large white patches on tail. Nocturnal; usually rests on ground or on a bough perched lengthwise during day. Best distinguished by call, a mellow *whip-poor-will.*
Habitat. Deciduous forests.

Nesting. Nest is a simple depression among leaves on forest floor. Eggs 2, whitish or cream-colored with brown and gray spots. Incubation 17–20 days, by female only. Young leave nest about 20 days after hatching.
Food. Large moths and other nocturnal insects caught in flight.

Blue-throated Hummingbird

Lampornis clemenciae

A blue-throated hummingbird extracts threads from a spider's web with which to bind her nest.

As nest builders, hummingbirds are true artists. Their neat little mounds of plant down, usually coated with lichens, mosses, or leaves, are gracefully sculpted and held together with spider or caterpillar silk. Typically, the inside is cozily lined with downy feathers or animal hair.

The female does the building. She may set her abode on a tree limb, plant stem, or twig, or perhaps on top of another species' nest. If she has built in past years, she often returns to a previous nest and adds another story or two. As soon as the platform is complete, she molds it with her belly, then proceeds to add the walls. These dwellings are always petite. For example, the nest of the blue-throat, one of the largest of all North American hummingbirds, measures just 2½ inches across and 2 inches high, and that of the tiny calliope is only 1½ inches across.

Her house built, the female lays in it two white eggs, about the size of small beans; when the nestlings hatch, she assumes the role of provider, thrusting her slender bill down their throats to feed them regurgitated insects. And heaven help the intruder that ventures near her brood or feeding ground, for she won't hesitate to attack even a huge hawk or a crow, darting at it with lightning speed and stabbing beak.

Recognition. 4½–5 in. long. Large for a hummingbird. Metallic green above, with 2 white stripes on face; tail blackish with large white patches at corners. Male has blue throat. Often glides briefly like a swift.
Habitat. Wooded canyons near Mexican border.
Nesting. Nest is a small cup of plant down and moss, held together with spider web and fastened to stem of a flowering plant, fern, or vine, often over running water. Eggs 2, white. Incubation by female only; period unknown. Nestling period unknown.
Food. Nectar and pollen from flowers; also captures insects in flight.

Ruby-throated Hummingbird

Archilochus colubris

The widest-ranging of all North American hummingbirds, the green and scarlet ruby-throat is the only one likely to be seen east of the Mississippi River. Like others in this family, it uses its long thin bill to probe flowers—especially those with red blossoms—for their nectar. A backyard feeder filled with sugared water will also attract this friendly and relatively tame visitor. But if the hummingbird had to depend solely on flowering plants or feeders it would quickly starve, for it has the highest metabolism of any warm-blooded vertebrate in the world, except perhaps the shrew (a tiny mouselike mammal). Just to stay alive, it must eat all day long.

Small insects—beetles, ants, aphids, gnats, mosquitoes, wasps—supply the protein in the hummingbird's diet. Spiders, too, are a favorite food. While gathering spider webs to bind her nest, the female may help herself to some of the insects dangling there and to the spiders as well.

Sap holes, drilled in trees by yellow-bellied sapsuckers, are also attended by ruby-throated hummingbirds, who share territory across the northern part of their ranges. It appears that the hummers time their arrival in spring to coincide with that of the sapsuckers, whose drillings provide the ruby-throats with food until flowers are plentiful.

It's not unusual for the ruby throat to fight for flower rights with a bumblebee.

Recognition. 3–3¾ in. long. Metallic green above, with no rusty tinge; underparts whitish. Male has ruby-red throat; tail blackish, shallowly forked. Female lacks throat patch; tail fan-shaped with white tips.

Habitat. Deciduous woodlands, orchards, and suburban areas.

Nesting. Nest is a cup of plant down and spider web, covered with flakes of lichen, on branch 5–20 ft. above ground. Eggs 2, white. Incubation 20 days, by female only. Young leave nest 20–22 days after hatching.

Food. Nectar and pollen from flowers; also small insects taken from flowers and spider webs.

Calliope Hummingbird

Stellula calliope

Weighing in at only one-tenth of an ounce, the petite calliope is the tiniest of hummingbirds and, indeed, of all the birds that nest in North America. But the male of this species transcends his small stature when he sets out to court a mate. Flying to a height of 100 feet or more, he dives toward the perched female and makes a short buzz as he pulls up nearby and begins to climb again, repeating the aerobatics until the female is duly impressed. It probably doesn't hurt his campaign that he is also good-looking—displaying a shimmering metallic green back, a throat streaked with iridescent red—and strong. One valiant little male was seen lifting a female who had been stunned when she hit a windowpane, grabbing her beak and managing to get her three feet off the ground before losing his grip.

With such tiny bodies, calliope hummingbirds face a challenge in surviving the low temperatures of their breeding places, which extend through the Rocky Mountains, the Sierra Nevada, and the Cascade Range. One way of coping is to become torpid at night—lowering body temperature, heartbeat, and respiration rate—to conserve energy. A nesting female, rather than go torpid, relies on the nest's insulation to reduce heat loss. And if she has placed her nest under an overhanging branch, she and her eggs are snugly protected from the elements.

The extraordinary feat of lifting a stunned female was once performed by a male calliope.

Recognition. 3–3¼ in. long. Very small and short-billed. Metallic green above; underparts whitish with buff on flanks. Male has red streaks on throat; female has dark spots on throat.
Habitat. Mountain meadows and coniferous forests in the West.
Nesting. Nest is a small cup of moss, pine needles, and bark, bound with spider web, on branch 2–70 ft. above ground. Eggs 2, white. Incubation 15 days, by female only. Young leave nest about 20 days after hatching. Female often builds on top of last year's nest; sometimes several are stacked together.
Food. Nectar, small insects, and spiders.

Broad-tailed Hummingbird

Selasphorus platycercus

A strange noise begins far away across the sunlit mountain meadow, where masses of wildflowers grow in the dappled shade of the pines. An odd, high-pitched whirring, it is impossible to ignore. As the sound moves closer, it seems to be coming out of thin air. Could it be a hallucination? But then the source materializes into a glittering speck of green and red, hovering by a flower. The sound is the wing-music of the broad-tailed hummingbird.

It was this rapid, buzzing wingbeat that first gave the hummingbird its name. So swiftly do the wings move—some 50 to 75 beats per second—the human eye sees only a blur of movement. Like a helicopter, the hummingbird can hover in midair as it gathers nectar from the heart of a flower. Remarkably, it can fly straight up, straight down, sideways, even backwards.

Though the hummingbird's incredible flying prowess would be enough for any bird to boast, this tiny dynamo is also endowed with some of the bird world's most splendid colors. With his metallic green back and ruby-red throat set off by white underparts, the broad-tailed could compete with the gaudiest avian jewel of the tropics. And he is not alone. Other hummingbirds—the calliope, the blue-throated, and the ruby-throated all come to mind—are equally polychromatic, every bit as handsome. If only they could be gathered together for a family portrait, it would be a dazzling vision indeed.

The unusual trilling sound of a male broad-tail in flight results from his tapered wingtips.

Recognition. 4–4½ in. long. Metallic green above; underparts mainly white. Male has ruby-red throat; tail rounded and blackish with some rusty. Female lacks throat patch; flanks rusty; tail broad with white tips. Wings of male produce loud trill in flight.
Habitat. Mountain forests and meadows.

Nesting. Nest is a small cup of plant down, bound with spider web, fastened to horizontal branch 1–15 ft. above ground. Eggs 2, white. Incubation 16 days, by female only. Young leave nest about 23 days after hatching.
Food. Nectar and insects from flowers; also catches insects in flight and on foliage.

Rufous Hummingbird

Selasphorus rufus

It is July in a mountain meadow high in the Colorado Rockies, and the rufous hummingbirds are out in force. The adult males—bright copper-colored like new pennies—engage in aerial dogfights over the meadow, while the more subtly hued females and young birds swarm around the wildflowers. It would seem that rufous hummingbirds are the most common nesting birds on this mountainside. But no: these are migrants. Though the calendar says this is midsummer, the rufous are already far from their nesting grounds in Montana and western Washington on their way to winter quarters in Mexico.

According to the conventional view, migratory birds of North America fly north in spring to a summer nesting place, then migrate south in the fall. But the rufous has its own rules: it flies northwest in late winter and southeast in summer, following completely different routes in each of these two seasons.

What is the advantage to this strategy? Picture the same mountainside in April, when patches of snow may still blotch the ground and the earliest flowers have not yet blossomed. At this time, the rufous is moving up the Pacific coastline, where the moderate climate is more hospitable and flowers are abundant in spring. As soon as their young are fledged, they head southeastward in time to catch the biggest bloom of the year in the mountain meadows. With these routes, the rufous hummingbird is assured a steady food supply during its annual travels.

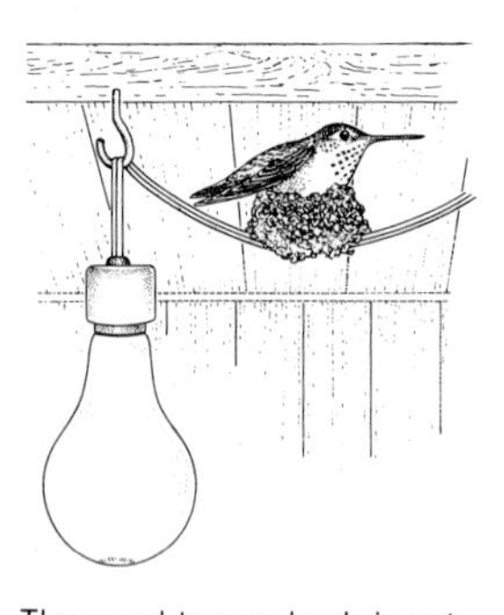

The usual tree or bush is not always a rufous hummingbird's choice for its nesting site.

Recognition. 3½–4 in. long. Male mostly rusty with white breast; throat orange-red. Female has rusty on flanks and in tail. Wings of male produce high-pitched trill in flight.
Habitat. Forests and forest clearings.
Nesting. Nest is a cup of plant down, moss, and bark, decorated with flakes of lichen, 5–50 ft. above ground in tree or shrub. Eggs 2, white. Incubation by female only; period unknown. Young leave nest about 20 days after hatching.
Food. Nectar and insects from flowers; also feeds on running tree sap.

Lewis' Woodpecker

Melanerpes lewis

From a perch astride a fence post a greenish-black bird with a deep crimson face and reddish belly sweeps forth to catch a mayfly on the wing. Judging by its graceful and steady movement through the air, the uninitiated might guess it's a small crow or possibly a kingbird. Actually, it's a Lewis' woodpecker, whose typical feeding habits during the summer are very un-woodpecker-like. Instead of hitching up the trunks of trees, then chipping off the bark or chiseling into the wood in search of beetles, the Lewis' spends most of its time hawking insects—a habit shared by only one other woodpecker, the red-headed.

The Lewis' has other characteristics that set it apart. Alone among our woodpeckers, it makes a habit of sitting on wires and other perches out in the open, and it is the only one with unmarked, dark wings and tail. Its flight too—unlike the undulating pattern of other woodpeckers—is a floating, effortless glide.

This unusual bird was named for Meriwether Lewis, who with William Clark in 1804–06 led an expedition from St. Louis to the Pacific coast to explore the area of the Louisiana Purchase. Their efforts produced a wealth of information about the region—not least, perhaps, the discovery of a singular woodpecker.

Though it prefers insects, the Lewis' woodpecker also shells and stores nuts for winter fare.

Recognition. 10–11 in. long. Mainly blackish, with gray collar and breast, red face, and pinkish belly. Wings broad and rounded. Flight direct, not undulating as in other woodpeckers.
Habitat. Open pine woodlands, groves, and areas with scattered trees.
Nesting. Eggs 5–9, white, laid in cavity excavated mainly by male 4–150 ft. above ground in dead branch. Incubation about 12 days, by both sexes (male incubates at night). Young leave nest 4–5 weeks after hatching.
Food. Insects picked from bark or caught in flight (seldom chisels wood like other woodpeckers); also berries and nuts.

Red-headed Woodpecker

Melanerpes erythrocephalus

If nest holes are scarce, a red-headed woodpecker may gain tenancy only by fighting for it.

Flag birds—as red-headed woodpeckers are sometimes called—are striking to behold, and their strong personalities match their bright coloration. They are often aggressive toward other woodpeckers or any avian visitor that approaches a nest tree or tries to raid a food cache. And in their feeding habits they are bold opportunists.

Like the Lewis' woodpeckers, they spend much of their time flycatching from exposed perches. But these gluttons seem to crave everything, from insects, spiders, earthworms, and birds' eggs, to nestlings, mice, nuts, berries, and corn. Some red-heads even find sustenance in tree bark, and many visit feeders in winter.

Inhabiting areas with well-spaced trees, red-headed woodpeckers are common across their extensive range east of the Rockies. Of late, however, there has been concern for their numbers, especially in the Northeast, where they have declined dramatically. One source of their problem is obvious: like bluebirds, red-heads face stiff competition for nest holes from starlings and other hole-nesters. Another difficulty is more subtle: red-heads like to nest in dead trees or large dead limbs, and in the habitats that they favor, dead or dying trees are often removed for firewood or to reduce the hazard of fire. To the red-headed woodpecker, however, dead wood is a necessity for life.

Recognition. 8½–9½ in. long. Adult black above; head red; underparts, rump, and wing patch white. Immature similar but browner; head brown; wing patch with 2 dark bands.
Habitat. Groves, open woodlands, farmlands, and shade trees.
Nesting. Eggs 4–7, white, laid in cavity excavated 5–80 ft. above ground in dead tree limb. Incubation about 2 weeks, by both sexes. Young leave nest about 27 days after hatching. Often 2 broods a year.
Food. Wood-boring insects and nuts; sometimes insects captured in flight.

Acorn Woodpecker

Melanerpes formicivorus

In the top of a tall tree, a company of clown-faced, black and white birds are raising a ruckus. Yammering at each other in scratchy voices, waving their wings for emphasis, they appear to be having a terrible quarrel. But in fact this is just an everyday conversation for acorn woodpeckers; they are actually among the most cooperative of birds.

Living in colonies of from 3 to 10 birds, they do everything by committee, and as a result can store enough of their winter staple—acorns—to feed them for several months. Choosing a handy tree, such as a sycamore or oak, the woodpeckers drill tiny holes and then gather acorns to stuff into the openings. Every bird shares in the work, and later in the bounty.

The spirit of cooperation even extends to the raising of young. When one pair lays its eggs, all birds in the group help with incubation, trading places several times an hour. After hatching, the young are fed by all the adults. With such cooperation, a nesting couple can raise two or three broods per season.

But this woodpecker society, built on an economy of acorns, can run into hard times if the acorn crop fails one year. For this reason, the largest and most stable colonies are found where there are half a dozen different kinds of oak trees: whatever the weather, at least one type is sure to produce a good harvest in any given year.

A typical granary tree for acorn woodpeckers will eventually be chock-a-block with nuts.

Recognition. 8–9½ in. long. Black above, with red cap and black mask through eyes; forehead, throat, and rump white. Usually in small groups.
Habitat. Oak woodlands, groves, and forested canyons of the Far West.
Nesting. Eggs 4–6, white, laid in cavity excavated 12–60 ft. above ground in dead oak or other tree. Incubation about 14 days, by both sexes. Young leave nest about 4 weeks after hatching. Often 2 broods a year.
Food. Insects, fruit, and acorns. Stores acorns in holes drilled in tree trunks, fence posts, and utility poles.

Golden-fronted Woodpecker

Melanerpes aurifrons

On the mesquite prairies of Texas, where golden-fronted woodpeckers live most abundantly, each pair appears to dig only one nest hole, which is then used for two or even several years thereafter. Away from the mesquite, in pecan groves and mixed-oak forests, each pair may dig two or more nest holes every nesting season. And in some areas, they drill so frequently into utility poles and fence posts that the damage they do is a serious matter.

But to a woodpecker, a pole or a post is simply a dead tree, an ideal place for digging out a dark little room to house eggs and nestlings. (Though occasionally woodpeckers will nest in a live tree, most prefer dead wood.) When a pair has chosen a nesting site, both male and female work on the excavation. They chip off the outer bark to the size and shape of the entrance hole, usually 1½ to 2 inches across, then drill straight back for a ways. From this entrance they then begin the real work of digging downward, chiseling out a cavity 6 to 18 inches, sometimes even 30 inches, deep. With the digging completed, they may make the bottom a little roomier and leave a few wood chips as a cushion for their eggs, but they add nothing else. The whole excavation usually takes from 6 to 10 days, but sometimes as long as three weeks. An astute observer may locate a tree containing a woodpecker's nest by looking for the telltale chips on the ground nearby.

When courting, the golden-fronted woodpecker sings his love song to the sky.

Recognition. 8½–10 in. long. Barred black and white above; face and underparts grayish tan; forehead and nape golden yellow; red patch at center of male's crown. White rump usually visible in flight.
Habitat. Dry woodlands, groves, and mesquite.
Nesting. Eggs 4–7, white, laid in cavity excavated 3–25 ft. above ground in dead trunk, limb, or fence post. Incubation about 14 days, by both sexes. Young leave nest about 4 weeks after hatching, but remain with parents after fledging.
Food. Insects, fruit, and acorns.

Red-bellied Woodpecker

Melanerpes carolinus

The red-belly, like most North American woodpeckers, has a firm body supremely adapted for a lifetime of climbing on and drilling into the trunks and limbs of trees. With short legs and long toes equipped with sharp, down-curved claws, it clings securely to trees and props itself erect with the stiff central feathers of its down-turned tail.

With its heavy, chisel-shaped bill, it chips insects from beneath tree bark, pecks holes to get at wood-boring beetles, slashes out chunks of wood while digging a nest hole, and beats out the rapid hammer strokes of its spring-time communications. To screen its lungs from all the wood dust it produces, its nostrils are conveniently covered by a small mask of fine bristly feathers. And to prevent brain damage from all that pounding, a strong neck, a thick skull, and a cushioning space between the heavy outer membrane and the brain itself act as special protectors.

But the most spectacular feature of a woodpecker is its unbelievably long, cylindrical tongue, which can be extended to an amazing length. The tip of this astonishing tongue is hard and pointed for spearing such insects as grasshoppers and beetles, and sticky with a gluey saliva for lapping up ants. In Florida, where they like to winter, red-bellies even use this adaptable instrument to draw the succulent juice and pulp from oranges.

The red-belly's enthusiasm for orange juice is not welcome news to Florida citrus growers.

Recognition. 8½–10 in. long. Barred black and white above; face and underparts grayish tan. Male has red crown and nape; female has red on nape only. White rump visible in flight.
Habitat. Open woodlands, farmlands, orchards, shade trees, and parks.
Nesting. Eggs 3–8, white, laid in cavity excavated 5–70 ft. above ground in dead trunk, limb, or utility pole. Incubation about 14 days, by both sexes. Young leave nest 24–26 days after hatching, but remain with parents after fledging.
Food. Insects, fruit, and seeds.

Northern Flicker

Colaptes auratus

Should you see a large bird with a striking white rump, brightly colored underwings, and spotted breast suddenly take flight from a treeless meadow, you would be looking at a northern flicker—a type of woodpecker that, compared with its tree-drumming kin, is a bit of an oddball.

Spending a great deal of time on the ground, the flicker searches for insects, fallen fruits, seeds—and above all for ants, of which it consumes more than any other North American bird. (Its tongue, which extends nearly three inches beyond its beak, is ideally suited to this purpose.) The flicker still uses trees for nesting, of course, but has also adapted well to farmlands, parks, and other more open habitats, where it may burrow into fence posts, utility poles, or barn sides, or under the eaves of houses. In the Southwest, it even makes use of the giant saguaro cactus.

Northern flickers occur throughout North America, but their appearance differs in various regions, and they are known by more than a hundred local names. In the East lives the yellow-shafted, with yellow under the wings and tail and a black mustache; in the West is the red-shafted, noted for its red underwings, undertail, and mustache; and in the Southwest lives the gilded, similar to the eastern bird but with a red mustache. Where their ranges overlap, different flickers sometimes interbreed, creating even more mixtures of their varied characteristics.

A northern flicker considers even house siding suitable dead wood for its nest hole.

Recognition. 10–14 in. long. Upperparts barred with black and tan; breast band black; rump white. Eastern males have black mustache, yellow flash in wings. Western males have red mustache, reddish flash in wings. Southwestern males have red mustache, yellow flash in wings. Females similar but lack mustache.

Habitat. Woodlands, deserts, and suburbs.
Nesting. Eggs 3–14, white, laid in cavity 8–100 ft. above ground in tree or cactus. Incubation about 12 days, by both sexes. Young leave nest 25–28 days after hatching.
Food. Ants and other insects; fruits.

Yellow-bellied Sapsucker

Yellow-bellied Sapsucker *Sphyrapicus varius*

Red-naped Sapsucker *Sphyrapicus nuchalis*

Ranging across much of the East and North, yellow-bellied sapsuckers are—like their Rocky Mountain counterparts, the red-naped sapsuckers—aptly named. For about a fifth of their diet consists of the sap obtained by drilling tiny holes in trees (insects and berries make up most of the rest). Sapsuckers are catholic in their tastes, tapping more than 250 species of trees and shrubs. Small mammals, hummingbirds, and several other species of woodpeckers also drink from sapsucker holes. And insects attracted to the flowing sweetness provide additional food for the birds.

The sapsucker's unorthodox feeding habits are facilitated by the structure of its tongue. Woodpeckers as a group are known for their unusually long, barb-tipped tongues, which allow them to reach deep into cavities to extract their insect prey. By contrast, sapsuckers have shorter tongues that are coated with fine hairs, instead of barbs, to help them lap up the sap.

The work of few other creatures is so easily noted. In late March or early April when the males arrive in their northern nesting grounds, it is a reassuring sight to see a sapsucker tree with its neatly tattooed rows of holes, and realize that spring will soon return.

Red-naped

Yellow-bellied

Drilling rows of holes, sapsuckers feed not only on the oozing sap but also on the insects it attracts.

Recognition. 8–9 in. long. Black and white above; wing patch white; face striped black and white; white rump visible in flight. Male yellow-bellied has red crown and throat; female has white throat. Male red-naped has red crown, throat, and nape; female has white or red throat.
Habitat. Forests; woodlands; aspen groves.

Nesting. Eggs 4–7, white, laid in cavity 10–45 ft. above ground in dead trunk or live aspen. Incubation about 13 days, by both sexes. Young leave nest 4 weeks after hatching.
Food. Sap from trees; also insects and berries.

Red-breasted Sapsucker

Sphyrapicus ruber

The sapsucker's long, finely bristled tongue is perfect for extracting sap from a tree.

Who would believe that this handsome member of the woodpecker family, with a flaming red head, yellow belly, and white patches on its raven-black wings, could have such an unsavory reputation? Yet the red-breasted and its cousins, the yellow-bellied, red-naped, and Williamson's sapsuckers, have long borne the blame for killing trees by girdling them with tiny drill holes.

There's no doubt that sapsuckers do drill shallow holes in bark and drink the sap that oozes forth and eat some of the delicate cambium (a light green layer of growth just beneath the bark). And row upon row of ¼-inch holes on a tree trunk may not be esthetically pleasing. But the extent of the harm sapsuckers actually cause is open to question.

If a tree is healthy it will quickly heal the tiny wounds, and will benefit from the sapsuckers' habit of eating insects that would otherwise inflict their own damage. But occasionally a sapsucker becomes enamored of one particular trunk and finally causes so much injury that the tree fails to recover. Sometimes, too, lumber cut from a sapsucker's tree shows scars caused by fungus in the drill holes. But in most areas, sapsuckers are simply not abundant enough to pose a serious threat—though such notoriety, once gained, is always hard to escape.

Recognition. 8–9 in. long. Black with white spots on back and wings; wing patch white; whole head and breast red. White rump visible in flight. Sexes alike. Immature more mottled; red areas tinged with brown; crown black.
Habitat. Forests, woodlands, orchards, and aspen groves of the Pacific Northwest.

Nesting. Eggs 4–7, white, laid in cavity 10–45 ft. above ground in live or dead trunk or limb. Incubation about 13 days, by both sexes. Young leave nest about 4 weeks after hatching.
Food. Sap from trees; also insects and berries.

Williamson's Sapsucker

Sphyrapicus thyroideus

The first naturalists to study American birds faced a formidable challenge in classifying them. One might puzzle over some drab little bird, only to discover that it was the female mate of a colorful songster. Even placing a species in the right family could be a problem.

But one family—the woodpecker—was easy to classify. Some of its members were called flickers or sapsuckers, but it was obvious they all belonged to the same clan of tree-climbing, hole-nesting, wood-pecking birds. And the females looked almost like the males, so it wasn't hard to sort out who belonged with whom.

In 1852, explorers out west found a new species: brown with a black chest patch, it was dubbed the black-breasted woodpecker. Soon afterward, another new one was discovered: patterned with black, white, and yellow, it was named Williamson's woodpecker to honor one of the explorers who first saw it. Some 20 years later, a naturalist found a black-breasted and a Williamson's in a Colorado grove, both of them feeding the young in the same nest. Clearly, they were a male and female of the same species—blissfully unaware that, acccording to the experts, they were supposed to look like each other.

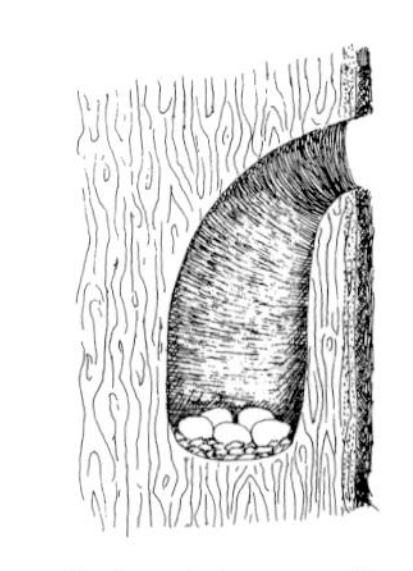

A marvel of aerial excavation, a sapsucker's nest may take up to two weeks to complete.

Recognition. 8½–9 in. long. Male mainly black; chin red; face with 2 white stripes; wing patch white; belly bright yellow; white rump visible in flight. Female barred with brown and black; head brown; belly yellow; rump white.
Habitat. Mountain pine forests.
Nesting. Eggs 3–7, white, laid in cavity excavated 5–60 ft. above ground in dead pine or aspen. Incubation about 14 days, by both sexes. Young leave nest 4–5 weeks after hatching.
Food. Sap from trees; also ants, wood-boring larvae, and other insects.

Hairy Woodpecker

Picoides villosus

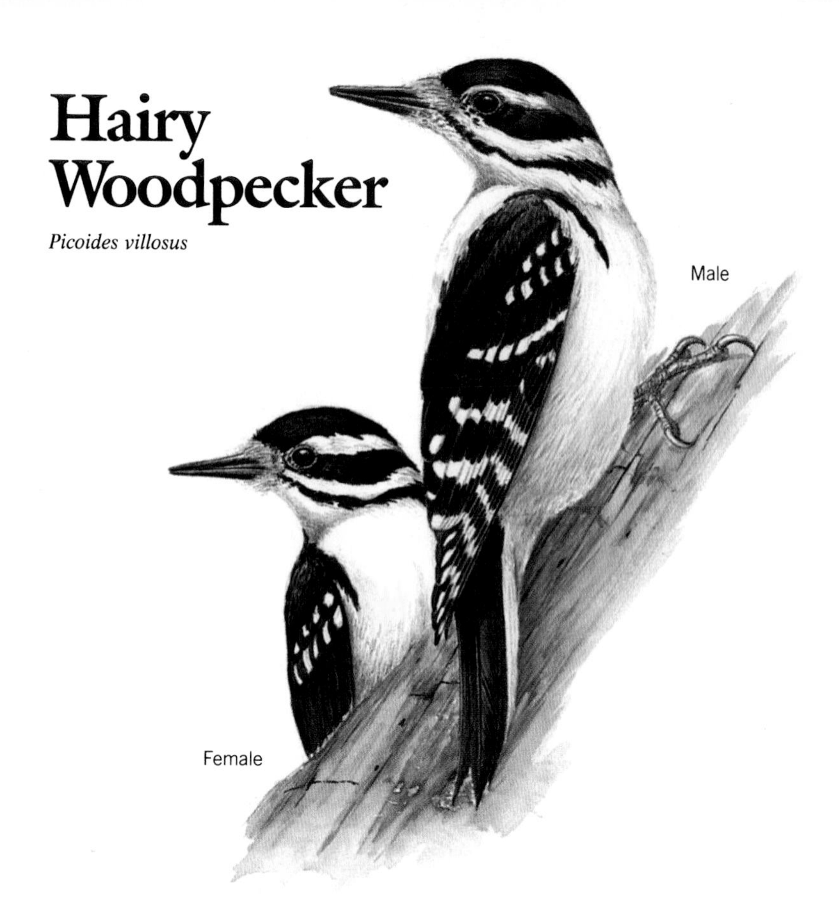

The long, slow courtship of the hairy woodpecker begins in deep winter with the male drumming rapidly on a favorite post to announce to the world at large that he has set up his territory, and to inform any females in the vicinity that he is looking for a mate. An interested female drums back. When they finally approach one another, after days or weeks of drumming, they perform special flight displays in which they strike their wings against their sides to produce a feathery clapping sound, or flutter their wings like butterflies, seeming almost to hover in the air.

At length, the pair bond is formed and the two mate and dig out a nesting hole, greeting each other with a quiet call each time they meet. After the three to six white eggs are laid on the bottom of the cavity, the two share incubation duties. The male sits on the eggs—and broods the young—through the night, and the female relieves him every morning after sunrise. They then alternate throughout the day, always changing duty with a ceremony. The relieving bird swoops in, alights just below the entrance, utters a soft greeting call while waving its head in front of the nest hole, then moves aside. The incubating bird answers, steps out of the nest, and flies slowly away. Thus they are united throughout the nesting season in this courtly fashion.

With all the sawdust it raises, a woodpecker needs the nostril bristles that act as filters.

Recognition. 8½–10 in. long. Larger than downy woodpecker, with longer bill. Upperparts black and white; face with black and white stripes; underparts pure white. Male has red nape patch. Call note a loud *peek!*
Habitat. Mature forests, orchards, and parks.
Nesting. Eggs 3–6, white, laid on bed of wood chips in cavity excavated 3–55 ft. above ground in dead limb of live tree. Incubation about 14 days, by both sexes (male at night). Young leave nest 28–30 days after hatching.
Food. Wood-boring insects, berries, and seeds.

Downy Woodpecker

Picoides pubescens

The smallest of our woodpeckers, the downies are prominent participants in the noisy activity of drumming (a rapid beating of the bill), which for woodpeckers is akin to a songbird's singing. Drumming notifies the bird community of the woodpecker's territory, tells intruders to "get out," and plays a part in courtship. A similar but slower action—tapping—is used by a mated pair for keeping in touch with each other. These activities are entirely distinct from the muted, irregular beats with which a woodpecker chips off bark to hunt insects or digs out a nesting cavity.

The drumming ritual has a distinct pattern for each species. The downy's is an unbroken roll, *tr-r-r-r-r-r-r-r,* lasting about two seconds and done so rapidly that the bird's trip-hammer head is only a blur. The hairy produces the same rapid roll but both louder and shorter, while sapsuckers add a few *tap-tap-tap*s to the end.

Giving an individual tone to the ritual, each woodpecker seeks out its own drumming object—a stub or a post or a limb with a certain resonance that pleases its ear. Sapsuckers and flickers love to drum on metal. But whatever its composition, every post is private property. Pairs of downy woodpeckers have three or four apiece in their territories, and they remain "his" and "hers," without question, for as long as their partnership lasts.

Near winter feeding stations, the downy woodpecker has been observed bathing in snow.

Recognition. 6–7 in. long. Smaller than hairy woodpecker, with shorter bill. Upperparts black and white; face with black and white stripes; underparts pure white. Male has red nape patch. Call note a soft *pick.*

Habitat. Open woodlands, orchards, and parks.

Nesting. Eggs 3–7, white, laid in cavity excavated 5–50 ft. above ground in dead trunk or branch, or in birdhouse. Incubation about 12 days, by both sexes. Young leave nest about 24 days after hatching. Sometimes 2 broods in southern states.

Food. Wood-boring insects, berries, and seeds.

Ladder-backed Woodpecker

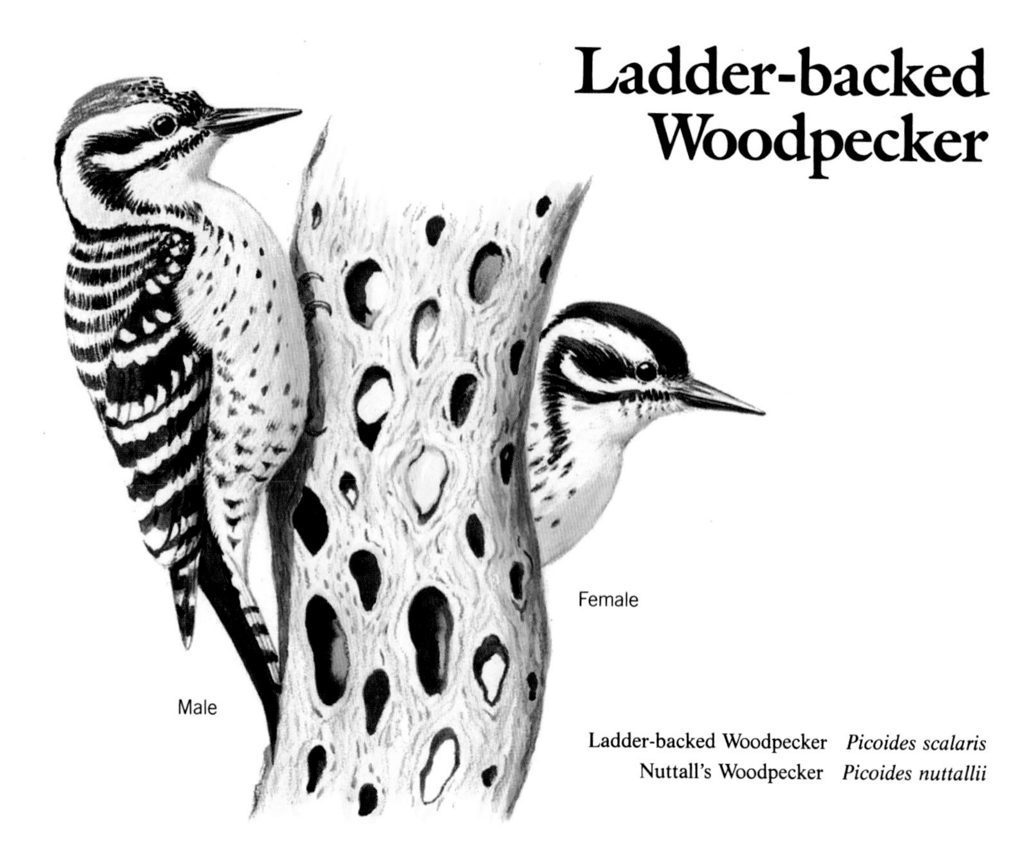

Ladder-backed Woodpecker *Picoides scalaris*
Nuttall's Woodpecker *Picoides nuttallii*

Ladder-backed Woodpecker

Nuttall's Woodpecker

Without trees, a woodpecker would seem as much out of place as a duck without water. Yet the little ladder-backed is able to make a living in the southwestern deserts where, in many parts, there is hardly anything resembling a tree. There is, however, something just about as good—the enormous saguaro cactus, which grows up to 60 feet tall and has columnar branches not unlike tree trunks. The ladder-backed, along with its cousins the Gila woodpecker and the gilded flicker, digs its nest in this giant, feeds on the insects that inhabit it, and even partakes of the fruit.

On the fringes of the more arid regions—in wooded canyons, mesquite thickets, and cottonwood groves—the ladder-back supports itself in the traditional woodpecker way, using its climbing and probing skills to explore every part of every plant. At times, both members of a pair may be seen foraging together, efficiently dividing up the terrain. Their uncanny ability to pinpoint the location of beetle larvae has long puzzled observers. One theory is that the birds listen for movement beneath the bark; another, that they can tell from their tapping where the insects lie, just as a carpenter locates a stud by thumping on a wall. Whatever their secret, the woodpeckers are highly successful in ridding trees of these unwelcome residents.

Recognition. 6½–7½ in. long. Small; barred above; face white with black stripes. Crown red in males, black in females. Ladder-backed has less black on face and upper back than Nuttall's. Best distinguished by range and habitat.

Habitat. Dry brushy areas and thickets. Nuttall's in less arid areas.

Nesting. Eggs 2–7, white, laid in cavity excavated 2–60 ft. above ground in tree or cactus. Incubation by both sexes; period unknown. Nestling period unknown.

Food. Wood-boring insects, caterpillars, and cactus fruit.

Red-cockaded Woodpecker

Picoides borealis

On a walk through an open stand of tall old pines in the Southeast, you might see signs of a red-cockaded woodpecker's work—small holes, below which a large wash of resin has flowed down the trunk. Many woodpeckers excavate in dead or dying trees, but red-cockades are unusually specific. They seek out living pines with red-heart disease, a fungus that affects the heartwood and undoubtedly makes the wood easier for the woodpeckers to remove. After digging out a hole to use for roosting or nesting, the red-cockade drills several small holes around the main cavity to release pitch. This sticky ooze apparently keeps his enemies in check, but it's a wonder that the bird himself manages to stay free of it.

In the southeastern pinewoods, red heart has long been a fairly common affliction of trees 70 years old or older. But today, most pines are cut before they reach that age, so the birds face a shortage of nesting sites. In addition, the open forests that red-cockades prefer are declining because of man's greater ability to control forest fires, which in the past kept the understory open and the trees well spaced. As a result of such well-intentioned management, red-cockaded woodpeckers have become an endangered species, and we could lose forever these colorful natives that are unique to our southland.

Recognition. 8–8½ in. long. Upperparts barred black and white; crown black; cheek white. Male's red "cockade" hidden behind eye and hard to see. Usually in small groups.
Habitat. Open pine forests.
Nesting. Eggs 2–5, white, laid in cavity excavated 18–100 ft. above ground in soft, decayed heartwood of living pine. Incubation about 13 days, by both sexes (male incubates at night). Young leave nest 26–29 days after hatching. Breeds in loose colonies in stands of tall pines; nest cavities may be used for many years.
Food. Wood-boring insects.

Three-toed Woodpecker

Picoides tridactylus

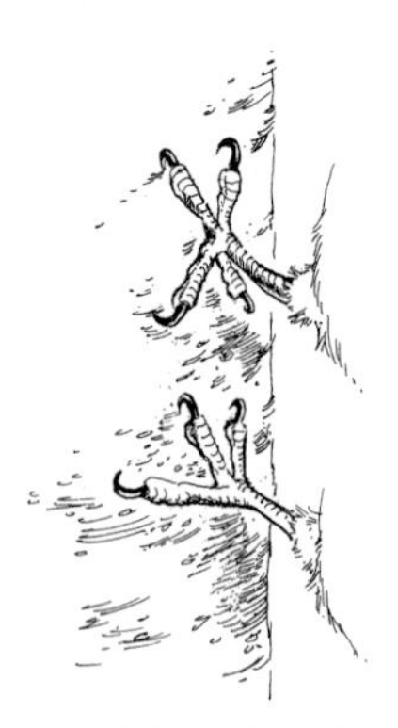

All toes point forward on three-toed woodpeckers; two forward, two back on four-toed types.

Most birds have four toes, with three fanned forward and one pointing backward. On woodpeckers the toes are arranged with two turned forward and two backward and are fitted with sharp-tipped, down-curved claws. Since these birds clamber about in trees, presumably their two-and-two arrangement gives them better balance and stronger support as they climb or stand on rough-barked and sometimes vertical surfaces.

Why, then, should two species of North American woodpeckers—the three-toed and the black-backed, both of which reside in forests of the Far North—have only three toes, all of them fanned forward? This is a condition they share with some seabirds and a few shorebirds, all of whom do their walking on rocks, sand, grass, or mud.

Whatever the reason for their unusual feet, the three-toed woodpeckers seem to have accommodated their lifestyle to the support their feet give them. Like other woodpeckers, they search for wood borers, but instead of drilling through the bark, they flake it off with their bills to expose the insects hidden there. They work only on dead and dying trees, seldom on healthy ones, and are exceptionally quiet for woodpeckers, often standing in one spot for minutes at a time—as if content to live their lives in a less frenetic manner.

Recognition. 8–9½ in. long. Black above with center of back barred black and white; underparts white; flanks barred with black and white; head black with white eyebrow and whisker mark. Male has yellow cap.
Habitat. Coniferous forests.
Nesting. Eggs 4–5, white, laid in cavity excavated 5–40 ft. above ground in trunk of dead spruce, fir, or birch, especially where flooding has killed trees. Incubation by both sexes, probably about 14 days. Nestling period unknown.
Food. Wood-boring insects, other insects and spiders, and berries.

Black-backed Woodpecker

Picoides arcticus

In the coniferous forests of southern Canada, black-backed woodpeckers cling to dead and dying trees with their three-toed feet and carefully flake away scales of bark until they reach the larvae of wood-boring beetles. But as the dead trees are gradually denuded, both beetles and birds must find new stands for continuing their life cycles.

Every winter at least a few black-backed woodpeckers wander farther south, especially into the northeastern United States, to feed in the pines, spruces, and balsam firs killed by fire. But some years they arrive in unusually large numbers. This irregular migration—ornithologists call it irruption—probably occurs as a search for new food sources. Such birds as snowy owls and northern shrikes seem to irrupt into the Northeast about every four years, when their food animals—voles and lemmings—are at a low point; goshawks and northern horned owls, too, appear about every 10 years, when the hares on which they feed reach periodic lows. Mystifyingly, the woodpecker irruptions occur in some years when northern winters are mild and the food abundant. One explanation: their numbers have so increased during these good years, they must expand their ranges. With ever-changing conditions, perhaps the colorful black-back will become a more regular resident in the northern woodlands of the United States.

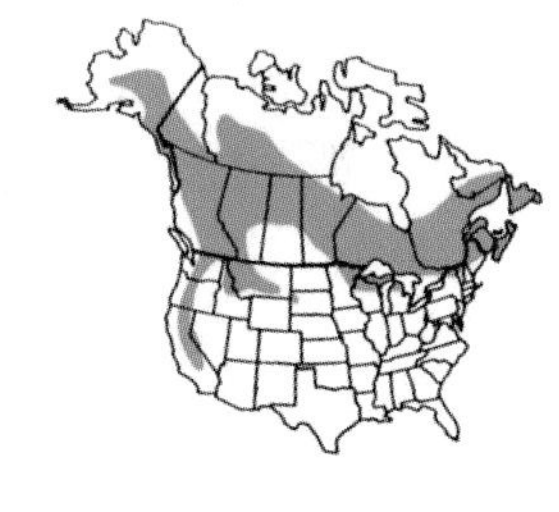

Chips flying, a pair of black-backed woodpeckers artfully excavate their nest hole.

Recognition. 9½–10 in. long. Back, wings, and tail black; underparts mainly white; flanks barred with black and white; head black with white whisker mark. Male has yellow cap.
Habitat. Coniferous forests.
Nesting. Eggs 2–6, white, laid in cavity excavated 2–15 ft. above ground, rarely much higher, in trunk or stump of dead spruce, fir, or birch. Incubation about 14 days, by both sexes. Nestling period unknown.
Food. Wood-boring insects, other insects and spiders, and berries.

Former range

Ivory-billed Woodpecker

Campephilus principalis

If these, our most magnificent woodpeckers, are not already extinct, they are so close to the edge that probably no measures could now save them. Creatures of the deep solitudes of our once immense southern forests, they were most likely never an abundant species. And because of their grandeur—the resplendence of the male's flaming crest and creamy white three-inch bill—men could not resist capturing them.

Chieftains of Indian tribes wore the crests and dagger-like bills as regal ornaments and prized the meat as a delicacy. Later, the advent of white frontiersmen reduced their numbers even more. As the great birds retreated to the most impenetrable of cypress jungles, man followed with axes, guns, and bulldozers and destroyed the creatures, scarcely knowing they existed.

But perhaps these birds were already doomed, not only by their beauty but by their own needs. They require vast woodlands, all suffering from disease, drought, or the ravages of fire, to supply their principal food of flat-headed borers that live beneath the bark of dead or dying trees. In all likelihood, today's smaller, better-managed forests could not sustain even a token population of ivory-bills.

Recognition. 20 in. long. Crow-sized, with long crest, red in male, black in female; pale yellow bill. Mainly black; face black; stripe down neck and large wing patches white.
Habitat. Mature swamp forests, mainly along large rivers.
Nesting. Eggs 1–4, white, laid in large cavity excavated 25–60 ft. above ground in tall, old tree. Incubation probably about 20 days, by both sexes (male incubates at night). Young leave nest about 5 weeks after hatching.
Food. Wood-boring beetle larvae exposed by shearing off slabs of bark.

Pileated Woodpecker

Dryocopus pileatus

With the virtual disappearance of ivory-billed woodpeckers, the title of largest North American woodpecker now goes to the magnificent pileated. To see these red-crested, crow-sized birds in flight across an open space, to watch them stripping slabs of bark and excavating huge holes (up to seven inches across) in a tree trunk, or to come upon a pair in excited courtship—these are privileged moments indeed.

Except that their bills are black, the pileateds look much like the glorious ivory-bills. But the resemblance is only superficial, for the two species are not closely related. And while ivory-bills have disappeared as their habitat has altered, pileateds have shown remarkable resilience.

When Europeans first arrived in North America, they found a land blanketed with mature woods that were the favorite haunts of pileated woodpeckers. During the 18th and 19th centuries, much of this great forest was cut, and pileateds became rare over much of their range. Then changes in lumbering practices and the abandonment of many eastern farms allowed much of the forest to regenerate. Pileateds still need large tracts of woodland, but they have adapted well to younger trees. And though they are quite shy, it is not unusual now to hear or see these majestic "log cocks" in many of their former haunts.

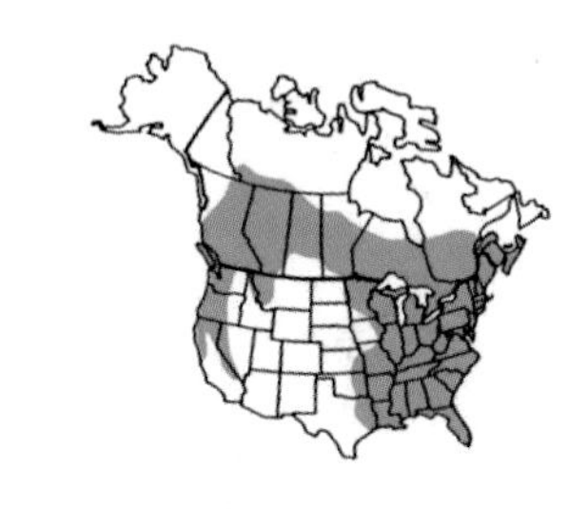

A pileated woodpecker once carried its eggs to a new site when its nest tree broke in two.

Recognition. 16–19 in. long. Crow-sized. Mainly black; crest red; face striped black and white; white stripe down neck; wing linings white. Male has red mustache.

Habitat. Mature forests with large trees.

Nesting. Eggs 3–8, white, laid in cavity excavated 15–70 ft. above ground in dead trunk or limb in stand of living trees, rarely in utility pole. Incubation probably about 18 days, by both sexes. Young leave nest after 22–26 days.

Food. Mainly ants and other wood-boring insects; also berries.

White-headed Woodpecker

Picoides albolarvatus

The white-head is seen drinking from streams or ponds more often than most other woodpeckers.

Boasting the only white head and throat among woodpeckers in North America, the white-headed woodpecker is easy to distinguish from any of its cousins that range among the conifers in the Pacific Northwest. Except for a sliver of white on the bird's wings and a dollop of red on the back of the male's head, its body is all black.

With such a contrasting color pattern, it stands to reason that this woodpecker would find it difficult to escape the keen eyes of predators. But in its evergreen world, it becomes almost invisible, hiding in plain sight against dark needles, mottled bark, white stubs of broken branches, and constantly changing patterns of light and shadow.

The camouflage is most surprising when sunlight dazzles on the bird's slender patches of whiteness; for that same sunlight also highlights pale streaks in pine bark, polishes shining needles, and makes a black body just another dark shading on the side of the tree. Even when the bird explores the light bark of young firs or wanders among gray, barkless stands of dead conifers, its white head somehow blends in with the gray trees and its black body becomes a knothole, a shadow, or a scar. Thus it goes about its business of gathering pine cone seeds and beetle larvae generally unseen and undisturbed.

Recognition. 9–9¼ in. long. Body, wings, and tail mainly black, with white crown, face, and throat, and white patch on wing, visible in flight. Male has small red patch on nape.
Habitat. Mountain pine forests.
Nesting. Eggs 3–7, white, laid in cavity excavated 3–50 ft. above ground in tall dead pine stump. May excavate several cavities close to each other before nesting in one. Incubation probably about 11 days, by both sexes. Nestling period unknown.
Food. Pine seeds and wood-boring insects.

Olive-sided Flycatcher

Contopus cooperi

High atop a weathered pine tree, standing like some proud burgomaster of the northern woodlands, the olive-sided flycatcher poses grandly. He's a portly fellow, whose olive-gray vest is stretched tight across an ample middle, while the seams of an embarrassingly shabby gray coat seem to have split at the sides. Large white patches peek through the breaches along both flanks. But the flycatcher seems oblivious to anything that might undermine his appearance. Shoulders back, head erect, he throws his order to the world: *Quick! three beers . . . Quick! three beers . . . .*

This burly bird seems loath to leave his favorite perch for long. After making a brief sortie in pursuit of a passing honeybee or dragonfly, or after aggressively driving an intruder from his territory, he habitually returns to the same spot. Even in migration, olive-sided flycatchers show a single-minded devotion to their hunting perches, a trait that is useful for separating flycatchers viewed at a distance from kingbirds or the slimmer, trimmer pewees.

With such regular habits and commanding vocalization, the olive-sided flycatcher is easy to locate in northern forest bogs or at high elevations in the burned-over areas it prefers. Observers who hope to see the bird in migration should note that it arrives later in the spring and departs earlier in the fall than most other migrants. But what else could be expected from such an important fellow?

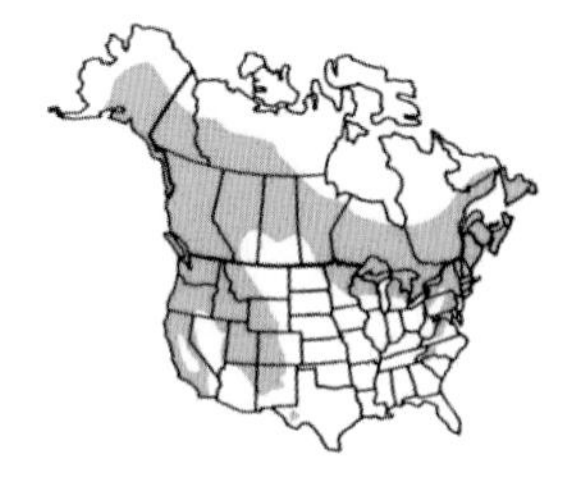

From a favorite perch the olive-sided flycatcher sallies forth only to catch a meal on the wing.

Recognition. 7–8 in. long. Large-headed and large-billed. Mainly dark olive-gray; dark flanks separated by white on center of underparts; sometimes has white tufts on lower back. Usually perches on exposed dead branch.

Habitat. Coniferous forests, flooded woodlands with dead trees, and eucalyptus groves.

Nesting. Nest is a cup of twigs, grass, and lichen, 5–70 ft. above ground in tree, usually conifer. Eggs 3–4, whitish or pink, spotted with brown and gray. Incubation about 17 days. Nestling period unknown.

Food. Flying insects.

Western Wood-Pewee

Western Wood-Pewee *Contopus sordidulus*

Eastern Wood-Pewee *Contopus virens*

Western Wood-Pewee

Eastern Wood-Pewee

To the casual observer, one flycatcher species hunts pretty much like another, darting out from a perch to chase down a passing insect. In fact, flycatcher techniques are subtly different, and where several species share a piece of habitat, their distinct hunting strategies serve to minimize competition for the area's food resources.

A biologist studying three flycatchers with overlapping territories in California found that western wood-pewees chose as perches the exposed branches on the periphery of trees—especially the top margin—while black phoebes preferred fences and the bottom tree fringes, and western flycatchers favored middle and lower interior branches. From their high stations, the wood-pewees launched attacks on insects flying in open airspace. Black phoebes also chased insects in the open, but mostly at lower levels, and they sometimes gleaned them from low foliage as well. Western flycatchers snatched flying insects in the more confined space of tree interiors and plucked prey from twigs and leaves nearly half of the time.

Observers in Virginia noted that eastern flycatchers parceled out foraging areas in much the same way and that differences in style led to variations in diet. While flies constituted more than half of a wood-pewee's food, they made up less than one-third of a least flycatcher's and only one-sixth of a great crested flycatcher's.

Recognition. 6–6½ in. long. Olive-brown above with no eye-ring; 2 pale wing bars; underparts whitish. Best distinguished by range and voice; western gives a shrill *fee-reet!*, eastern a plaintive *pee-a-wee?*
Habitat. Woodlands and groves.
Nesting. Nest is a cup of stems and plant fibers 15–75 ft. above ground on horizontal branch. Eggs 2–4, whitish, speckled and blotched with brown. Incubation about 12 days, by female. Young leave nest 15–18 days after hatching.
Food. Flying insects; berries.

Great Crested Flycatcher

Great Crested Flycatcher *Myiarchus crinitus*
Brown-crested Flycatcher *Myiarchus tyrannulus*

No eastern woodlot would be complete without this prominent flycatcher. Its staccato chattering and demanding *wheeep?* filter through the dense summer forest. Sometimes, yellow underparts or chestnut flashes peeking from spread wings and tail hint at the bird's presence, high in the canopy. Squeak like a mouse in distress and the "cresty" may respond with a torrent of scolding notes.

Before eastern woodlands knew the sound of axes, the great crested flycatcher was probably a bird of the deep forest. There, in the limbs of ancient trees, it found the cavities it sought for its nest. But as the forests fell, the bird adjusted to orchards, thinned woodlots, and, more recently, artificial birdhouses. (The closely related brown-crested flycatcher of the southwestern deserts nests in abandoned woodpecker holes found in saguaro cactus.)

The nests of great crested flycatchers are flexible affairs, built with whatever material is at hand—leaves, sticks, bark, roots, and other forest oddments. According to lore, the bird's habit of weaving a shed snake skin into its nest is a trick intended to discourage predators. But if so, then nest-robbers are equally daunted by cellophane, plastic, and wax paper; more and more, these man-made items are turning up in nests. In any case, whether by trickery or more direct means, this "tyrant" flycatcher's defense of its nest is formidable.

Great Crested Flycatcher

Brown-crested Flycatcher

Recognition. 8–9 in. long. Olive-brown or gray-brown above; wings and tail with rusty tinge; belly yellow. Great crested has pale lower mandible, loud *wheep!* call. Brown-crested has all-black bill, sharp *whit!* call.
Habitat. Deciduous woodlands; brown-crested also inhabits cactus deserts.

Nesting. Nest is a mass of twigs and grass, and usually a snake skin, in a tree cavity 5–70 ft. above ground (5–30 ft. above ground in brown-crested). Eggs 3–8, buff, blotched with brown and purple. Incubation 13–15 days. Young leave nest about 15 days after hatching.
Food. Flying insects and berries.

Willow Flycatcher *Empidonax traillii*
Alder Flycatcher *Empidonax alnorum*

Willow Flycatcher

Willow Flycatcher

Alder Flycatcher

In a streamside alder thicket a slim, olive-colored flycatcher jerks its head back and tosses an assertive challenge to the world: *fee-bee-o.* In an overgrown pasture, not far away, a bird that could pass for the alder flycatcher's reflection issues a challenge of its own: *fitz-bew.* By their songs and their habitats the alder and willow flycatchers avoid direct competition and draw the fine line that separates one species from another.

John James Audubon first described the willow flycatcher along the Arkansas River. He named the bird after Dr. Thomas Traill, founder of the Royal Institution of Liverpool. It may be that from his travels in Canada, Audubon also knew the alder flycatcher, a bird he called the short-legged pewit flycatcher. But the differentiation was apparently overlooked or ignored, and the two flycatchers were regarded as different races of a single species until as recently as 1972.

But these two birds are distinct. The more northerly ranging alder flycatcher is somewhat greener and shorter-billed than the willow and boasts a more prominent eye-ring. In the hand, the birds can be distinguished by using a complicated formula involving feather measurements. But the best way to tell them apart is by using the technique the birds use themselves: listening to what they say and noting where they say it.

Recognition. 5–6 in. long. Olive-green above, with 2 pale wing bars, vague eye-ring, whitish underparts. Best distinguished by voice: willow gives a dry *fitz-bew,* alder a burry *fee-bee-o.*
Habitat. Willow and alder swamps; willow flycatcher also nests in drier, brushy areas.
Nesting. Willow's nest is a cup of grass and bark 1–6 ft. above ground. Alder's is a compact cup of plant down and fibers 5–20 ft. above ground. Eggs 2–4, whitish, buff, or pink, speckled with reddish brown. Incubation 12–15 days. Young leave nest 2 weeks after hatching.
Food. Flying insects; berries.

Hammond's Flycatcher

Hammond's Flycatcher *Empidonax hammondii*
Dusky Flycatcher *Empidonax oberholseri*

High atop a spruce clinging to the flanks of the northern Rockies, a Hammond's flycatcher sits and waits. Wings and tail flicking, feet clamped upon its perch, the bird fairly vibrates like a slingshot at full draw. Suddenly it darts forth. The wide flycatcher bill opens and closes with an audible snap—and a flying ant disappears within. Then, as if tethered by invisible elastic, the feathered projectile snaps back to its launch point, where it sits, fidgeting nervously, and waits for another victim to pass.

The Hammond's is a bird for the heights, foraging in the higher branches of tall western evergreens. In the warmer, southern reaches of its nesting range, the bird's love of high places may carry it to elevations above 10,000 feet—a penchant that helps to distinguish it from the dusky flycatcher. This near-twin of the Hammond's shuns high places and favors drier, more open woodlands. Not quite so nervous as the Hammond's, dusky flycatchers seem able to hold their wing-flicking in check, but even the strongest-willed dusky must surrender to a periodic twitch of the tail.

The high-strung Hammond's must be anxious about being separated from its nesting territory, too. In the spring it arrives uncommonly early—often migrating with those vanguards of spring, the hardy kinglets. In the fall it lingers longer than most other flycatchers—far longer than might seem prudent for a bird that catches insects for a living.

Hammond's Flycatcher

Dusky Flycatcher

Recognition. 5–5½ in. long. Dull olive above, with 2 pale wing bars, pale eye-ring, whitish underparts. Best distinguished by habitat and call note: Hammond's gives a sharp *pick,* dusky a soft *whit.*
Habitat. Mature mountain forest; dusky prefers more open, brushy areas.

Nesting. Hammond's nest is a cup of bark, fibers, and rootlets 6–60 ft. above ground. Dusky's is a cup of grass and stems 4–7 ft. above ground. Eggs 3–4, whitish or yellowish, usually unmarked. Incubation 12–15 days. Young leave nest 18 days after hatching.
Food. Flying insects.

Pacific-slope Flycatcher

Pacific-slope Flycatcher *Empidonax difficilis*
Cordilleran Flycatcher *Empidonax occidentalis*

Beneath the forest dome, morning is still a rumor. Yet one bird has already wakened to greet the dawn. Over the sound of the nearby stream can be heard a murmured, monotonous *bz-zeek trip seet! . . . bz-zeek trip seet!* But by daybreak the bird has changed his tune. *Pawee!* he insists loudly, *pawee!* Sure enough, we do see him, perched on a branch at the edge of the glen, and chances are this flycatcher will call attention to himself all day. During nesting season, once he has begun his dawn chorus, nothing short of nightfall seems to stop him—unless some insect passes temptingly close, or another bird breaches the privacy of the glen. Like many of their flycatcher cousins, Pacific-slope and cordilleran flycatchers (until 1989 considered a single species, the western flycatcher) have a low tolerance for intruders. They can easily put much larger birds to rout, and the females will join the fray whenever their young are threatened.

Once tranquillity is restored to the glen, the female will return to the nest. It may be situated, as tradition has it, on the end of a forked branch or among the roots of a windblown tree. But these flycatchers also adopt abandoned woodpecker holes and cavities on cliffs, stream banks, or abandoned buildings. The male will take up where he left off—snapping up passing insects, boldly proclaiming his presence, and frustrating intruders brash enough to cross a flycatcher's threshold.

These hardy flycatchers will call almost any place home, even the underside of a bridge.

Recognition. 5½–6 in. long. Olive-green above, with 2 pale wing bars, bold eye-ring; underparts with yellowish tinge. Call of Pacific-slope flycatcher *pawee.* Call of cordilleran *wee-eee.*
Habitat. Woodlands and forest.
Nesting. Nest is a cup of fresh moss, stems, and grass in crevice in bank, among upturned tree roots, or up to 30 ft. above ground on branch or in woodpecker hole. Eggs 3–5, whitish or cream with brown and lavender spots. Incubation about 14 days. Young leave nest 14–17 days after hatching.
Food. Flying insects; also berries and seeds.

Acadian Flycatcher

Acadian Flycatcher *Empidonax virescens*
Least Flycatcher *Empidonax minimus*

As a bird watcher's identification skills grow, there are formidable challenges to be confronted and conquered. Learning the sparrows requires patience; mastering the profusion of fall warblers is an anguishing rite of passage; and sorting out small sandpipers is an enduring trial. But for pure consternation, all other groups pale next to the scientific genus *Empidonax.* These small, enigmatic, gray-green flycatchers can bring tears of frustration to the most skilled birder's eyes. Imagine the puzzlement of early ornithologists whose task it was to determine where the fine lines are that separate species. One lingering reminder of that early confusion is the misnamed Acadian flycatcher, a southern member of the clan.

Initially, ornithologists recognized just one small flycatcher with wing bars and an eye-ring—from a bird collected in Acadia, as Nova Scotia was once called. The specimen was dubbed *Empidonax acadicus,* and for years the subtle differences between it and other birds within this genus went unnoticed. Over time, it was realized that several different species do exist, distinguished by their songs, nesting practices, and habitats, rather than by appearance. But it was a southern flycatcher, which ranges no closer to Nova Scotia than Connecticut, that inherited the name of the original bird from far-off Acadia. Its scientific name was changed to *Empidonax virescens,* but the common name, Acadian flycatcher, stubbornly persists to this day.

Acadian Flycatcher

Least Flycatcher

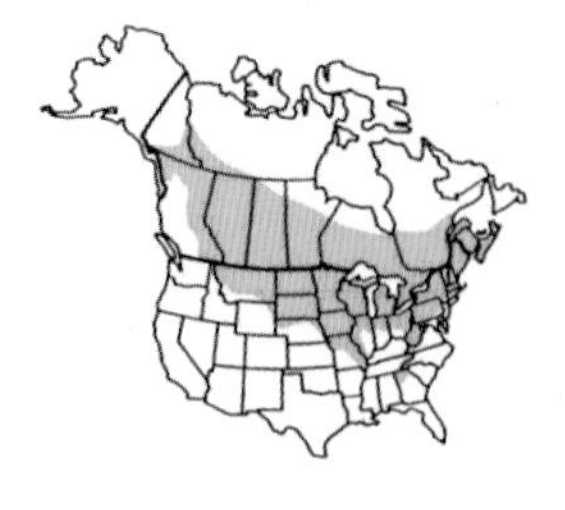

Recognition. 5½–6½ in. long. Olive above, with 2 pale wing bars, bold eye-ring, whitish underparts. Acadian is yellowish below; call a sharp *peet-seet.* Least is grayer; call a sharp *chebec.*
Habitat. Rich deciduous forests. Least also nests in groves and orchards.
Nesting. Acadian's nest is a hanging cup of stems and twigs, 8–25 ft. above ground. Least's is a deep cup of fibers, 2–60 ft. above ground. Eggs 3–6, spotted with brown (Acadian) or plain white (least). Incubation 14 days. Young leave nest 2 weeks after hatching.
Food. Flying insects; berries.

Gray Jay

Perisoreus canadensis

The gray jay is a bold bird, equipped for survival in the harsh North Woods. No stranger there to humankind, it has many names—camp robber, moose bird, and meat bird among them—each reflecting the bird's relationship to people and its environment. Indians call it *wiskedjak,* from which "whiskey-jack," one of its commonest names, is taken. In Algonquin-Ojibwa lore, the sly Wiskedjak is the gray jay in human form. In real life, the large chickadee lookalike both vexes and entertains. For if pilfering defines the family, then the whiskey-jack is the quintessential jay. Campers soon learn that all food is the fearless bird's prey; indeed it may boldly alight on a plate or a frying pan to filch a tasty morsel. Hunters, too, know that the sound of a gun will bring the gray jay, eager for moose or deer meat. The Indians sum up by declaring that whiskey-jack will eat anything, including moccasins and fur caps.

The bird's opportunistic behavior is not purposeless. It is storing food and fattening itself against the long northern winter, when it may have to eat lichens or fir needles, and when its biological clock will impel it to begin nesting in late February or early March. With snow on the ground and subzero temperatures keeping it close to eggs in a feather-lined nest, the gray jay will need secret stores, perhaps tidbits of bacon and hash-browned potatoes as well as the berries and nuts stored away during summer's halcyon days.

A campground scavenger, the gray jay brashly invites itself to dine with woodland visitors.

Recognition. 10–13 in. long. Gray back, wings, and tail; crown, cheeks, and throat white; back of head and line through eye black; underparts whitish. Immature grayish brown with whitish whisker mark. Tame; travels in small parties.
Habitat. Coniferous forests to treeline.
Nesting. Nest is a cup of twigs, bark, plant fibers, and moss, 4–30 ft. above ground in tree. Eggs 2–5, pale gray, pale green, or white, spotted with brown and gray. Incubation 16–18 days, by female only. Young leave nest about 15 days after hatching.
Food. Insects, fruit, mice, and birds' eggs.

Scrub Jay
also known as
Florida Scrub Jay

Scrub Jay *Aphelocoma coerulescens*
Mexican Jay *Aphelocoma ultramarina*

Two ornithologists, studying scrub jays in central Florida, found to their surprise that many nests were attended not only by the breeding pair but by other jays as well. In typical years, about half the breeding pairs have "helpers at the nest," generally one or two, but occasionally five or six. These aides, usually the pair's offspring from a previous year, help keep the nest clean, feed the young, and guard against other jays and predators. This apprenticeship usually lasts a year or two, but some helpers, mostly males, carry on for three, five, or even six years.

But why would these birds, sexually mature at one year, defer pairing to serve as helpers? The reason probably is overcrowding. Florida scrub jays inhabit isolated sandy tracts with low vegetation, especially scrub oaks and palmettos. Virtually all territories are occupied year-round by long-lived jays. Helpers, it seems, are simply practicing nesting while biding their time to jump at the first vacancy.

The much larger population of western scrub jays breeds conventionally without helpers, but the closely related gray-breasted, or Mexican, jay has taken cooperative breeding even further than the Florida scrub jay. This species lives in permanent clans comprising close relatives. A clan of 10 or 15 birds typically establishes two nests, both of which are attended by all members.

Recognition. 11–13 in. long. Blue above; back gray; throat and upper breast whitish with blue necklace; rest of underparts off-white. Travels in small parties. Gray-breasted jay similar but much duller, with dark cheek patch, no blue necklace.
Habitat. Open woodlands and brushy areas.
Nesting. Nest is a bulky mass of twigs 2–12 ft. above ground in bush or low tree. Eggs 2–7, variously marked with red and brown (plain grayish green in gray-breasted). Incubation 16–19 days, by female. Young leave nest about 18 days after hatching.
Food. Nuts, seeds, fruits, and insects.

Blue Jay

Cyanocitta cristata

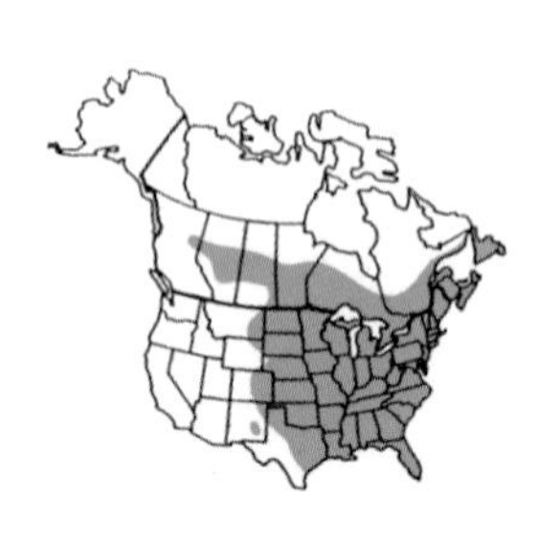

Strident, steely notes fracture the day's quiet: *Jay! Jay! Jay!* Other birds scatter at the warning notes, and the screaming blue jay crash-lands on an abandoned bird-feeding table, claiming the viands all for itself. Thus grows the jaybird's reputation as a noisy rascal, a buffoon with a darker side, a bully whose bright blue coat identifies it as one of the most common birds of eastern woodlands.

Yet the blue jay, like others of its family, is a study in contrasts. Known for its harsh voice and its mimicry of the wild cries of hawks, the blue jay also charms listeners with its own seldom-recognized soft and low song, or with its musical imitations of other songbirds. Although eminently visible because of its relatively large size and its vivid blue, black, and white coloration, the jay is a camouflage artist and can quickly disappear behind the branch of a tree, even in winter's barren landscape. And though its feeding habits can lead it to prey on mice or the young of other birds, the male blue jay is a picture of domestic tranquillity as he gently proffers his mate a kernel of corn, while she sits patiently on the nest they built together.

Though blue jays become secretive around the nest, they can be found in close proximity to humans year-round. This relative tameness, plus their constant curiosity, makes them ideal birds for observation and for stimulating exercises in separating fact from stereotype.

Boisterous and brazen, blue jays routinely bully their way to the feeding table.

Recognition. 11–12½ in. long. Crest, back, wings, and tail bright blue; wings with white spots and black bars; tail feathers with white tips; face and underparts whitish; necklace black. Noisy; often travels in small parties.

Habitat. Deciduous, mixed, and coniferous forests, gardens, and parks.

Nesting. Nest is a bulky cup of moss, twigs, and leaves, 5–50 ft. above ground in crotch of tree or on branch. Eggs 3–6, olive, pale blue, or buff, spotted with brown and gray. Incubation 16–18 days, usually by female. Young leave nest 17–21 days after hatching.

Food. Nuts, seeds, fruits, and insects.

Steller's Jay

Cyanocitta stelleri

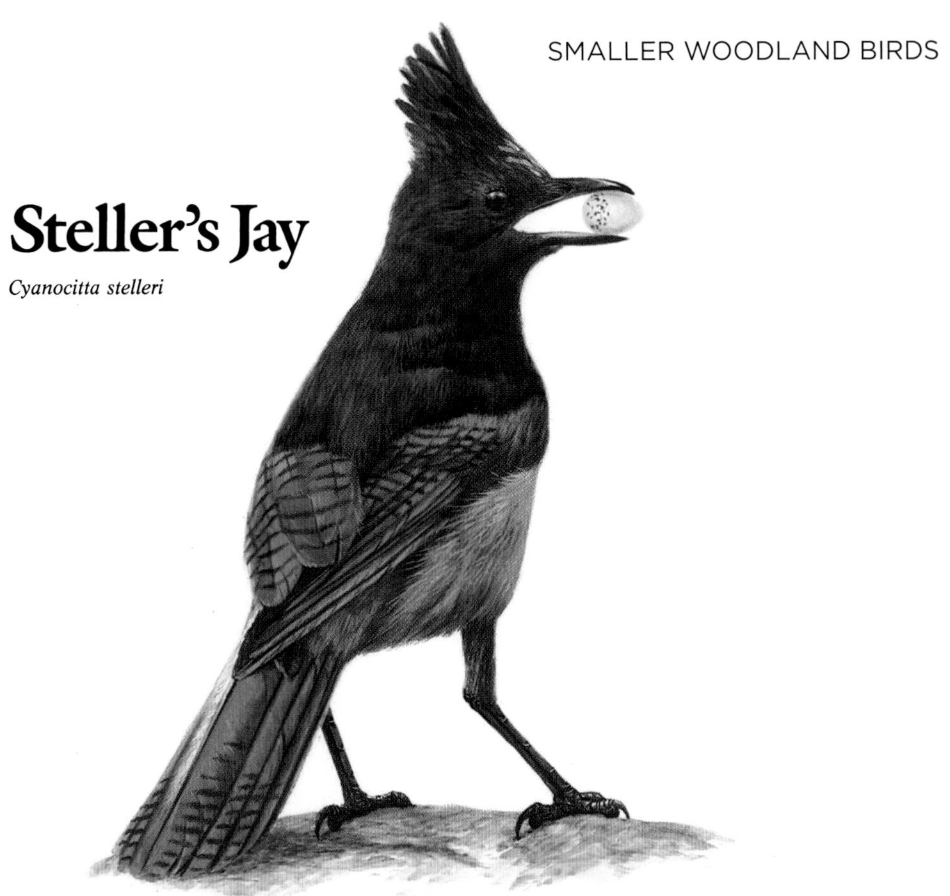

High on the mountain slopes of the West, amid forests of cone-bearing evergreens, the Steller's jay lives a life much like that of its eastern relative, the blue jay. Brightly colored and conspicuous, clad in aristocratic azure and sable, it is North America's largest jay—and certainly among the noisiest, with a voice described by one observer as "raucous as a fishwife's." A reverberating, repetitious *Shaack! Shaack! Shaack!* is its most familiar cry, delivered with unfailing energy and accompanied by a display of the long, dusky crest that rises whenever its owner is agitated—which, in true jay fashion, is most of the time.

Like many of its cousins, the Steller's is also an accomplished mimic, especially of the calls of birds of prey. It saves its own softly warbled song for a sequestered spot seldom within hearing of humans. Indeed, once paired for mating, Steller's become quite secretive until the young are fledged and the family begins to forage for food with renewed assertiveness. They capture insects proficiently and have a special fondness for acorns and pine seeds. But the Steller's tastes are eclectic, and they have long provoked ire, human and avian alike, for their frequent raids on grainfields, orchards, and the nests of other songbirds.

Skittish in the woods, Steller's jays can become positively audacious in campgrounds.

Recognition. 12–13½ in. long. Dark and crested. Belly, wings, and tail deep blue; head, crest, back, and breast blackish. Call a loud *Shaack! Shaack! Shaack!*
Habitat. Coniferous forests.
Nesting. Nest is a sturdy cup of sticks and twigs lined with mud, 8–16 ft. (rarely to 100 ft.) above ground, on branch or in crotch of conifer. Eggs 3–5, pale blue or greenish blue, lightly spotted with brown. Incubation about 16 days, by female only. Nestling period unknown.
Food. Nuts, seeds, fruits, and insects.

Black-capped Chickadee

Black-capped Chickadee *Poecile atricapillus*
Carolina Chickadee *Poecile carolinensis*

Black-capped Chickadee

Carolina Chickadee

When nesting is over and the young are on the wing, black-capped chickadees form flocks of eight or a dozen birds, which will roost and forage together until spring. Finding food in the winter is often tough, and hunting for it in groups increases the chances for success. As a band of chickadees flits among the trees and shrubs searching for insect eggs and pupae, they keep an eye on one another. When an individual discovers a tidbit, its fellows not only renew their search with enthusiasm, but concentrate on the particular niche where the discovery was made. In this way, new food-source bulletins are continually disseminated throughout the company.

So many eyes provide extra notice of danger as well. The first chickadee to spot a predator gives a warning note and the whole flock freezes, then utters thin, ventriloquial notes. The predator, confused by disembodied calls coming from every direction and no direction, is typically unable to initiate a hunt. When it moves on, an "all's well" note brings the flock back to life.

Birders scrutinize fall and winter chickadee flocks, knowing that other species, such as titmice, nuthatches, kinglets, warblers, and creepers, often travel with them. While the accompanying birds don't seem to tap into the chickadees' network of food information, they appear to profit from their predator-defense system.

Recognition. 4¾–5½ in. long. Gray above; cap and bib black; cheeks white; flanks pale buff. Black-capped has narrow white edges on feathers at bend of wing; Carolina does not. Both usually seen in pairs or small groups.

Habitat. Deciduous, mixed, and coniferous forests, gardens, and parks.

Nesting. Nest is a loose cup of plant fibers, moss, and feathers at bottom of tree cavity or in birdhouse, 1–10 ft. above ground. Eggs 5–10, white with brown spots. Incubation 11–13 days. Young leave nest 14–18 days after hatching.

Food. Insects, seeds, and berries.

Mountain Chickadee

Poecile gambelli

Conservationists bemoan man's continuing destruction of standing deadwood in pursuit of everything from firewood and profits to beautification. The loss of dead limbs makes it difficult for cavity-nesting birds to find breeding accommodations, a problem exacerbated by fierce competition from those two aggressive hole-nesters, the house sparrow and the European starling.

Whether a bird is a primary cavity-nester, such as a woodpecker, which drills its own holes, or a secondary cavity-nester, such as a bluebird or swallow, which depends on existing cavities, its nest site problems may not end with the drilling or the finding. The cavity may have to be defended, as a fascinated observer learned one June morning in the White Mountains of Arizona.

A pair of mountain chickadees, returning to their nest hole high in a quaking aspen with food for their young, found it expropriated by two violet-green swallows, and the four birds went at each other for more than an hour. They fought in the air, raking with their claws; they tumbled about on the ground, cuffing and pecking. Early in the fracas, the swallows were ahead, wrenching one chickadee's wing and jerking the other's leg when they tried to enter the cavity. But in the end, the plucky chickadees won the day and reclaimed the nest, leaving the frustrated swallows to search elsewhere for a hole in which to raise their family.

Little but a white eyebrow sets the mountain chickadee apart from its black-capped cousin.

Recognition. 5–5½ in. long. Gray above; cap and bib black; narrow eyebrow and cheeks white; flanks gray. Often found in pairs or family groups.
Habitat. Mountain coniferous forests.
Nesting. Nest is a loose cup of plant fibers, moss, feathers, and fur of small mammals, excavated at bottom of tree cavity or in birdhouse, 6–16 ft. (rarely as high as 80 ft.) above ground. Eggs 5–10, white with brown spots. Incubation about 14 days. Young leave nest about 20 days after hatching.
Food. Insects, seeds, and berries.

Boreal Chickadee

Poecile hudsonicus

When summer wanes in the Far North and the myriad warblers and other seasonal visitors head south, only the quintessential birds of the spruce-fir forest remain. One of these is the boreal chickadee, a subtly colored relative of the more familiar black-capped. The boreal's personality generally matches its plumage: It is often more retiring and quiet than its curious and vocal cousin.

Boreal chickadees range across Canada and into Alaska where, among willows or stunted spruces near the tree line, lives a similar chickadee that has a grayer crown, longer tail, and larger white cheek. However, it has an odd name—Siberian tit. Why isn't it, too, called a chickadee?

Such are the vagaries of naming birds that similar North American and European species do not always have similar names. Siberian tits, which occur across northern latitudes from Norway to Siberia as well as in Alaska, *are* chickadees, but they and other European birds of this type have always been called tits. When their North American kin were first named, the term "chickadee" was deemed more poetic—or onomatopoeic, to be exact. The Old and New Worlds have yet to agree on other names as well. Thus an American birder visiting Europe for the first time might be confused to find that loons are divers, mergansers are goosanders, brant are brent, murres are guillemots, jaegers are skuas, a robin there is not like our robin here, and their blackbird is a thrush!

An uncommonly tame boreal chickadee perches on a man's finger to feed from his palm.

Recognition. 5–5½ in. long. Brownish above; cap brown; bib black; cheek patch white; ear patch gray; flanks rusty. Usually seen in small groups of same species.
Habitat. Coniferous forests.
Nesting. Nest is a loose cup of moss, lichens, bark, and plant down on floor of tree cavity, either natural or excavated by parents, 1–10 ft. above ground. Eggs 4–9, white, speckled or spotted with brown. Incubation 11–16 days, by female. Young leave nest after about 18 days.
Food. Seeds and insects.

Chestnut-backed Chickadee

Poecile rufescens

The most colorful members of their family, chestnut-backed chickadees live on the western fringe of the continent, especially in coniferous woods. Like other chickadees, these sprightly birds often travel in bands of half a dozen or more, as well as in mixed flocks with kinglets, nuthatches, creepers, warblers, other chickadees, and juncos. There are many reasons why birds of different species move about the woods in groups. But how can so many insect-eaters share the same area and often the same tree or limb without competing with each other?

The answer is a loosely knit, multispecies flock called a bird guild, in which each member avoids competition by seeking food in a slightly different way. Some forage high in the trees, others low. Creepers and nuthatches explore chiefly on trunks—though one searches by going up, the other by going down—whereas other birds prefer the main limbs, and still others spend their time near the tips of branches. Some birds prefer one insect type to another, or large insects to small. And many have a preferred foraging method: Creepers and nuthatches poke under bark and in crannies, chickadees glean from the branches and twigs, while kinglets inspect foliage or snatch a flying insect. In this way guild members use their habitat to the fullest, while minimizing the disadvantages of sharing it.

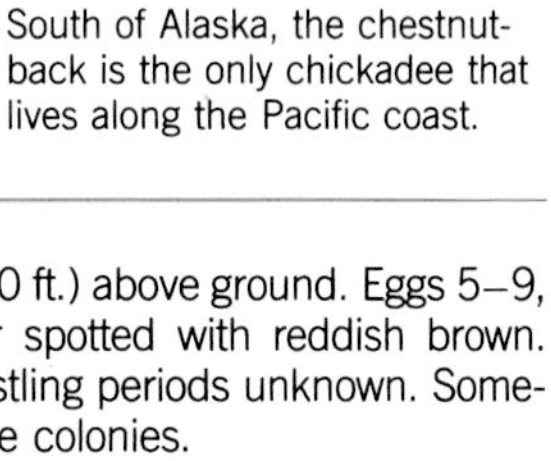

South of Alaska, the chestnut-back is the only chickadee that lives along the Pacific coast.

Recognition. 4½–5 in. long. Rich rusty brown on back; cap dark brown; bib black; cheek patch white; flanks rich rusty. Forages in small groups.
Habitat. Coniferous forests, gardens, and parks.
Nesting. Nest is a loose cup of moss, hair, fibers, and plant down, placed at bottom of tree cavity, either natural or excavated by parents, 1–20 ft. (rarely as high as 80 ft.) above ground. Eggs 5–9, white, speckled or spotted with reddish brown. Incubation and nestling periods unknown. Sometimes nests in loose colonies.
Food. Insects, seeds, and berries.

Bridled Titmouse

Baeolophus wollweberi

Birds are no respecters of international boundaries. The bridled titmouse, which spills over into Arizona and New Mexico but flourishes predominantly in Mexico's western highlands south almost to Guatemala, is one of them. With little variation, it prefers habitats with elevations of 5,000 to 7,000 feet.

Extremely sensitive, as is the entire family, to "trespassers," it will object with a harsh sputter—but from hiding. Elusive, it should be sought in the early mornings before the fiery southern sun heats its ovens. A slowly repeated *hoo hoo-hoo,* imitative of its feared and hated predator, the northern pygmy-owl, may well produce a bristling horde of bridle-faced titmice, along with other small birds joining in to mob the owl and drive it away.

Acrobatics while feeding are characteristic of all titmice. Hiking in the shade of oaks and sycamores, with the strong spicy scent of juniper in the air, birders have found them hanging upside down and swinging from side to side to feed on a bud. One moment the bird is gleaning from leaf and limb, the next it is concentrating on the tree trunk, where it probes and even peels back bits of bark to reveal a spider in hiding or a cluster of insect eggs.

Whatever its preoccupation, the beady black eye is ever alert, and we can almost hear this Hispanic member of the family muttering: "You may not know where I am, but I always know where *you* are!"

Dauntless in self-defense, a bridled titmouse will attack almost any would-be predator.

Recognition. 4½–5 in. long. Small, crested, and gray; face patterned with black and white; bib black; crest bordered with black; underparts whitish. Usually found in small parties.
Habitat. Woodlands of oak, pinyon, or juniper.
Nesting. Nest is a loose cup of cottonwood down, stems, leaves, and grass, placed at bottom of natural tree cavity or woodpecker hole, 4–28 ft. above ground. Eggs 5–9, white, speckled or spotted with reddish brown. Incubation and nestling periods unknown.
Food. Insects.

Plain Titmouse

Baeolophus inornatus

The size and shape of a bird's bill are determined by the food it must take to survive. The titmouse family is endowed with short, sharp bills designed for probing, picking, and pecking, not for hacking woodpecker-fashion.

On a springtime chaparral-covered slope a tufted little gray bird backs out of a hole in a yucca plant. The bird, like other members of this family, builds its nest in a tree cavity or even behind a loose flap of bark. But if dispossessed by the original owner or a predator it will make do even with a hole in a cactus!

Gleaned from leaves and wrinkles in rough bark, fat grubs, caterpillars, and plump, wormlike pupae are carried by the parents in their bills—necessary protein for the hungry chicks clamoring within.

The arrival of any newcomer, be it a strange bird, a coyote, or a human, will set off a flurry of excitement among small birds—led usually by the plain titmouse, who flits about nervously and scolds in a loud voice. This quickly attracts other, even better screamers and shouters such as the common raven, who is welcomed to their side in the fray, but as soon as it subsides is once again a predator and an enemy. All the small birds, once nesting is done, will flock together for feeding tours.

Each bird has at least two Latin names, and some have three. This plain bird's second name in Latin is *inornatus.* It means, appropriately, "unadorned."

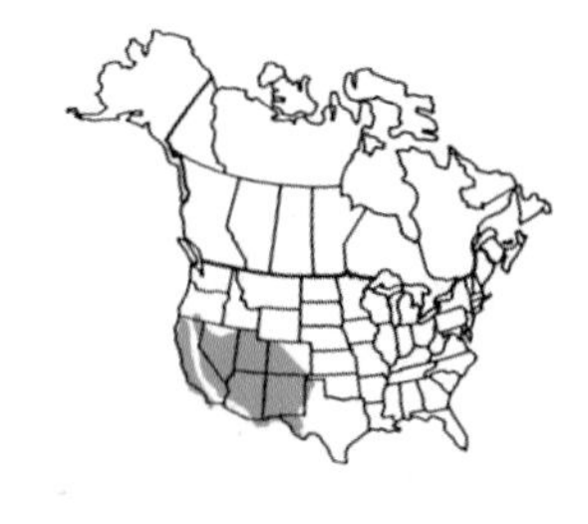

Poison oak seeds are a favorite of plain titmice, more eclectic diners than their bridled cousins.

Recognition. 5–5½ in. long. Uniformly gray, paler below, with no bold markings; crest held erect. Feeds quietly and alone.
Habitat. Woodlands of oak, pinyon, and juniper; gardens and parks.
Nesting. Nest is a loose cup of moss, grass, stems, hair, and feathers, placed at bottom of tree cavity or in birdhouse, 3–35 ft. above ground. Eggs 3–9, white, unmarked or faintly speckled with reddish brown. Incubation 14–16 days, by female. Young leave nest about 3 weeks after hatching.
Food. Insects, seeds, and berries.

Tufted Titmouse

Baeolophus bicolor

A sunflower seed wedged into a bark crevice will make a handy snack for this tufted titmouse.

It is petulant while nesting in spring, peevish when providing provender to the young in summer, pert though exhausted in fall—yet the tufted titmouse will very likely be the perkiest bird at a backyard feeder in winter.

Like the two other species of titmice in North America, the tufted has a highly nervous temperament but, on the whole, a friendly and sociable disposition, at least once the young are fledged and gone. In spring it seeks out bottomlands and wet forests. There it finds an abundance of insect life and scouts out a split or a hole in a gnarled old tree, preferably oak, in which to nest and raise its brood.

By autumn this bird may be found on higher ground in the scarlet and gold deciduous forests, tasting the acorn mast and oak galls but still searching out insects and their eggs. And often, along with chickadees, it will be acting the part of maître d', leading a mixed flock of migrating warblers, kinglets, and vireos to choice tables set with the finest cuisine in the neighborhood.

In winter, when the birdbath holds solid ice and there is an almost holy silence over the land, the tufted titmouse turns to seeds and nuts, especially beechnuts and acorns. And it is this plucky little bird that announces himself at your chilly window feeder. *Peter! Peter! Peter!* The call is sharp and clear and rings over the snowed-in hills and fields like the sound of a beaten anvil.

Recognition. 6–6½ in. long. Gray and crested; underparts whitish; flanks buff; forehead black on eastern birds; whole crest black on Texas birds. Black eye conspicuous on gray face. Tame and noisy.
Habitat. Deciduous and coniferous forests; gardens and parks.

Nesting. Nest is a cup of moss, bark strips, and hair, placed at bottom of tree cavity, in birdhouse, or in other hole 3–90 ft. above ground. Eggs 5–8, white, speckled with brown. Incubation about 14 days, by female. Young leave nest about 18 days after hatching.
Food. Insects, fruits, and seeds.

Verdin

Auriparus flaviceps

The most adaptable birds are, not surprisingly, the most widespread. Those that can accommodate to a wide variety of foods and nesting sites, and to the constant presence of man, simply do better than their more finicky brethren. Still, there are specialists.

In country where the sun is greeted by the clatter of cactus wrens, where each day the greater roadrunner scurries busily from shrub to shrub like the town mayor organizing a last-minute fiesta—here in our great southwestern desert lives the verdin, a very special specialist.

Only in recent years has the verdin been found to be unrelated to titmice and chickadees and shifted from the Paridae family to the family Remizidae. Its closest relatives are now thought to be the penduline tits of the Old World.

For such a tiny bird it builds an extraordinarily large, dome-shaped nest, usually in the safety of the prickly mesquite. It is used both for breeding and roosting, although often the bird will construct smaller nests for roosting through the bone-chilling cold of winter nights.

For much of the year this ash-gray bird with the yellow head is the color of the mesquite in which it nests. Oddly, it is said the bird has no need for water. Evidently getting what little it needs from its diet of insects and seeds, the verdin has never been seen to drink or bathe!

A nest bristling with thorny twigs protects the verdin and its young from swooping predators.

Recognition. 4–4½ in. long. Mainly gray, palest on underparts; head yellow; bend of wing with small reddish patch. Immature all dull gray. Usually forages alone.
Habitat. Streamside woodlands, brushy deserts, and mesquite thickets.
Nesting. Nest is a ball-shaped mass of thorny twigs with entrance at side, attached to branch or placed in dense cactus, 2–20 ft. above ground. Eggs 3–6, pale blue or pale green, spotted with reddish brown. Incubation about 10 days. Young leave nest about 21 days after hatching. Sometimes 2 broods per season.
Food. Insects and berries.

Bushtit

Psaltriparus minimus

Dangling in plain sight from the fork of a branch, a bushtit's nest hangs down 10 inches or more.

Most birds travel in flocks; bushtits are part of a troupe. To this company of roving acrobats the entire West is the Big Top, and any bush or tree will do for a stage. When moving from tree to tree, they seem to play follow the leader, yet each one tries to outdo the other. Performers all, they are masters of tumbling and trapeze artistry.

Arriving in a fruiting mulberry tree, they spread out into an antic formation—one on a leaf swinging to and fro, one hanging upside down from a twig, one swinging in circles on a piece of fruit—all as they feed while seemingly waiting for applause. The air is filled with gentle chips, and one senses they are saying, "Here I am. Where are you?" The contact-calling is constant. A squeaking sound will bring them very close, for they are unafraid of man. But at the slightest hint of a captor's approach, they commence a loud chorus of tinkling bell-like sounds that confuses, distracts, and often disconcerts the predator.

As soon as the young are fledged the family joins up with all the other families in the area. From then on, they will feed together, so it is not unusual to see some 30 or 40 birds moving as a unit from one feeding spot to another. But, amazingly, in contrast to the behavior of some other species, there is never an argument, never a clash, never a conflict between bushtits over the choicest aphid-covered leaf or the cushiest perch from which to swing while feeding on seed. The leaf cleaned, the bird will flaunt its tail, flick its wings, and fly to another bit of Eden.

Recognition. 3¾–4½ in. long. Small, short-billed, and long-tailed; mainly gray. Birds in some areas have brownish cap. Female has yellow eyes, male black eyes. Some near Mexican border have black mask. Forages in flocks.
Habitat. Oak woodlands and scrubby thickets.
Nesting. Nest is a hanging, gourd-shaped mass of twigs, moss, rootlets, and leaves with entrance near top, attached to fork of twigs 6–25 ft above ground in tree. Eggs 5–13, white. Incubation 12 days, by both sexes. Young leave nest about 2 weeks after hatching.
Food. Insects and spiders; some berries.

White-breasted Nuthatch

Sitta carolinensis

Rightside up on the top of a limb or upside down on the bottom, white-breasted nuthatches dart about so erratically that they look more like small windup toys than live birds. But when, suddenly, they call a nasal *Yark, yark* and scamper down the trunk headfirst, they seem like little mice, defying gravity by pausing here and there to reach into crevices and probe under broken bark, picking out insects with their thin little turned-up bills.

These chunky little birds have stubby legs so perfectly placed on their bodies, they are always in absolute balance no matter what their position. Their toes are unusually long and so are their down-turned claws. The tiny hooks at the tips catch easily into the slightest roughness, allowing the birds to run helter-skelter as they search for insects, insect eggs, and larvae nestled in the bark.

Although nuthatches and woodpeckers share the role of gleaning insects from the trunks and larger limbs of trees, their different styles permit a happy division of the bounty. Woodpeckers, braced back on their tails, hop *up* the tree, while nuthatches move *downward*, spotting whatever beetles and other insects might have been missed by the woodpecker. True to his name, the nuthatch also enjoys nuts and can be seen in winter foraging for acorns, hickory nuts, and other seeds.

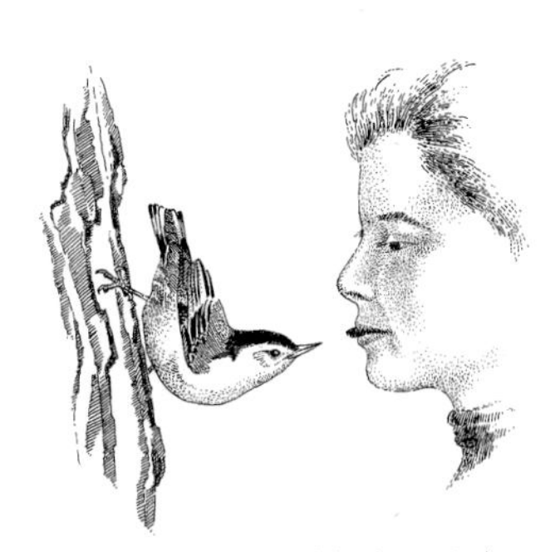

This tame, agile white-breasted nuthatch snatches a seed from the lips of a friendly birder.

Recognition. 5–6 in. long. Stocky, long-billed, and short-tailed; back blue-gray; crown black; face and underparts white; flanks rusty. Creeps down tree trunks.
Habitat. Deciduous forests, suburban areas, and parks.
Nesting. Nest is a mass of bark strips, hair, and feathers at bottom of tree cavity or in birdhouse, 15–50 ft. above ground. Eggs 5–10, white, spotted with brown, red, and gray. Incubation 12 days, by both sexes. Young leave nest about 14 days after hatching.
Food. Nuts, seeds, insects, and fruits.

Red-breasted Nuthatch

Sitta canadensis

Wherever it nests, a red-breasted nuthatch smears pine pitch around the entrance hole.

Like all the nuthatches, red-breasts build their nests inside a cavity—either an abandoned woodpecker hole or one that they whittle themselves in a dead tree. They spend six or seven weeks building the nest, incubating the eggs, and then brooding and caring for their young.

From the very beginning of nest building until the young are ready to leave, the parents exhibit an unusual behavior: they take droplets of resin, or pitch, from balsam fir, spruce, or pine trees and smear it all around the entrance of their nesting cavity. Carrying the droplets in the tips of their bills, they place them on the bark next to the nest entrance, then sweep their bills back and forth to spread it. By the time nesting season is over, the entrance may be surrounded, for two inches or more, with a quarter-inch of the sticky, glistening substance.

No one knows why the nuthatches do this or what use it may serve—unless the stickiness prevents ants from entering the nest. Many ornithologists believe it is an atavistic, no longer useful, trait. Some of the red-breasts' relatives also do strange and, as far as we can see, useless things around their nest entrances. White-breasted nuthatches, for example, stuff bits of fur into cracks and crevices close to the entrance hole, and pygmy nuthatches do the same kind of caulking. We don't know why; it's just another wonderful mystery about our feathered neighbors.

Recognition. 4½–5½ in. long. Small, stocky, and short-tailed; back blue-gray; crown black, with white eyebrow; black stripe through eye; underparts rusty. Creeps down tree trunks; forages at pine cones.
Habitat. Coniferous forests.
Nesting. Nest is a mass of bark strips, grass, and plant fibers, at bottom of tree cavity or in birdhouse, 5–100 ft. above ground. Often smears pitch around entrance. Eggs 4–7, white, finely spotted with brown. Incubation 12 days, by both sexes. Young leave nest 18–21 days after hatching.
Food. Mainly conifer seeds and insects.

Pygmy Nuthatch

Sitta pygmaea

If chickadees and nuthatches are the small boys of the woods, as George Gladden once wrote, then pygmy nuthatches are the typical smallest boys in the gang—noisy and irrepressible, as they exaggerate every swashbuckling quality of the bigger guys. They dash through the tops of junipers, pinyons, and pines of our western states in a pandemonium of incessant callings and twitterings. Just like other nuthatches, they run upside down all over trees, but they dash about faster, probe for insects in the bark with quicker, almost inattentive darts, and seem to be just generally more haphazard and hastier then their kin.

Their tails are shorter than those of other nuthatches, and they certainly cannot use them as braces the way woodpeckers do when drilling into trees. Yet the pygmy nuthatches, balancing on their two well-placed legs and held in place by their sharply hooked claws, manage to do a great deal of formidable pecking when digging out their nesting cavities and when chipping away the bark under which some highly palatable insects are hiding.

Even their roosting habits are exaggerated. While other nuthatches, and woodpeckers too, spend the night in their own roosting holes, with one or, at most, two birds in any one hole, the pygmy nuthatches sleep communally. In fact, one of the largest roosting groups ever reported consisted of 100 birds in a single cavity!

Peppy and independent by day, pygmy nuthatches huddle snugly in roosting holes at night.

Recognition. 3¾–4½ in. long. Very small, stocky, and short-tailed; back blue-gray; crown dull brown; eyeline blackish; underparts plain whitish. Forages in flocks in foliage, less often on tree trunks.
Habitat. Coniferous forests.
Nesting. Nest is a mass of pine-cone scales, plant down, leaves, and feathers, at bottom of tree cavity 8–60 ft. above ground in dead stub. Eggs 4–9, white, finely spotted with reddish brown. Incubation about 16 days, mainly by female. Young leave nest about 22 days after hatching.
Food. Insects, spiders, and pine seeds.

Brown-headed Nuthatch

Sitta pusilla

Because their bills are not adapted to strenuous drilling through bark and wood as the chisellike bills of woodpeckers are, nuthatches must be content with picking insects and larvae from open cracks and crevices in the bark with their thin, upturned bills. Even though they can *hear* other insect larvae munching away on the sweet, moist cambium beneath the bark, they cannot break through that tough layer to reach them.

But occasionally the brown-headed nuthatch *does* reach them, for this bird has uncannily learned how to use a tool. It first finds a small, very hard piece of bark lying on the ground or hanging loose on a tree, where it pulls and probes until it breaks the chip free. With this sturdy chip held firmly in its bill, the brown-headed nuthatch attacks the bark above the hidden larvae. Using the tool as both a hatchet and a lever, the bird strikes from the side and pries from beneath until it has exposed the fat, rich grubs. Then it drops the bark and enjoys the feast. Since it has no way of holding onto the tool while it eats, it will have to find the original chip or another for its next attack on bark that shields some coveted insects.

But just using a tool is a remarkable thing for this small bird to do! Once we thought that only man had this capability. Then we learned that a few exotic animals used primitive tools. Now we know that our own southeastern bird, the brown-headed nuthatch, is a tool user, too.

Lacking the woodpecker's sturdy beak, a brown-headed nuthatch uses a wood chip to find larvae.

Recognition. 4–5 in. long. Very small, stocky, and short-tailed; back blue-gray; crown brown, often with whitish spot on nape; eyeline blackish; throat whitish; breast pale buff; flanks grayish. Forages alone or in pairs.
Habitat. Pine forests.
Nesting. Nest is a mass of stems, bark strips, grass, and wood chips, at bottom of cavity in tree or utility pole, or in birdhouse, 2–50 ft. above ground. Eggs 3–9, white, spotted and blotched with brown. Incubation about 14 days, by both sexes. Young leave nest about 18 days after hatching.
Food. Insects, spiders, and pine seeds.

Brown Creeper

Certhia americana

In spite of its coat of mottled brown feathers, the brown creeper does not look like a bird as it spirals up the trunk of a tree, lisping quietly to itself. Its short legs are folded so closely beneath its body that its underfeathers just brush the bark. With its long, stiff, down-curving tail pressed against the tree for balance, and its long, slender, down-curving bill just skimming the bark, it actually looks like a nearsighted and not-quite-bright tree mouse as it searches the cracks and crevices for insects, their eggs, and larvae.

The brown creeper begins its climb very near the ground at the base of a tree, usually a large one. It creeps swiftly upward with peculiar jerking movements, going around and around the trunk in a loose spiral until it almost reaches the top. Then it just lets go and lands on the base of the next tree. It does not fly down but seems to flutter, as though it were a dry brown leaf on an autumn wind.

John Kieran said that the brown creeper is "one of the easiest birds to overlook," and it must be so. For though it breeds all across southern Canada and the northern United States and winters clear down to the Gulf states, comparatively few people have ever seen the bird. Those who have usually see only one, feeding all alone. It's quite a surprise to find on occasion two or three of them frolicking in the merry company of chickadees, kinglets, and downies, as sociable and alert as any of the band.

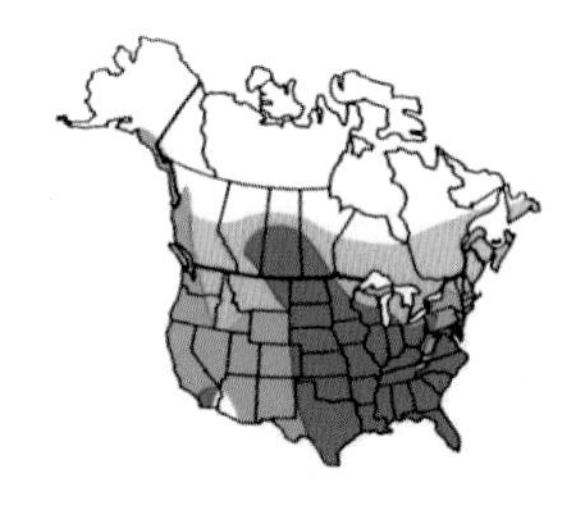

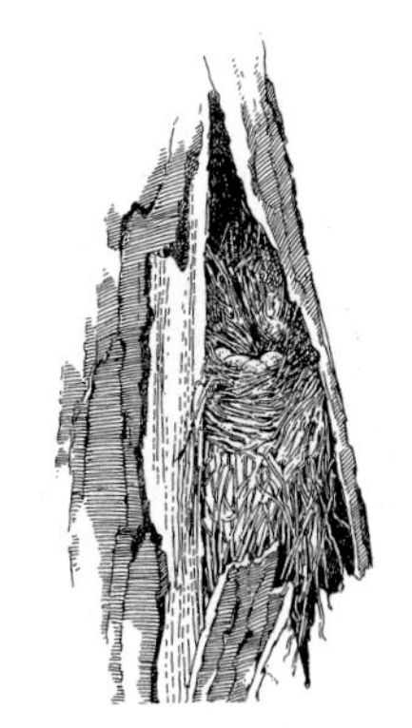

The brown creeper's nest is a hammock tucked snugly beneath a wedge of bark.

Recognition. 5–5¾ in. long. Small and slim; bill thin and curved; tail long and stiff; streaked brown above; underparts whitish, often with buff tinge on flanks. Spirals up tree trunk, then flutters to bottom of next tree. Call note a wiry *seep.*
Habitat. Forests, woodlands, and parks.
Nesting. Nest is a cup of moss, bark strips, and twigs, placed behind loose strip of bark on tree trunk or sometimes in tree cavity, 5–15 ft. above ground. Eggs 4–8, white, lightly spotted with brown. Incubation about 15 days. Young leave nest about 14 days after hatching.
Food. Insects and spiders gleaned from bark.

Carolina Wren

Thryothorus ludovicianus

Baskets, mailboxes, and even pockets of old overcoats provide nesting sites for Carolina wrens.

In a voice so loud and rollicking that it eclipses the songs of all other birds in the neighborhood, the Carolina wren announces its presence in woodland tangle or dooryard shrubbery. *Wheat-eater, wheat-eater, wheat-eater, wheat!* he clamors, and, without a break, switches suddenly to *Giddyap!* or *Tea-party* or *It's-raining!* all in the same cadence and pitch. And he sings practically all day long, the whole year round, no matter what the weather. But he can scold, too, and in this the female joins him, bobbing in excitement, jerking their tails, irately chucking any trespasser out of their territory.

Carolina wrens are southern birds, never migrating, spending year after year on the same range. But unknown pressures often send their youngsters northward. So long as the winters are mild, they survive to build resident populations. Then comes a bitter winter, severe and pitiless. Since Carolina wrens are not migratory, they do not instinctually turn southward. They stay, and they perish. Even as far south as Maryland, all the boisterous Carolina wrens have died in a single winter.

But the process of northern expansion begins all over again. After several years of vacant niches, one spring morning *Wheat-eater, wheat-eater, wheat-eater, wheat!* rings once again through northern woodlands.

Recognition. 5¼–6 in. long. Mainly rich rusty brown, with bold white eyebrow, white chin, and buff-tinged underparts.
Habitat. Woodlands, thickets and undergrowth near water, and garden shrubbery.
Nesting. Nest is a cup of grass, stems, and bark, 1–10 ft. above ground in hole in tree or in birdhouse. Often nests in cavities in man-made objects. Eggs 4–8, white or pinkish, spotted with brown. Incubation about 14 days, by female. Young leave nest about 14 days after hatching. 2 or 3 broods per season.
Food. Insects and other small animals; some fruits and seeds.

Bewick's Wren

Thryomanes bewickii

Our most common birds have lived on earth far longer than *Homo sapiens* has, thanks largely to their skill in adapting to various changes and challenges—the impact of predators, fluctuations in food supply, drought, flooding, and other climatic events. But there is one type of change that cannot always be adapted to: loss of a species' habitat, whether woodlands cleared for farming, fields used for housing, marshes drained, valleys filled in, or any other landscapes reshaped by the hand of man.

It appears that the Bewick's wren in the eastern United States is the victim of such activity, for its numbers there have declined sharply. But there is good news too: the Bewick's is the most common wren in the West and seems to be holding its own quite well. Favoring open fields with cover from low-growing brush, it usually nests in a cavity at the foot of a tree or a crevice between rocks low in a wall—though it can make do with an opening in the dilapidated seat of an abandoned tractor.

Alert and inquisitive, a Bewick's can be roused to action by almost any squeaky sound and will scurry from bush to bush rendering its distinctive "irritation buzz," which some have likened to an old-fashioned Bronx cheer. Its chief competitor in the West is its cousin and archrival, the house wren, which normally departs for warmer regions in winter. The Bewick's, in contrast, is a stay-at-home whose intimate knowledge of the territory gives it a vital advantage over those who must start from scratch each spring.

Tame and chipper, the Bewick's wren is happily at home among humans and their dwellings.

Recognition. 5–5½ in. long. Slender and long-tailed; brown above and on crown; eyebrow white; face brown; underparts whitish; tail feathers with white tips.
Habitat. Open, shrubby woods, clearings, thickets, and suburbs.
Nesting. Nest is a mass of sticks, moss, and leaves, placed in any natural or man-made cavity. Eggs 4–11, white, speckled with brown and lilac. Incubation about 14 days, by female. Young leave nest about 2 weeks after hatching. Usually 2 broods per season.
Food. Insects and spiders.

House Wren

Troglodytes aedon

For many North Americans, spring hasn't really "sprung" until this high-strung little brown bird refurbishes the birdhouse and graces the backyard with its bubbly, full-throated song. Any house wrens that don't opt for a lease on the nest box will seek out either a natural cavity or a woodpecker hole in a tree, preferably an oak. However, true to its common name, the house wren generally remains close to human dwellings.

Don't be misled by its small size and comfortably domestic habits, though. A "plain Jane" with no distinctive marks to set it apart from other members of its family, the house wren distinguishes itself with the temperament of a terrier. Spirited and aggressive, this bird will not hesitate to drive a downy woodpecker from a just-completed nest hole. (Fortunately, most woodpeckers hack out more than one hole, for comfort, convenience—and insurance.)

All members of the wren family feed on insects and spiders; this one finds them in low trees, brush, and hedgerows. Out foraging, the house wren often crosses paths with Carolina and Bewick's wrens, and the resulting friction at times appears to be more than normal competition intensified by a gradual loss of habitat. In fact, the bitter, grudging antipathy that seems to exist—especially between the house and Bewick's wrens—cannot help but evoke images of the Hatfields and the McCoys.

A swarm of wasps nesting in a bird box can prove lethal to an unwary wren.

Recognition. 4½–5¼ in. long. Small and stocky; back, wings, and tail gray-brown, with fine bars on wings and tail; underparts dull whitish. Birds in mountains of southern Arizona rustier, with buff throat.

Habitat. Open woodlands, wooded canyons, farmyards, gardens, and parks.

Nesting. Nest is a mass of sticks, twigs, and feathers, placed in any natural or man-made cavity. Eggs 5–9 white, finely spotted with brown. Incubation 13–15 days, by female only. Young leave nest 12–18 days after hatching. Often 2 broods per season.

Food. Insects and spiders.

Winter Wren

Troglodytes troglodytes

This tiny songbird doesn't look for tornadoes, it just makes the most of them. Seeking out trees knocked down by violent storms, the winter wren builds twig nests among their upturned roots. Or, if no uprooted trees are available, other dark, impenetrable places will do. Brush heaps and wood piles are also favorite homesites of this nervous little bird who creeps through forest windfalls like a furtive deer mouse.

Ever in motion, head bobbing and stubby tail cocked, the male winter wren announces his discovery of a nesting spot with a burst of song—a fervent babble of trills, warbles, and gurgled tinklings—that is a marvel of northern woodlands. Lasting several seconds and containing more than 100 separate notes, the medley surprises many listeners, who can't believe that such a small bird is capable of such varied, vigorous expression.

And the song is seductive as well. Lured by its cascading notes to the male's territory near a wooded swamp or forested ravine, a female winter wren may discover her prospective mate has built several concealed nests. Often only one nest is used, the others serving as dummies or decoys. However, if another female appears and food is sufficient to sustain an additional brood, a decoy nest may become a real home as the male turns polygamist and takes a second mate.

A simple cavity in the ground may offer enough security for the winter wren's large, dense nest.

Recognition. 4–4½ in. long. Small, stocky, and short-tailed; upperparts dark brown; underparts paler, with dark bars on belly and flanks; eyebrow pale. Secretive; creeps through low vegetation like mouse; bobs up and down when excited or curious.

Habitat. Coniferous forests and thickets.

Nesting. Nest is a large mass of twigs, grass, and moss, concealed among roots of overturned tree, in cavity in stump or bank, or under bark. Eggs 4–7, white, spotted with brown. Incubation 14–17 days, by female only. Young leave nest about 19 days after hatching.

Food. Insects.

American Dipper

Cinclus mexicanus

The stream hurtles wildly along its bed in a western mountain canyon. Churning white with foam, the roaring water is almost deafening. The air is wet with cold spray. Here, in the heart of this evergreen forest, is the solitude that the American dipper seeks.

On a protruding rock in midstream stands a chunky dark gray bird. Every few seconds it launches into deep knee bends—the "dipping" from which it takes its name. Suddenly, like a New England dory streaking to a rescue, the bird plunges into the stream and disappears.

One of nature's true anomalies, the dipper is a land bird that spends its life in, on, and *under* water! Its long toes, neither lobed nor webbed, were meant for perching, so the dipper propels itself underwater with its short wings, in effect "flying" while it swims. Although it can catch insects on the surface, the dipper spends most of its time walking the bottom, snagging small fish and probing for flatworms and snails. Walking its beat, it is kept warm and dry by an undercoating of down.

Dippers usually build their bulky, moss-covered nests on the underside of rock overhangs, sometimes even beneath waterfalls. Here they enter their homes by flying through the fall's cascading current—a daredevil maneuver that not only foils would-be predators but also evokes applause from anyone lucky enough to witness it.

This most aquatic songbird can swim down 20 feet to walk a river bottom in search of food.

Recognition. 7–8½ in. long. Chunky and short-tailed; slate-gray, with brownish tinge on head. Forages along streams, diving under water to capture food; call a loud *bzeet!*
Habitat. Rushing mountain streams.
Nesting. Nest is a bulky, domed mass of moss and fine grass, with entrance at side, built overlooking water on rocky ledge, under bridge, or sometimes on rock in stream. Adults moisten all nesting material except lining before adding it to nest. Eggs 3–6, white. Incubation 15–17 days, by female only. Young leave nest about 25 days after hatching.
Food. Aquatic insects and snails.

Blue-gray Gnatcatcher

Blue-gray Gnatcatcher *Polioptila caerulea*
Black-tailed Gnatcatcher *Polioptila melanura*
California Gnatcatcher *Polioptila californica*

Twisting, turning, rising, and diving, gnatcatchers pursue their elusive prey in rapid flight among the treetops. Whether these little birds are darting from twigs flycatcher-fashion to snatch gyrating gnats in midflight or hovering above blossoms to nab a hidden insect, their long tails accentuate every move. Cocked in alert anticipation when the birds perch, opening and closing in flight, these stiff plumes seem almost to throw the birds off balance when a vagrant breeze catches them the wrong way.

Despite their fussy and seemingly frail appearance, gnatcatchers—the blue-grays and their black-tailed and California cousins—are not afraid of anything. They attack blue jays and even crows who encroach on their territory. And it never occurs to them to be secretive about their nests. Every activity is accompanied by conversations in their distinctive high-pitched nasal call, whether they are off gathering nesting materials or are actually on the nest itself. While the female lays her eggs, incubates them, and broods the young, the male stands nearby singing with all his might. When his own turn to tend the eggs or young arrives, he sits right with them and continues his warbling song at the top of his voice. It is not his fault that this song is so soft it can scarcely be heard a hundred feet away!

Recognition. 4–5 in. long. Tiny and slender; upperparts blue-gray; underparts white; tail long, black, with white outer feathers. Male black-tail and California have black cap; little white in tail; best differentiated by range.
Habitat. Open woods and streamside thickets.
Nesting. Nest is a compact cup of grass, bark strips, and fibers, 2–70 ft. above ground in tree. Eggs 3–6, pale blue, spotted with reddish brown or unmarked. Incubation about 13 days, by both sexes. Young leave nest 10–12 days after hatching.
Food. Insects and spiders.

Ruby-crowned Kinglet

Regulus calendula

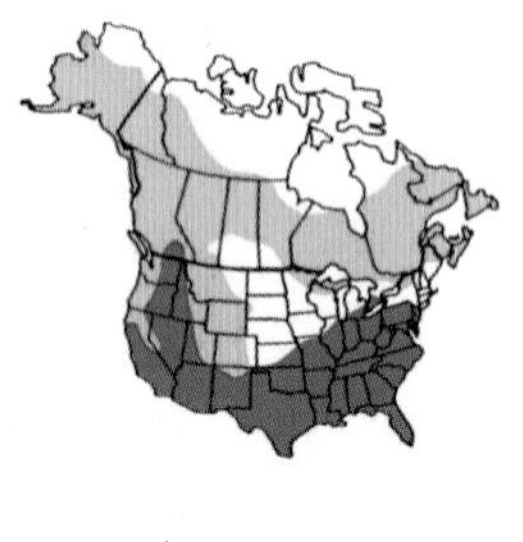

In typical kinglet fashion, the ruby-crown cushions its nest with soft bark and feathers.

Some birds make a vivid first impression. But not ruby-crowned kinglets. Possessed of grayish heads, olive backs, and whitish breasts, these fluttery mites appear drab at first glance. And their diminutive size doesn't help. Barely four inches long, they rank among North America's smallest birds, along with golden-crowned kinglets, winter wrens, marsh wrens, and hummingbirds. The truth is, if these active insect-eaters didn't flick their wings in nervous bursts while hopping from limb to limb, or hover like hummingbirds while snatching aphids off branch tips, we might not spot them at all.

First appearances, however, can be deceptive. There is good reason for this apparently colorless bird's colorful name. Concealed beneath the male's grayish head feathers is a bright red "flash patch," which remains hidden while the bird feeds, flies, and sleeps. But come courtship season, these crests flare like flames when two males face off to battle for territory, heads bent forward and crown feathers erect. Hopping at each other, calling excitedly, the combatants avoid contact while comparing colors, and feisty temperaments, to determine which bird will prevail.

Concealed flash patches are also common in other species. Red-winged blackbirds, for example, have crimson wing epaulets that remain hidden at rest but flare during courtship and territorial defense.

Recognition. 3¾–4½ in. long. Tiny and greenish, with 2 white wing bars and bold white eye-ring. Male has tuft of red feathers on crown, usually concealed. Flicks wings nervously; call note a scratchy *didit.*
Habitat. Coniferous forests; thickets in winter.
Nesting. Nest is a round mass of moss and lichens with entrance at top, hanging from twigs 2–100 ft. above ground in conifer. Eggs 5–11, white or cream-colored, spotted with brown and gray. Incubation probably about 2 weeks, by female only. Nestling period unknown.
Food. Insects and spiders; some berries.

Golden-crowned Kinglet

Regulus satrapa

What is there about evergreens this bird can't ignore? Perhaps it's the insects—the aphids, scales, and budworms—that feed golden-crowned kinglets both summer and winter. Or perhaps it's the shelter that keeps cold winds at bay. Whatever the allure, golden-crowned kinglets and evergreen stands are inextricably linked. Nesting among them in springtime northern forests, flitting through them on wintering grounds to the south, these acrobatic songbirds never seem to be far from needles and pine cones—be they natural or planted by humans.

In the wilds of Quebec, a spruce bog may attract kinglet pairs, while farther south, in the northern United States, kinglets are drawn to groves originally planted by the Civilian Conservation Corps in the 1930's. Formed by the federal government to provide jobs during the Depression, the CCC planted thousands of spruce and pine seedlings across the northeastern states. As these young trees matured, golden-crowned kinglets flocked tc the dense stands they created, and the birds' population burgeoned. When later the trees were harvested, the kinglets moved on, seeking native stands or other, younger, plantations that timber growers had planted. Adapting to an altered landscape, their numbers ebbing and flowing with the growth and harvest of evergreens, golden-crowned kinglets have tailored their lives to the human presence.

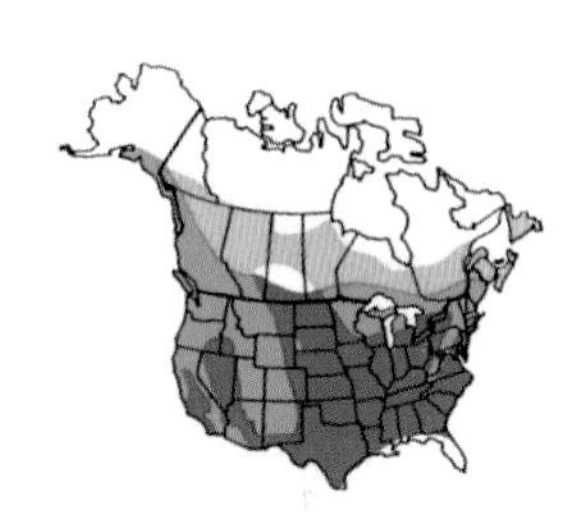

Calmer than it looks, the golden-crown flicks its wings busily while hopping among branches.

Recognition. 3¼–4¼ in. long. Tiny and greenish, with 2 white wing bars; crown orange (yellow on female) with 2 black stripes at sides; eyebrow and underparts whitish. Flicks wings nervously; call note a thin *see-seep.*
Habitat. Coniferous forests; some other forests during winter.
Nesting. Nest is a round mass of moss and lichens with entrance at top, attached to twigs 30–60 ft. above ground in conifer. Eggs 5–10, white or cream-colored, spotted with brown and gray. Incubation probably about 2 weeks. Nestling period unknown.
Food. Insects.

Townsend's Solitaire

Myadestes townsendi

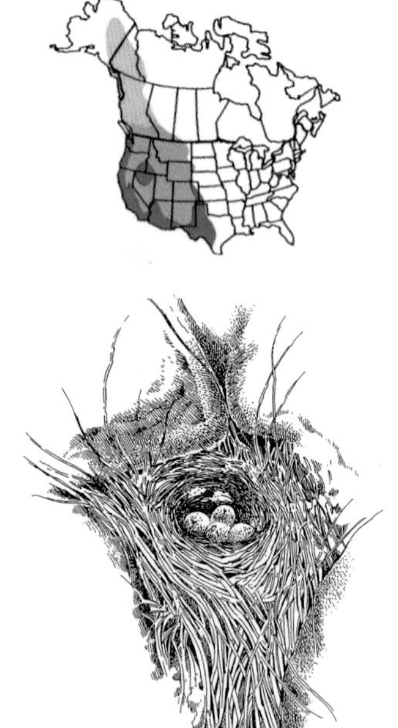

Puzzlingly, the Townsend's builds an "apron" of grasses and pine needles at the edge of its nest.

A bird's song and its call have different sounds and different meanings, as a few moments spent listening to a Townsend's solitaire will quickly show. Blessed with the musical mastery typical of its family, this mountain thrush sings a sweet, loud, prolonged song that melodically lilts through trills and slower-paced warbles as it rises and falls in volume. But this virtuoso's call, heard monotonously at all hours, is a short, unpleasant squeak during the day, like the sound of a rusty gate. At night it becomes an equally sour but deeper, muffled chant.

The Townsend's solitaire seems to have only these two call notes, which is typical of many songbirds, who almost never have more than four or five. Each call note carries a different message, some for its mate, some for its fledglings, some even for other species. It may be a call to assemble, or a warning, or a message about food, water, danger, or any number of things unknown to human beings. Although the meaning may be complicated, the call itself is simple.

Birds' songs, in contrast, have elaborate phrasing, but their practical use is just to announce the male's presence on his territory, to warn off intruders, and to attract a mate. The songs' beauty, however, is undeniable, and experts haven't ruled out the possibility that birds may sometimes sing for the sheer joy they take in their own musical achievement.

Recognition. 8–9½ in. long. Slender and long-tailed; upperparts gray; underparts paler; wing patch buff; eye-ring white. Perches upright much like flycatcher.
Habitat. Mountain coniferous forests.
Nesting. Nest is a loose cup of twigs, grass, rootlets, and pine needles, concealed on ground among tree roots, in shallow cavity in bank, or in crevice on rocky slope. Eggs 3–5, pale blue, pinkish, or whitish, heavily speckled with brown. Incubation and nestling periods unknown.
Food. Berries and other fruits; insects, spiders, and pine seeds.

Veery

Catharus fuscescens

To most people the veery is an unreal creature, a mysterious voice spiraling downward from remote and starry distances, filling the darkening woods with rippling, heavenly music. It has been called the voice of a disembodied spirit holding within itself all the sweetness and wildness of the forest, and a vocal will-o'-the-wisp drifting beyond touch through the nighttime woods.

Yet the veery is not a phantom. It is a real live thrush, tawny and only faintly spotted, the most slender and graceful thrush of all. It lives in the woods and swamps, where it scratches through the leaf mold to feed on insects and nibbles on fruits and berries. Its bulky nest is usually on the ground, and if the earth is damp, the veery will build a thick platform of dead leaves to insulate the thin shreds of bark, grasses, and mosses it assembles as a cradle for its delicate, pale blue eggs.

Shy and elusive, the bird is real enough, but seldom seen, for it slips quietly away through the undergrowth if anyone comes near. And so the veery remains a mysterious disembodied voice that sings a seemingly impossible duet with itself in the darkness—an alto and a soprano twining a silver thread of tremulous music that spirals downward through the moonlight.

Although a ground feeder, the veery can take such agile prey as butterflies on the wing.

Recognition. 6½–7¾ in. long. Upperparts from head to tail rich rusty brown on eastern birds, darker brown on western; underparts whitish; breast sparsely spotted, tinged with buff in East. Feeds mainly on ground.
Habitat. Deciduous and mixed woods; also willow thickets in West.

Nesting. Nest is a firm-walled cup of leaves, grass, bark, and stems, concealed on or near ground at base of shrub or weeds. Eggs 3–5, pale blue and unmarked. Incubation 10–12 days. Young leave nest about 10 days after hatching.
Food. Insects, spiders, earthworms, and fruit.

Gray-cheeked Thrush

Catharus minimus

The berries of American spikenard attract even the most skittish gray-cheeks.

Alaskans and northern Canadians may be excused for thinking that of the 150 thrush varieties living in the Western Hemisphere, they possess the sweetest singer of all in the gray-cheeked thrush. But they may have an argument on their hands: many prefer the deeper, flutelike tones of the hermit thrush to the slightly more nasal song of its longer-legged, gray-cheeked cousin. However, most ornithologists agree that for singing endurance, the gray-cheeked thrush is unique. Often during the long days of summer in its northernmost range, this bird will break into song at the sun's first appearance, then continue to carol his way, almost nonstop, through 20 hours of daylight. He may sing on a tree limb (where trees exist) or while on the wing above windswept stretches of green tundra.

No other thrush migrates the great distances—from 6,000 to 8,000 miles—that the gray-cheeked does. Some begin their journey from points as deep in South America as Peru or the immense rain forests of the Amazon, and their northward thrust takes them through the eastern half of the United States. But even in migration, most of it nocturnal, the gray-cheeked thrush seems to find time for singing. One ornithologist has described the bird as a small troubadour who makes frequent stops to fill short musical engagements for all who are willing to pause and listen.

Recognition. 6¼–8 in. long. Upperparts dull brown or olive-brown from head to tail; underparts whitish; vague eye-ring; cheek patch grayish; boldly spotted breast. Feeds mainly on ground.
Habitat. Coniferous forests and thickets of willow and alder.
Nesting. Nest a cup of grass, mud, bark, and leaves, 6–20 ft. above ground in small tree; sometimes on ground. Eggs 3–6, pale blue or greenish, lightly dotted with brown. Incubation about 14 days, by female only. Young leave nest 11–13 days after hatching.
Food. Insects, spiders, earthworms, and fruit.

Swainson's Thrush

Catharus ustulatus

In summer, across North America, the beautiful songs of five very similar birds fill our forests—five thrushes, often heard but infrequently seen. Indeed, we most often spot them not on their singing grounds but while they are migrating, appearing briefly in a park or garden. Then we reach for our field guides and search for small details of eye-rings, back color, breast spotting, or behavior that will answer our question: Is it a Swainson's thrush, or a hermit, gray-cheeked, or wood thrush, or a veery?

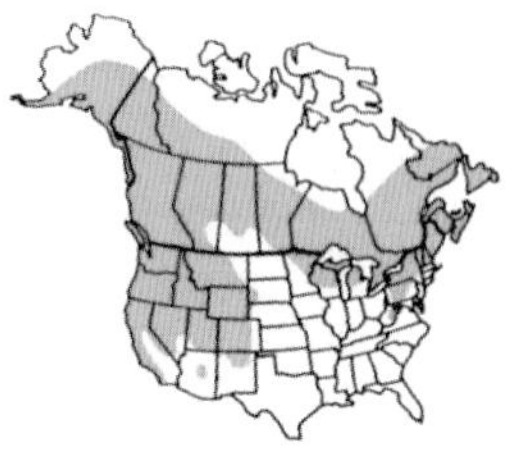

One or another of these five brownish thrushes inhabit wooded areas throughout the United States and Canada, and in some places two or even three species may nest in close proximity. Though confusingly similar to us, they are distinct species and will not interbreed.

The barriers to hybridization are complex. Certainly appearance is important, but it is not the only characteristic that defines a species. Baltimore and Bullock's orioles, for instance, have more obviously different looks than the Swainson's and gray-cheeked thrushes, yet they are in fact two varieties of a single species, the northern oriole. Among our thrushes, nonvisual cues may be equally or more significant: For example, each species has a distinctive song and call—an acoustic "signature"—that the wrong mate will not respond to. And the way in which the male displays and otherwise courts the female is a distinguishing factor. Undoubtedly each female prefers her own kind.

Young thrushes venture out of the nest very early, complicating the task of parental care.

Recognition. 6½–7½ in. long. Upperparts from head to tail olive-brown in East, rusty brown in West; underparts whitish; eye-ring buff; breast buff with bold spots. Usually seen on ground.
Habitat. Coniferous forests, willow thickets, and aspen groves.
Nesting. Nest is a compact cup of twigs, leaves, and plant fibers, 2–20 ft. above ground, close to trunk in small tree. Eggs 3–5, pale blue spotted with light brown. Incubation 11–14 days, by female only. Young leave nest after 10–14 days.
Food. Insects, spiders, earthworms, and fruit.

Hermit Thrush

Catharus guttatus

One of the shyest birds, the hermit thrush favors either deciduous or coniferous woodlands, where it finds the protein-rich food it needs and the privacy it craves. With its spotted chest, the hermit resembles its slightly larger cousin, the wood thrush, but can be distinguished by the reddish tail that it holds erect when first at rest, then drops gradually from its cocked position.

The hermit thrush announces all its songs with one long-held, indescribably lovely note, followed by marvelous improvisations. John Burroughs, the great naturalist, once described its singing at dusk: "Listening to this strain on the lone mountain, with the full moon just rounded on the horizon, the pomp of your cities and the pride of your civilization seemed trivial and cheap," he wrote.

This "shy and hidden bird" also figures prominently in a literary tribute to a great man—Abraham Lincoln. In the poem "When Lilacs Last in the Dooryard Bloom'd," Walt Whitman wrote movingly of the just-slain president. For Whitman, the forgiving mildness as well as astonishing strength of Lincoln's character could be symbolized by two of America's most welcome annual occurrences—the blooming of fragrant lilacs and the unforgettable music of the hermit thrush. "O liquid and free and tender," Whitman wrote of the bird's song, " . . . he sang the carol of death, and a verse for him I love."

Despite their shyness, hermit thrushes are quick to defend their nests against intruders.

Recognition. 6½–7½ in. long. Crown, back, and wings dull brown or olive-brown; tail rusty; underparts whitish; breast with bold spots. Raises and slowly lowers tail; usually seen on or near ground.
Habitat. Forests, bogs, and pine barrens.
Nesting. Nest is a bulky cup of twigs, strips of bark, grass, and moss, concealed on ground or 2–5 ft. above ground in tree. Eggs 3–6, pale blue, unmarked or dotted with brown. Incubation about 13 days, by female only . Young leave nest about 12 days after hatching.
Food. Insects, spiders, earthworms, and fruit.

Wood Thrush

Hylocichla mustelina

While the winds of March are still rattling windows and driving the rain, those who can wait no longer to hear one of nature's most beautiful sounds begin watching woodlands from the Dakotas southward to the Gulf states. Usually, if the woods are deciduous and moist, they will not have long to wait. Nearby a wood thrush will suddenly begin to sing—a song so lovely, so unforced, that one distinguished naturalist declared "with their silvery, bell-like notes, they can sing their way into anyone's heart."

Some naturalists believe that the wood thrush sings because the sight lines in wooded areas are so short, a male must constantly be declaring his whereabouts to his mate, who spends as much as 90 percent of her time sitting on her clutch of eggs. In that nest may be scraps of white rag or paper, tucked there, so some authorities believe, to help the mother bird's spotted chest blend more readily into the nest itself. In there, with her rufous head tilted skyward, she also sends forth a cascade of sound that is the equal of any produced by a master flutist in a symphony orchestra.

Although the wood thrush is primarily a forest dweller, it has become an increasingly welcome sight in suburban areas where there are sufficient shrubs and trees for the bird to conceal itself from enemies. Gardeners give it special welcome, for this thrush is an avid eater of such garden pests as beetles, cutworms, and snails.

Within minutes of her young's hatching, a mother wood thrush may start to clear away the shell.

Recognition. 7½–8½ in. long. Stockier than other brown thrushes. Upperparts rusty brown, brightest on head; tail dull brown; underparts whitish with bold black spots.
Habitat. Woods, swamps, suburbs, and gardens.
Nesting. Nest is a sturdy cup of twigs and grass reinforced with mud, 6–50 ft. above ground in tree. Eggs 2–5, pale blue or greenish blue, unmarked. Incubation about 14 days, by female only. Young leave nest about 13 days after hatching. Sometimes 2 broods per season.
Food. Insects, spiders, earthworms, and fruit.

Varied Thrush

Ixoreus naevius

Ever belligerent, the varied will drive other birds from a feeder even when not hungry itself.

A sense of the primeval permeates the older coniferous forests of the Pacific states, favorite haunts of the varied thrush. On still, misty mornings, droplets of moisture fall from the needles and leaves of great trees to soak rotting logs and rocks on the ground below. Suddenly an eerie, quavering whistle arises, followed by several more at different pitches. These sounds seem the perfect accompaniment to this somber, damp scene—like a melancholy pipe organ tuning up in a natural cathedral.

The varied thrush's whistle is the quintessential sound of this drizzly habitat, but compared to the beautifully patterned songs of other thrushes it is unexpectedly simple. And there are visual surprises as well: a bright eyebrow, a complex orangey wing pattern, and across the breast of the male, a broad black band—certainly distinct from our spotted woodland thrushes or the ubiquitous robin.

Though it lives only in western North America, the varied thrush may be most closely related to a large group of Asian thrushes; perhaps it is a relic from a long-ago expansion of this group. Such expansions have occurred more successfully in some groups than others. Thrushes have diversified to fill nearly all parts of the world. Consider, by contrast, the wrens: 60 species occur in the New World, but only one—the winter wren—inhabits the Old.

Recognition. 9–10 in. long. Male clear gray above, with orange wing bars; eyebrow orange; underparts rusty; breast band black. Female similar but duller, especially above; breast band gray or lacking. Forages in trees or on ground.
Habitat. Coniferous forests.
Nesting. Nest is a bulky cup of twigs, leaves, and stems, often reinforced with mud, 6–20 ft. above ground in tree. Eggs 2–5, pale blue, lightly spotted with brown. Incubation about 14 days, by female. Nestling period unknown.
Food. Insects, spiders, earthworms, nuts, seeds, and fruit.

Wrentit

Chamaea fasciata

It is the quintessential homebody among North American birds. While shorebirds, warblers, and other migrants make round-trip journeys of thousands of miles each year, a wrentit, once established, rarely wanders from the acre or two that constitutes its brushy territory. And after a male attracts a female, the pair mate for life.

Neither wrens nor tits but one-of-a-kind relatives of Old World babblers, wrentits daily make the rounds of their territory, feed on insects, bathe, care for their young, and ward off intruders. Nearly constant companions, mates can sometimes be seen in a bout of mutual preening, each inspecting and arranging the other's plumage. By night they huddle together against the chill air. "A pair sit side by side," one observer wrote, "facing in the same direction and so near together that they appear as a single ball of feathers from which tails, wings, and feet protrude."

These secretive birds are heard far more often than they are seen in the chaparral or brushland where their bouncing-ball song is a common refrain. They are loath to cross open spaces. Perhaps because of this reluctance, their range—which extends up the West Coast from Baja California—stops at the Columbia River, which to a wrentit must seem a formidable barrier indeed.

Mated wrentits help maintain their close pair bond by carefully preening each other's plumage.

Recognition. 6–6½ in. long. Small and long-tailed. Grayish brown; eye pale. Skilled at staying out of sight; loud *yip-yip-yip-yrrrr* call usually heard more often than bird is seen.
Habitat. Dense brush, chaparral, thickets, and garden shrubbery.
Nesting. Nest is a compact cup of grass, plant fibers, and bark bound with spider web, 1–15 ft. above ground in dense shrub. Eggs 3–5, pale blue. Incubation about 15 days, by female only. Young leave nest about 16 days after hatching.
Food. Insects, spiders, and berries.

Long-billed Thrasher

Toxostoma longirostre

In the intense heat of southern Texas, the long-bill places its nest in deep, cool shade.

But for occasional variations in coloring and a few other details, a long-billed thrasher could easily be mistaken for the brown thrasher of eastern North America. The long-bill's face is grayer, its eye orange instead of yellow, and its bill a bit longer, but these are minor points amid many similarities. For years, in fact, the two were thought to be the same species. Within the United States, however, long-bills live only in south Texas, and if they and the brown thrashers overlap there at all, it is only in winter.

Essentially, the long-bill is a Mexican bird. Cattle grazing and the activities of settlers in south Texas over the past 300 years have changed much of that area from grassland to chaparral. It may be these changes that allowed long-bills to expand north of their former range and come into contact with wintering brown thrashers.

In a sense this is a grand natural experiment, one that ornithologists would love to re-create with many other birds. When two closely related species suddenly come in contact, will they share the same habitat or occupy separate ones? Will they compete for survival or make use of different resources? These are difficult questions, but the thrashers seem to be sorting them out. Long-bills nest and forage in the chaparral, while wintering browns favor woodlands along rivers—an accommodation that permits both species to survive and flourish in the same small area.

Recognition. 10–12 in. long. Slender and long-tailed; dull brown above with 2 white wing bars; underparts whitish with black streaks; eyes orange; bill curved. Forages mainly on ground.
Habitat. Dense woodlands, mesquite chaparral, and streamside thickets.
Nesting. Nest is a compact cup of thorny twigs, 4–8 ft. above ground in dense thicket or clump of cactus. Eggs 2–5, blue-green or greenish, speckled with reddish brown. Incubation and nestling periods unknown. Sometimes 2 broods a season.
Food. Insects, spiders, and berries.

Brown Thrasher

Toxostoma rufum

In late spring or early summer, anytime during that brief period when brown thrashers are courting, walk along a brushy border, up a bushy slope, or down a winding lane perfumed by hedges of multiflora rose. Listen for an eloquent song and, letting eyes follow ears, scan sapling tops for the lone singer. There the brown thrasher will be—conspicuous for once on his lofty perch, in rapturous song. With body trembling, head upraised, and bill wide open, he seems to pour his very soul into a paean of praise to his territory, his mate, his nest, and his nestlings-to-be. This impassioned singing will cease soon after eggs are laid, and it stops immediately if he realizes he is being observed.

A secretive bird, the brown thrasher is most commonly glimpsed as a cinnamon-brown shape escaping into the thorny tangles it chooses for its home. There it forages for food on the ground, stroking its bill from side to side, scattering leaves—and suggesting to some a possible origin of its name. Others believe the name derives from the Middle English word for thrush. The thrasher, of course, is no thrush; it is a mimic, albeit with a limited repertoire compared to the mockingbird's or catbird's. It also tends to sing in couplets, voicing each phrase twice—though with so beautiful a song, the notes are well worth repeating.

In courtship, the female brown thrasher picks up a twig and approaches the male as he sings.

Recognition. 10–12 in. long. Slender and long-tailed; rusty brown above with 2 white wing bars; underparts whitish with brown streaks; eyes yellow. Forages mainly on ground.
Habitat. Brushy woodlands, garden shrubbery, and parks.
Nesting. Nest is a bulky cup of twigs, leaves, grass, and bark, from ground to 15 ft. up in dense thicket or brush pile. Eggs 2–5, pale blue or whitish, finely speckled with brown. Incubation 12–14 days, by both sexes. Young leave nest about 13 days after hatching. Sometimes 2 broods per season.
Food. Insects, spiders, and berries.

Gray Catbird

Dumetella carolinensis

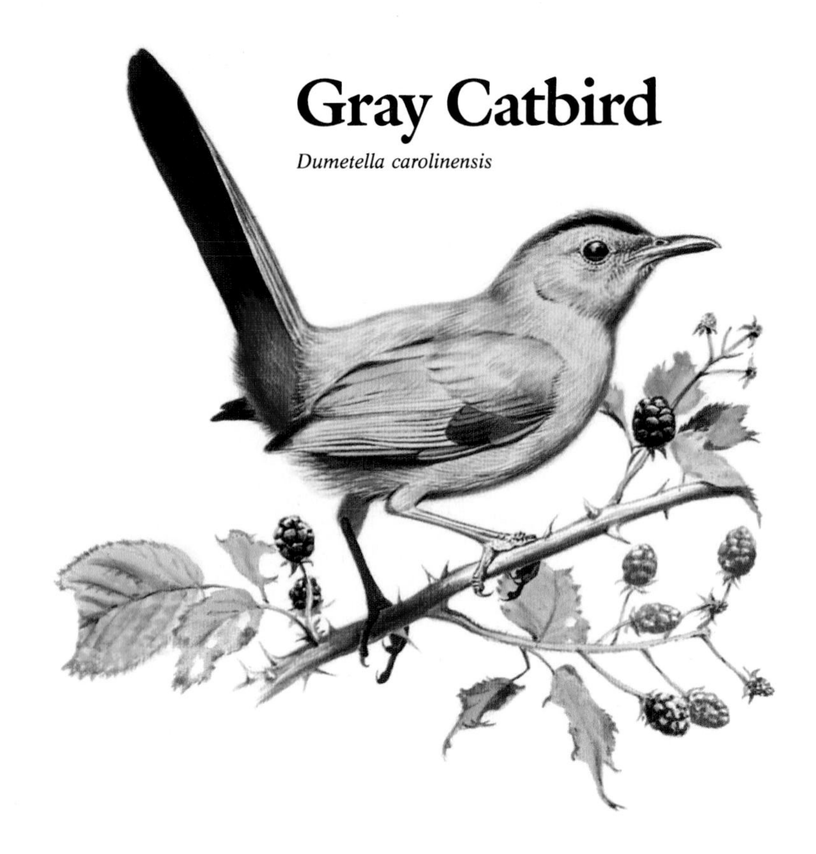

Catbirds begin incubating before a clutch is complete, so their eggs hatch at different times.

Watch and listen: the gray catbird, inconspicuous at first glance, invites closer acquaintance. As a mimic it ranks second only to the mockingbird, and it is an excellent songster in its own right. The male sings throughout the breeding season, mingling musical renditions of all the birdsongs in the area with an amusing cacophony of other sounds—shrieks, squeaks, whistles, cackles, as well as the cat's *meeeow* from which the species derives its name.

Wherever a tangle of shrub or briers is preserved near a house, catbirds will take up lively residence in it as readily as they would in a remoter site. A squabble in the garden signals that the resident catbird is driving off a gray-feathered intruder of its own kind. Yet the catbird is amazingly tolerant of other species. It has been known to share a nest-bush with a family of shrikes or robins. And should an orphaned bird appear, catbirds will readily adopt it, so strong are their caretaking instincts. In fact, these fascinating birds seem to keep an eye on everything. Just glance toward a rustling in the summer shrubs and you may see a bright gray bird with bright black eyes, watching you watch it. Discovered, the catbird retreats, mewing an alarm; recovered, it may fly off for a dip in the birdbath or resume its outpouring of song, regaling the ear and cementing the bond between human and bird.

Recognition. 8–9¼ in. long. Mainly gray, darkest on wings and tail; cap black; patch under tail dark rusty. Usually stays close to dense cover.
Habitat. Woodland thickets, garden shrubbery, and parks.
Nesting. Nest is a rough cup of twigs, stems, leaves, and grass, 3–10 ft. (rarely to 60 ft.) above ground in dense shrub, tree, or tangle of vines. Eggs 2–6, unmarked greenish blue. Incubation about 13 days, by female only. Young leave nest 10–15 days after hatching. Often 2 broods a season.
Food. Insects, spiders, and berries.

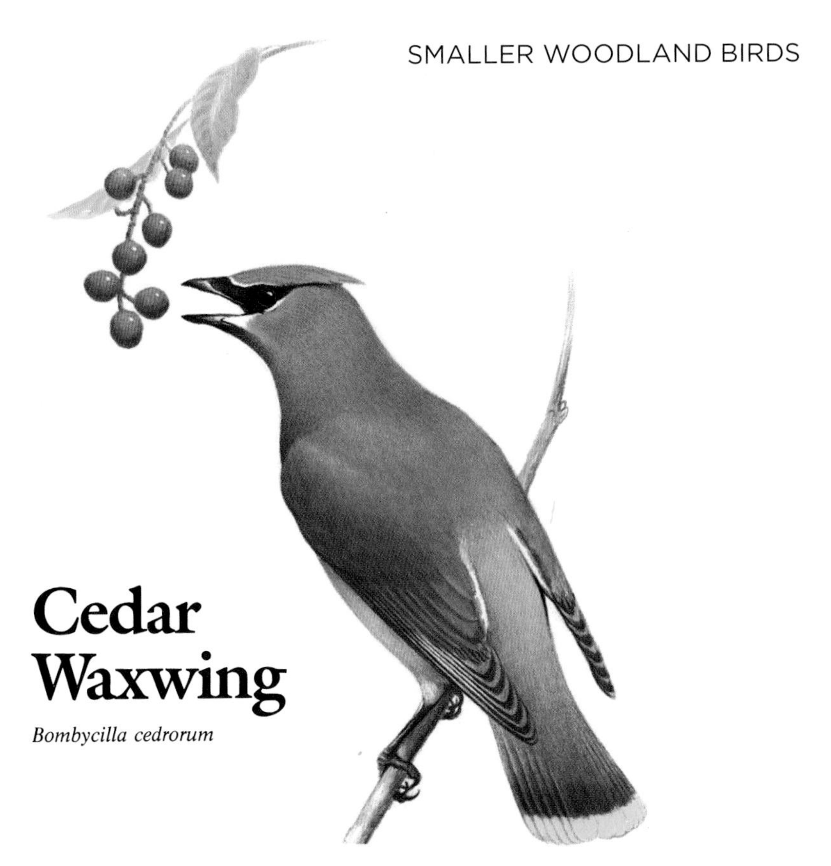

Cedar Waxwing

Bombycilla cedrorum

We are lucky that cedar waxwings seldom travel singly. For if one of these elegant birds provides a visual feast, a flock is a gourmand's delight. Culinary terms are easy to associate with the cedarbird, a sociable creature who likes to eat berries—lots of berries. A fruiting tree becomes a banquet hall for the fattening waxwings, who may swarm like bees over the pendant fruit. Their hearty appetites can be their undoing, however, when overripe fruit ferments. More than one backyard bird watcher has been astounded by the sight of tipsy cedarbirds, feathers and coordination askew, tumbling from trees and having to sober up before they can fly again.

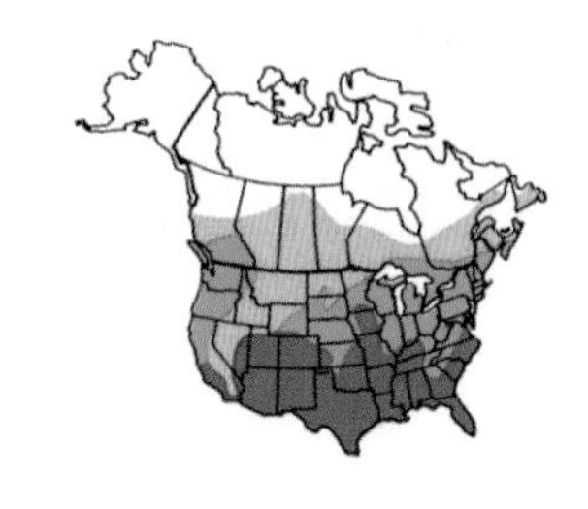

Fortunately, these interludes are not typical of the species' domestic routine, which is marked by a relatively high degree of social behavior. At times waxwings can be seen lined up in a row along a branch, passing berries from beak to beak. Courting pairs sometimes play what seems to be a game, passing flower petals back and forth, the only apparent aim being to surprise each other. Among the few North American songbirds to nest in colonies, cedar waxwings accept the presence of other breeding pairs nearby. After breeding season ends in late summer, they flock in even greater numbers—the better for us to admire their beauty, and certainly the better for them to rove in search of food.

Waxwings often pass a berry back and forth along a row of birds before one bird finally eats it.

Recognition. 6½–8 in. long. Adult brown and crested; mask black; yellow band across tip of tail. At close range, inner wing feathers show red "wax" tips. Immature similar but streaked below. Travels in flocks.

Habitat. Open woodlands, orchards, gardens, parks, and shade trees.

Nesting. Nest is a bulky cup of twigs, stems, and grass, 6–50 ft. above ground in tree. Eggs 3–6, pale blue or grayish dotted with black. Incubation 12–14 days, by female only. Young leave nest about 18 days after hatching. Several pairs often nest close together.

Food. Berries, flower petals, and insects.

White-eyed Vireo

Vireo griseus

Like other vireos, the white-eye is fearless at the nest, especially when the eggs are about to hatch.

One memorable account of the white-eyed vireo tells of an early ornithologist who approached a male bird brooding eggs. Unable to scare off the little creature, which was furiously pecking at his fingers, he lifted it bodily from its nest. Released, the bird promptly returned and trumpeted its full, defiant song. This fearlessness toward humans, along with a distinctive voice, typifies the diminutive white-eyed vireo, which may be one of the smallest of its clan but is clearly not inhibited by that fact.

The white-eye's rather unvireolike song consists of a series of from three to nine loud notes. The refrain often begins with a *chick* or a *tick* and almost always includes a *whee* or a *wheeyo.* The melody may be repeated countless times before a variation begins: *Chick! ticha wheeyo chick! Chick! ticha wheeyo chick! Chick! ticha wheeyo chick!* The white-eye broadcasts its song in a syncopated sort of rhythm, drawing a listener to the brushy thickets it inhabits. There the olive-gray and white bird may be hard to see, perfectly camouflaged in a leafy, shadowy tangle.

But let the oberver remember the bird's character and pause for a time: this vireo will creep forth to within arm's reach, sizing up the situation with pale, yellow-spectacled eyes. If alarmed, it may deliver a scolding or simply disappear into protective cover with an impertinent flick of its tail, saving its song for a likelier prospect in olive drab.

Recognition. 4½–5½ in. long. Grayish green above with 2 white wing bars; spectacles yellow; eyes white; underparts whitish with pale yellow tinge on flanks.
Habitat. Thickets in deciduous woods and dense, brushy pastures.
Nesting. Nest is a small cup of bark strips, grass, fibers, and spider web, suspended from forked twig, 1–8 ft. above ground in sapling or shrub. Eggs 3–5, white, dotted with brown and black. Incubation 12–15 days, by both sexes. Nestling period unknown. Southeastern birds raise 2 broods a season.
Food. Insects, spiders, and berries.

Bell's Vireo

Vireo bellii

John Graham Bell (1812–89) is undoubtedly less well-known today than the vireo that bears his name. To naturalists of the 19th century, however, he was familiar as one of the pioneering amateurs whose contributions were vital to the growth of American ornithology. Among those efforts were Bell's participation in John James Audubon's great Missouri River expedition of 1843, during which he collected the vireo that Audubon named for him.

An innovator in the art of taxidermy, Bell operated a shop in New York City that served as an unofficial clearinghouse of information for many of the country's leading bird authorities—and as a classroom for a budding 13-year-old naturalist named Theodore Roosevelt. Bell was, in addition, mentor, confidant, friend, and companion to many who discovered the bird riches of North America. He collected a number of previously unknown species while on expedition in California with John Cassin, who honored him by giving the scientific name *Amphispiza belli* to one of the new finds, the sage sparrow.

To Bell also goes the dubious credit for having collected the last Labrador duck known to science, on Long Island, New York, in 1875. In the course of his career Bell became one of the country's leading collectors of bird specimens, which were preserved both for scientific study and for sale to wealthy enthusiasts. When he died in 1889, however, sentiment against the unregulated collecting of birds and bird eggs was running high. An era was coming to an end.

Although tiny, the Bell's takes on larger prey than do other vireos, even a bulky grasshopper.

Recognition. 4½–5 in. long. Small; olive-green above with 2 pale wing bars; underparts tinged with yellow; eye-ring and eyebrow whitish. Southwestern birds grayer above, lack yellow tinge below.
Habitat. Streamside woods and thickets.
Nesting. Nest is a small cup of bark strips, grass, fibers, and spider web, suspended from forked twig, 1–10 ft. above ground. Eggs 3–5, white, sparsely dotted with brown and black. Incubation 14 days, by both sexes. Young leave nest after about 12 days. Often 2 broods a season.
Food. Mainly insects and spiders; berries.

Solitary Vireo

also known as

Blue-headed Vireo

Vireo solitarius

Before mating, a male solitary may start two or more nests; the female chooses the final site.

Frost glazes the southern woodlands. The January dawn is stark and quiet. No leaves rustle, no twigs snap, no bird utters a new-day note, for there are no false starts in the waiting game. The wind whispers, leaves rustle, twigs snap. The new day warms. Mockers, towhees, and cardinals equivocate. Now? Now! The day erupts in a spate of bird sounds—titmice, chickadees, wrens, robins, nuthatches, and others. Each sallies to its own food source, to glean, probe, chisel, and scratch to make a winter living.

On the periphery a small bird moves in, a latecomer. It pauses for a tidbit of beetle, then snatches a borer in its stout, hooked bill and swallows it whole. It is a neat bird. In the sunlight its head looks slaty blue, and bold white spectacles frame its alert, dark eyes. Suddenly agitated, it stops its exploration, crouches forward on the pine bough behind the cones and needles, and flings a repertoire of epithets at those who come too close. It is a solitary vireo. By writ of nature, it is alone.

Except during the nesting season, when it is the first of many migrants to arrive in northern nesting grounds, this vireo is indeed somewhat of a hermit. When most others of the clan have left to winter in Central America and Cuba, many solitaries remain in the southernmost United States. Each is a loner, a quiet afterthought to the flocks of more gregarious birds that roam the woodlands in winter.

Recognition. 5–6 in. long. Olive-green above with 2 white wing bars; head slate-gray with bold white spectacles and throat; flanks tinged with yellow. Rocky Mountain birds duller and grayer.
Habitat. Coniferous forests.
Nesting. Nest is a cup of bark strips, grass, rootlets, and spider web, suspended from forked twig, 3–40 ft. above ground, usually in small conifer. Eggs 3–5, white, sparsely dotted with brown and black. Incubation about 11 days, by both sexes. Nestling period unknown.
Food. Insects, spiders, and berries.

Yellow-throated Vireo

Vireo flavifrons

Widely acclaimed as the handsomest of the vireos, the yellow-throat is also an accomplished architect, constructing a hanging nest whose intricate design and summer-long durability would do credit to any hummingbird, pewee, or gnatcatcher. The task of building it involves both female and male, usually working in the upper levels of a tree toward the end of a branch. For as long as a week, using only their bills and feet as tools, the pair instinctively choose the right materials and make all the right moves.

The result is a beautiful cup—deeply rounded, with thick walls, an incurved rim, and a lining woven of fine grasses and pine needles. All have been stretched and shaped to meet the highest vireo standards. Supports and outside walls, frequently rearranged after interim inspections, are a lacework of moss and fibers, stippled with lichens and bonded with spider silk—the cement of bird architects. Lichens and spider silk not only bind the nest, but also help to conceal it from predators and provide a rudimentary type of weatherproofing.

Birds are conditioned by the lessons of nature to build the nest that suits them best. But when the building is done with the artistry the yellow-throated vireo brings to this most practical task, it is hard to think of the know-how as something that just comes naturally.

In late summer and fall, the yellow-throat adds berries to its insect diet before migrating south.

Recognition. 5–6 in. long. Olive-green above with 2 white wing bars; spectacles, throat, and upper breast bright yellow.
Habitat. Deciduous forests.
Nesting. Nest is a thick-walled cup of bark strips, grass, fibers, and spider web, decorated on outside with bits of lichen and moss, suspended from forked twig 3–60 ft. above ground. Eggs 3–5, white, heavily marked with brown and lilac. Incubation about 15 days, by both sexes. Young leave nest 14 days after hatching.
Food. Insects and berries.

Hutton's Vireo

Vireo huttoni

Some birds seem to have been designed mainly to make life more complicated for bird watchers. Many a birder visiting the West Coast has mistaken a Hutton's vireo for the more widespread ruby-crowned kinglet. True, the male kinglet has a crimson patch on its crown that the vireo lacks, but this sliver of color is difficult to see, and in all other respects the two birds are so similar that only careful scrutiny will sort them out.

It might be assumed that, given such similarities—not to mention a shared taste for insects and preference for tree-nesting—Hutton's vireos and ruby-crowned kinglets would compete directly with each other for food. But though they often travel together in winter flocks in oak woodlands, their feeding strategies are different enough to allow them to coexist reasonably well. (The Hutton's real competitor, in fact, may be another member of these woodland groups, the chestnut-backed chickadee.)

Of course, determining exactly how much two similar bird species compete is a difficult task, for an observer may miss key details that distinguish them. We do know that over the long run, only one species can occupy a particular niche of habitat. If chestnut-backed chickadees and Hutton's vireos in a given woodland area are hunting the same prey in the same manner, one must finally prevail and the other will have to cultivate new eating habits—or else, in the parlance of the Old West, get out of town.

Among most vireos, males and females share in the incubation and care of their young.

Recognition. 4½–5 in. long. Small; olive-green with 2 white wing bars; eye-ring white, incomplete. Flicks wings nervously. Very similar to ruby-crowned kinglet but has larger head and bill, moves more slowly.
Habitat. Oak woodlands; thickets.
Nesting. Nest is a small, deep cup of lichens, bark strips, grass, fibers, and spider web, suspended from forked twig, 5–35 ft. above ground. Eggs 3–5, white, sparsely marked with brown. Incubation about 15 days, by both sexes. Young leave nest about 2 weeks after hatching.
Food. Insects, spiders, and berries.

Warbling Vireo

Vireo gilvus

Those who know the warbling vireo agree: its outstanding quality is its voice. "A ripple of melody threading its way through the mazes of verdure, now almost absorbed in the sighing of the foliage, now flowing released on its grateful mission." So wrote the eminent naturalist Elliott Coues over a century ago.

That "ripple of melody" is a languid or melodious warble, often ending (though only briefly) on a rising, strongly accented note. Some accounts that have the bird singing *brig-adier, brig-adier, brigate* are an injustice to the vireo and not much help to the birder.

The warbling vireo sings—quite sweetly and all day long—usually from a sun-dappled hiding place in a deciduous tree, even from the nest. It is not a cooperative bird. Many know the song, but few the singer. Whatever ruses, ingenious or absurd, bird watchers resort to in their quest for just one good look, the elusive vireo is typically one step ahead or one leaf behind. In its wake it leaves only subtle hints. It has nothing more to reveal—no wing bars, wattles, spectacles, tail spots, or bright colors. Those who identify their first warbling vireo often do so by a process of elimination. Such a moment of bird-watching truth may be an empty victory. Better to follow the vireo's lead, relying almost entirely upon the ripple of melody that communicates it is there, in the treetop, behind the leaves.

This young cowbird fills the nest of a warbling vireo, which has no defense against such intruders.

Recognition. 5–6 in. long. Grayish olive above with no wing bars; eyebrow whitish. Stays out of sight in tall trees; best recognized by drowsy, warbling song, similar to canary's.
Habitat. Streamside forests and woodlands; shade trees and suburban areas.
Nesting. Nest is a small cup of bark strips, grass, fibers, and spider web, suspended from forked twig, 4–60 ft. above ground in bush or tree. Eggs 3–5, white, sparsely dotted with brown and black. Incubation 12–14 days, by both sexes. Young leave nest 14 days after hatching.
Food. Insects, spiders, and berries.

Red-eyed Vireo

Vireo olivaceus

This bird has a formidable enemy—the brown-headed cowbird, an enormously successful parasite that has made the red-eyed vireo one of its prime victims. The cunning cowbird lays one or more eggs in the vireo's nest, foisting incubation and other parental duties upon the red-eye more often than it does on almost any other species. And the little surrogate pays dearly: each cowbird it raises costs the life of one or more of its own offspring. Rare is the vireo who thwarts the cowbird, as some birds do, by roofing over the intruder's eggs before laying its own.

The cowbird surely contributed to the population crash of the red-eyed vireo in the middle of this century, as did the widely used insecticide DDT. Equal blame could fall on those who cleared the forests of the eastern United States, and on those who today are denuding the tropical wintering grounds of red-eyes and scores of other species.

Yet for all this the bird's spirit is hardly glum. As a vocalist, the red-eyed vireo is by all odds our endurance champion: on a single day one was recorded singing more than 22,000 individual songs. Does the red-eye have anything to sing about? Right now the species is holding its own in the eastern United States. But for every woodman who spares that tree, there is another harvesting large tracts of forest. And the brown-headed cowbird continues to proliferate at a breathtaking pace, raising questions that only time will answer.

Where no pesticides have been sprayed, red-eyes nest in tall shade trees in suburban areas.

Recognition. 5½–6½ in. long. Olive-green above with no wing bars; crown gray, bordered by black stripes and white eyebrows; eyes red; underparts whitish. Sings in deliberate phrases.
Habitat. Deciduous forests, shade trees, gardens, and parks.
Nesting. Nest is a thin-walled cup of bark strips, grass, fibers, and spider web, suspended from forked twig, 2–60 ft. above ground. Eggs 3–5, white, sparsely dotted with brown and black. Incubation 11–14 days, by female only. Young leave nest about 12 days after hatching.
Food. Insects and berries.

Blue-winged Warbler

Vermivora pinus

The scientific name may be *Vermivora pinus,* but evidently the blue-winged warbler does not read Latin, because it has no connection with pine woods at all. Second-growth woodlands and overgrown fields afford a natural home for this bird, whose head and underparts gleam like buttercups in sunshine. Methodically, unhurriedly, the blue-winged warbler ambles through thick foliage, picking off a caterpillar here, gleaning a spider there. At times it may dangle upside down like a chickadee or titmouse, but this matter-of-fact feeder seems to have none of the flash of other members of the warbler tribe.

Where the ranges of blue-winged and golden-winged warblers overlap, interbreeding occurs, and the young inherit traits of both parents. The classic hybrid types have even been given their own names—the Brewster's warbler and the less common Lawrence's warbler. These hybrids are fertile and may interbreed with either parent species, or with still other hybrid birds.

With so much mixing and matching, a basic question occurs: are blue-winged and golden-winged warblers really different species at all? The question may not be asked for long. Where ranges overlap, blue-wings generally increase and golden-wings decline, the victims of genetic swamping. Eventually, golden-winged warblers may simply disappear, their identity diluted to the vanishing point within the blue-wings' genetic pool.

Blue-wings not only nest on the ground, but occasionally hunt for food there as well.

Recognition. 4½–5 in. long. Blue-gray above with 2 white wing bars; crown and underparts yellow; line through eye black. Brewster's warbler, hybrid between blue-wing and golden-wing, similar, but wing bars often yellow, underparts more or less white.
Habitat. Open, brushy woodlands and thickets.

Nesting. Nest is a conical cup of bark strips, grass, and hair on ground, woven among weed stems, or hidden under shrub. Eggs 4–7, white, finely spotted with brown and gray. Incubation 10–12 days, by female only. Young leave nest about 10 days after hatching.
Food. Insects and spiders.

Golden-winged Warbler

Vermivora chrysoptera

Most female warblers do all the incubating, but males help to feed the young after the eggs hatch.

From its perch among the taller saplings that flank a field, the golden-winged warbler might be mistaken for a chickadee. Gray above, white below, its throat nicely bibbed, the golden-wing resembles a chickadee down to its habit of swinging upside down to gather caterpillars from the underside of leaves. But when the male golden-wing moves away from the shadows, the masquerade is over. Sunlight exposes the yellow crown and shoulders. Even the black mask etched across the bird's face can't disguise its identity now.

In the early 19th century, when John James Audubon and Alexander Wilson were pioneering the study of ornithology, the felling of eastern hardwood forests was creating the patchwork fields and briery landscape that golden-winged warblers favor. But over time, untended fields revert to mature woodland, and eventually the golden-wings must look elsewhere for nest sites. Wherever they nest, they usually arrive in May to begin the process of homesteading and courtship. June is a busy time, filled with parenting and the defense of nestlings, and in July clearings bustle with the clumsy first flights of newly fledged young. But before August, the desultory song of the chickadee-mimic ceases and the clearings are abandoned. And by September the birds have returned to Central America, where they make their home for the better part of each year.

Recognition. 4½–5 in. long. Male gray above with crown and 2 wing bars yellow; face mask and bib black; underparts white. Female duller. Lawrence's warbler, hybrid between this and blue-winged, similar but underparts yellow.
Habitat. Open, brushy woodlands and thickets.
Nesting. Nest is a bulky cup of grass and bark strips on ground, woven among weed stems, or hidden under shrub. Eggs 5–7, white or cream-colored, speckled with brown and gray. Incubation about 10 days, by female only. Young leave nest about 10 days after hatching.
Food. Primarily insects.

Tennessee Warbler

Vermivora peregrina

The oaks are in flower, and overhead in the forest's canopy cold-numbed insects come to life under the touch of the morning sun. It's too late now for spring to be halted by a single cold snap. The May sun is too strong, the season too far along. Darting about and snatching insects amid the outermost branches are dozens of small, energetic birds—green above, white below, a pencil-fine line drawn across each eye. Now and again, one of the frenetic treetop feeders pauses and sends a shower of notes earthward, loud, emphatic notes followed by a cascading trill that reveals its identity as a Tennessee warbler.

In many locales this welcome song heralds spring's conquest of winter, though nowhere does the Tennessee warbler seem quite so common as in the Mississippi Valley, its principal spring migration route. It was here in 1832, along the Cumberland River in Tennessee, that Alexander Wilson discovered (and misleadingly named) this treetop-loving species, which winters in South America and nests in Canada. On its first journey south in the fall, an immature Tennessee warbler may feed in low, weedy tangles, but after a winter in the tropics it will return with irrepressible zest to forage and make music in the heights.

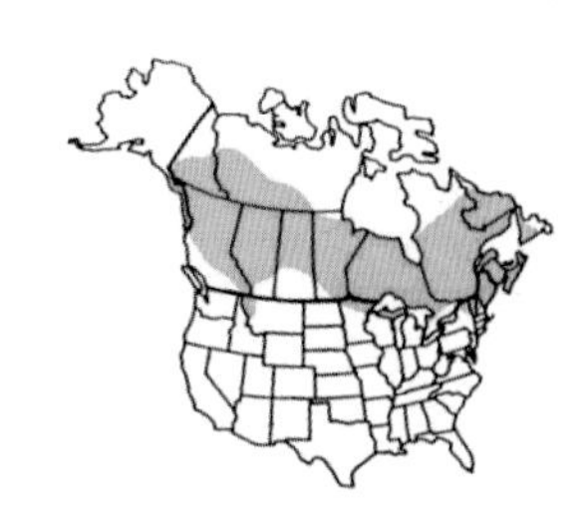

Recognition. 4–5 in. long. Male olive-green above; crown clear gray; eyebrow white with darker line through eye; underparts whitish. Female similar but duller.
Habitat. Forests and wooded bogs.
Nesting. Nest is a solid cup of grass and rootlets, sunk into sphagnum moss or built on drier ground at base of shrub or clump of grass. Eggs 5–7, white or cream-colored, speckled with brown. Incubation probably about 12 days, probably by female only. Nestling period unknown.
Food. Insects, spiders, berries, and seeds.

Orange-crowned Warbler

Vermivora celata

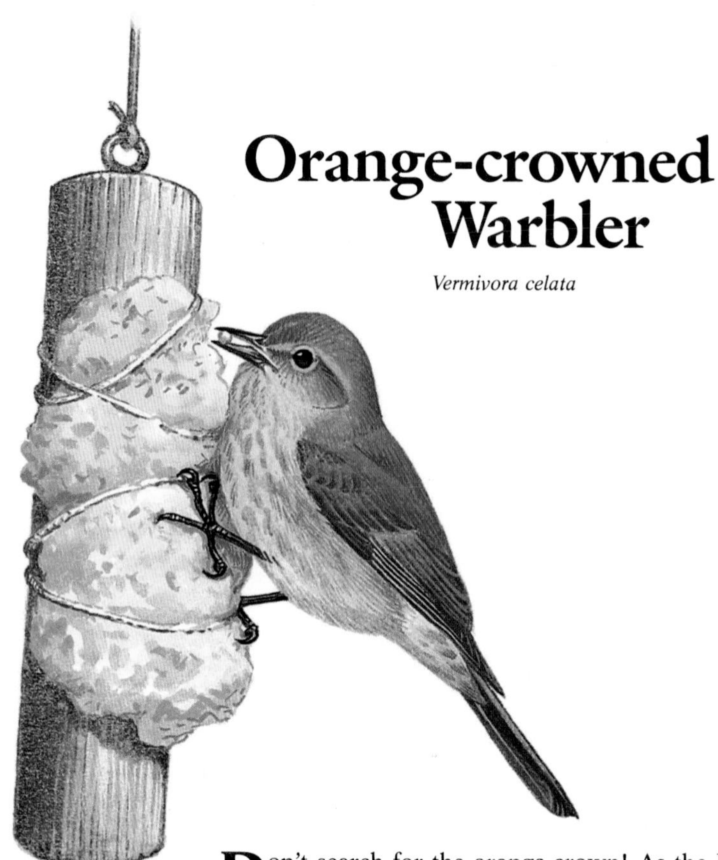

Don't search for the orange-crown! As the Latin name *celata* implies, it is hidden. Virtually invisible in the field, the tiny rust-colored patch may be absent entirely on female and immature members of the species. But as a field mark, the crown is hardly necessary anyway. This nondescript LGB ("Little Green Bird" in bird watchers' shorthand) betrays its identity by its very manner.

Hyperactive! That, in a word, is the orange-crowned warbler. A feathered bundle of quicksilver, it hops the length of a goldenrod stalk, darts from branch to branch, drops to the ground, springs to the top of a bush, disappears into a weedy tangle—then stretches its neck to the breaking point, wings aflutter, to reach some tasty morsel.

Birders, however, need not crane their own necks to watch. The orange-crown, unlike its cousin the Tennessee warbler, shuns the heights at all times. Its nest, situated along an aspen-crowded stream or concealed amid sparse alpine vegetation, will be on or very near the ground. And whether nesting or in migration, the bird keeps a low profile, foraging in thickets and weedy fields. Hardier than many other insect-eating birds, orange-crowns migrate later than most warblers and are among the few that winter in North America. They are common in the southwestern United States, and some of them on the East Coast may linger as far north as Connecticut.

Whether nesting or wintering, orange-crowns sip nectar at flowers like hummingbirds.

Recognition. 4½–5½ in. long. Dull olive-green, paler and tinged with yellow below (brighter on western birds); feathers below tail always yellow; orange-rusty patch on crown seldom visible.
Habitat. Brushy woodlands, aspen groves, and dense thickets.
Nesting. Nest is a cup of grass, bark strips, and leaves, hidden under shrub or on bank, or 2–4 ft. above ground. Eggs 3–6, white, spotted with reddish and brown. Incubation and nestling periods unknown.
Food. Insects and berries.

Nashville Warbler

Vermivora ruficapilla

Alexander Wilson, who discovered this species near Nashville, Tennessee, in 1811, saw but three specimens in his lifetime. John James Audubon, during his long and well-traveled career, chanced upon only four. Yet by 1879 another eminent ornithologist, John Krider, reported the bird to be "very abundant as far as Minnesota." What had happened? A great deal, as Americans moved west, clearing forests, cultivating the land, then abandoning it. Today, the Nashville warbler is a common resident of second-growth northern woodlands and spruce bogs, its ground nest hidden in grass beneath low brush or nestled within some snug, mossy nook.

In its primary haunts, below a shaded canopy, the bird's gray-green plumage is muted. Were it not for a bright, active song and an unquenchable restlessness, the Nashville would hardly warrant any attention at all. But catch one in the open, in a shaft of sunlight leaking through leaves, or against the raven-green of a wall of spruce, and its plumage comes alive. Yellow underparts gleam, the olive back ripples with hidden color, and the pearl-gray head makes a perfect complement to the overall design.

During migration, the Nashville warbler may be found over much of North America in a variety of settings. It can even be found in the parks and gardens of the city after which it was named, though it appears there only in transit, passing through with the changing of the seasons.

More than most other warblers, the Nashville eats caterpillars and helps combat the gypsy moth.

Recognition. 4½–5 in. long. Back, wings, and tail olive-green; head gray, with white eye-ring; throat and rest of underparts bright yellow; rusty patch on crown seldom visible.
Habitat. Open, brushy woodlands and bogs; birch and aspen groves.
Nesting. Nest is a cup of moss, grass, fur, and stems, on ground among weeds, thick shrubs, or in clump of moss. Eggs 4–5, white, spotted with brown. Incubation about 12 days, by female only. Young leave nest about 10 days after hatching.
Food. Insects.

Virginia's Warbler

Vermivora virginiae

In a family of birds that boasts names like Tennessee warbler, Nashville warbler, and Kentucky warbler, Virginia's warbler would seem to be just another southern belle. Actually, the bird rarely gets closer to the state of Virginia than Texas. It was named for Virginia Anderson, wife of the army surgeon who discovered the bird at Fort Burgwyn, New Mexico, in 1858. Prof. Spencer Baird, secretary of the Smithsonian Institution, described the new species for science in 1860 and graciously honored the wife of the bird's discoverer in designating both its common and scientific names.

This western warbler is common to the southern Rocky Mountains and western basin. From the tips of yellow pines in Colorado foothills to the scrub oaks that cling to near-vertical walls in mountain canyons, male Virginia's warblers fill mornings in June with accelerating bursts of song. Slight and spry, nervous and shy, the bird forages close to the ground and seems most at home in the most impenetrable thickets. Such reticence might suggest that the bird was endowed with great beauty, but the Virginia's warbler is ordinary enough—gray with yellow and white underparts and staring white spectacles. Many observers catch no more than a glimpse of a furtive gray bird already en route to the next twig before the one it just claimed has stopped vibrating. Some may see even less—a momentary flash of yellow as the bird turns tail and disappears, a slight gray shadow fading into a pale maze.

So attentive are Virginia's warblers to their young that both parents often bring food at the same time.

Recognition. 4–4½ in. long. Mainly gray; eye-ring white; rump and feathers below tail yellow. Male has lemon-yellow patch on breast, faint or lacking in female; reddish crown patch visible only at close range.
Habitat. Oak and pinyon woodlands and brushy hillsides near streams.

Nesting. Nest is a cup of moss, grass, bark strips, and rootlets, hidden on ground among dead leaves or in tuft of grass at base of shrub or small sapling. Eggs 3–5, white, finely dotted with brown. Incubation and nestling periods unknown.
Food. Insects.

Northern Parula

Parula americana

Some ornithologists speak of a nest site "search image," an inherited picture of the kind of place—treetop, cliff face, barn rafter—where a particular species typically nests. According to this theory, an ovenbird, for example, doesn't *choose* to nest on a woodland floor, but simply matches an inborn blueprint with the appropriate piece of the world. If so, what the northern parula must visualize is a cross between witch's hair and a string mop.

In the North, parulas nearly always nest in hanging tufts of *Usnea,* tree lichens commonly called old-man's beard, moss beard, or hanging moss. In the South, their nests are found in hanks of Spanish moss, the epiphyte, or air plant, that (thanks to Civil War movies) has become an enduring part of our vision of antebellum cotton plantations. In the middle states, where neither moss beard nor Spanish moss grows, northern parulas sometimes build conventional open cup nests or bag nests like those of orioles. But whenever anything straggly and hanging turns up, a parula is likely to discover it. Even a clump of conifer twigs or high-water detritus hanging from a tree limb over a stream or river will do.

Hal Harrison, a photographer, found one parula nest in a tattered piece of burlap lodged in an oak tree in the mountains of West Virginia. Needless to say, such nests are difficult to locate. Harrison says his wife is pretty good at finding parula nests in *Usnea.* "She looks," he says, "for the beard with the 'tennis ball' in the bottom."

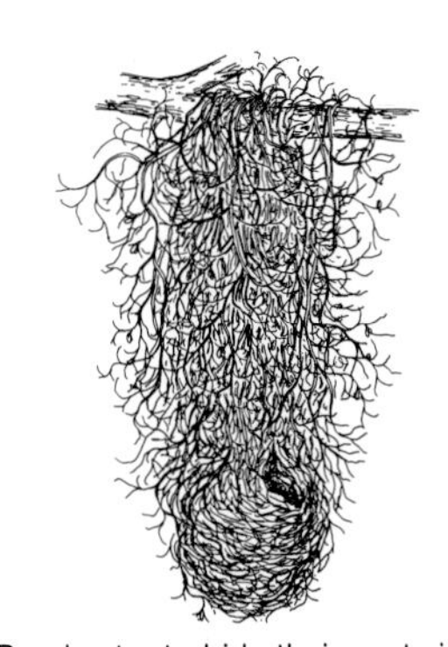

Parulas try to hide their nests in Spanish moss, but often the ball-shaped structure is still visible.

Recognition. 4–4½ in. long. Blue-gray above with 2 white wing bars; eye-ring white and broken; throat and breast yellow. Male has black and rusty breast bands, lacking in female.
Habitat. Woodlands, especially near water.
Nesting. Nest is a loosely woven cup of Spanish moss or *Usnea* lichen hanging from branch, 4–60 ft. above ground. Eggs 3–7, white or cream-colored, spotted with brown. Incubation 12–14 days, by female only. Nestling period unknown.
Food. Insects and spiders.

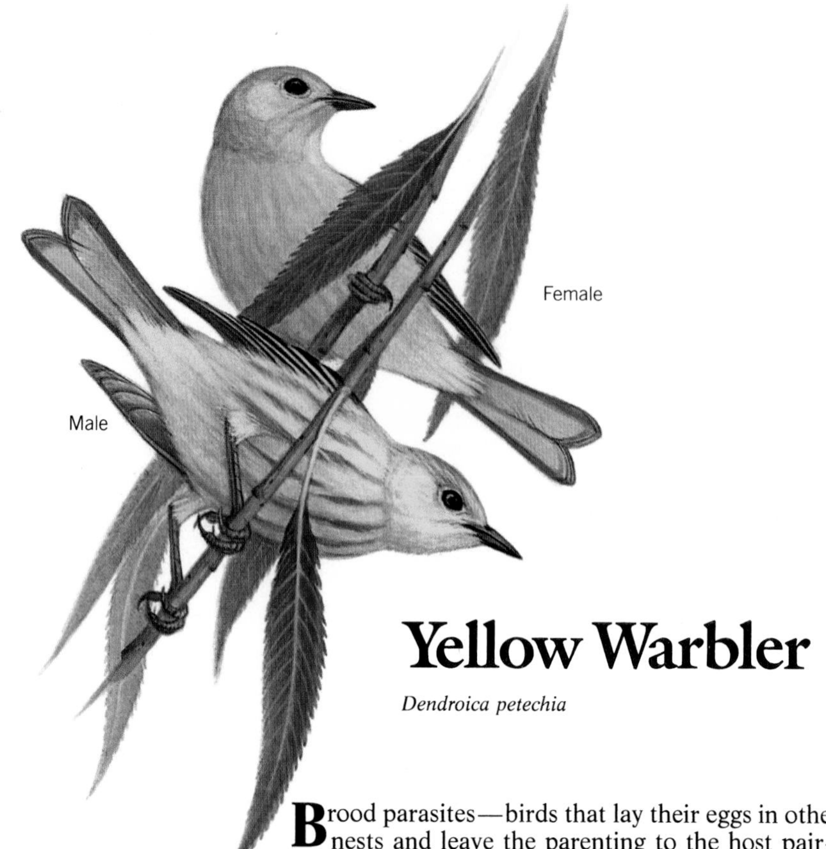

Yellow Warbler

Dendroica petechia

Brood parasites—birds that lay their eggs in other birds' nests and leave the parenting to the host pair—often seem to have favorite victims. The brown-headed cowbird is no exception. A study made near Pontiac, Michigan, noted 18 eggs possibly laid by one cowbird in a single nesting season—all deposited in yellow warblers' nests.

There are two kinds of host species, accepters and rejecters. Accepters dutifully incubate cowbird eggs and raise the young, often at the cost of their own offspring's lives. Rejecters treat cowbird eggs as foreign objects; larger birds simply chuck the eggs out of the nest, while smaller ones typically abandon the nest and build anew.

Yellow warblers are a sophisticated accepter-rejecter mix. An Ontario study found that when a cowbird egg was laid in a nest containing one or no warbler egg, it was nearly always rejected. Either the warblers deserted or the female built a new nest floor over the egg(s) and started a new clutch. But if the cowbird egg was laid in a nest holding two or more warbler eggs, the hosts usually accepted it. Probably the behavioral switch is adaptive. When a warbler clutch nears completion before a cowbird egg is laid, the warbler young have a head start. This may offset the double cowbird advantage of a shorter incubation period and larger chick size, which often enables the cowbird young to monopolize food and so threaten the parents' own offspring with starvation.

Yellow warblers may reline a nest if a cowbird egg is laid there; the record is a nest with six layers.

Recognition. 4½–5½ in. long. Bright yellow, tinged with green above; 2 wing bars yellow. Male has bold reddish streaks on breast, faint or lacking in female.
Habitat. Streamside woods and thickets, swampy thickets, and garden shrubbery.
Nesting. Nest is a cup of silky plant down, grass, and spider web, fastened to crotch of small sapling or shrub. Eggs 3–6, pale bluish or greenish, spotted with brown and gray. Incubation about 11 days, by female only. Young leave nest 9–12 days after hatching.
Food. Insects and spiders.

Chestnut-sided Warbler

Dendroica pensylvanica

Conservationists rightly decry the destruction of wildlife habitat, but it's difficult to imagine any sort of environmental alteration that wouldn't benefit *some* forms of life—admittedly nobody's favorites, perhaps.

The history of 19th-century America, as written by woodland birds, might be called The Destruction of the Eastern Forests; as authored by open country birds, it would be titled The Opening of New Land, for many species benefited from the clearing of mature woods. For example, the cowbird found new pastures in the East, while the relatively scarce American robin proliferated thanks to that new garden feature, the lawn.

The chestnut-sided warbler was another notable beneficiary of deforestation. Apparently John James Audubon saw but a single chestnut-sided in a lifetime of tramping the countryside. But even before his death, more and more of the bird's *pleased, pleased, pleased to meetcha* song was being heard as land cleared for agriculture provided burgeoning acreage of its preferred breeding habitat: scruffy open country partially covered with brush and brambles and dotted with small trees. Ironically, the clean farming techniques and reforestation of recent decades have probably diminished the chestnut-sided's numbers once again.

The chestnut-sided warbler often resorts to "fly-catching," shooting into the air after a passing insect.

Recognition. 4½–5½ in. long. Male has yellow crown, dark rusty stripe along side of breast and flanks; face patterned in black and white; underparts white. Female similar but duller. Fall immature yellow-green above, whitish below.
Habitat. Brushy woodlands and pastures.
Nesting. Nest is a cup of bark strips, stems, fibers, and plant down, 1–5 ft. above ground in fork of sapling or shrub. Eggs 3–5, white or cream-colored, spotted with brown and purple. Incubation about 13 days, by female. Young leave nest 10–12 days after hatching.
Food. Mainly insects; also berries and seeds.

Magnolia Warbler

Dendroica magnolia

Migrating overland at night was part of the cycle of nature long before man learned to put bricks and mortar together. But night flying has grown increasingly dangerous for birds because of the proliferation of tall structures and confusing lights across the landscape. A harbinger of things to come was noted more than half a century ago by an observer who spent one foggy night in the torch of the Statue of Liberty and watched as songbirds streamed by like a "swarm of golden bees." The silence was broken by intermittent thuds as birds "struck the light with terrific impact," many of them suffering fatal injuries.

Television transmitting towers present a special hazard. Close monitoring showed that a single tower in northwest Florida killed nearly 2,000 birds during each of three autumn migrations. One of the hardest-hit species was the magnolia warbler; after just one night, 106 were found dead. Such wholesale carnage is most common when tall, illuminated structures are shrouded in foggy weather, which diffuses the light and blots out any other points of reference. In 1954, drizzle blanketed an area from Canada to the Gulf of Mexico during three nights of heavy songbird migration. On the mornings after, surveys in 25 localities identified some 10,000 victims—only a fraction of the total. Of those, nearly one-tenth were magnolia warblers; how many more of their ill-fated kin perished on those three terrible nights will never be known.

With its tail fanned, a magnolia warbler is easy to spot as it forages busily for small insects.

Recognition. 4½–5 in. long. Male has black on sides of head and back; crown gray; underparts yellow with bold black streaks; eyebrow, wing bars, and patches on tail white. Female similar but duller, with less white on wing.
Habitat. Coniferous forests.
Nesting. Nest is a loose cup of twigs and grass, saddled to horizontal branch of conifer 1–35 ft. above ground. Eggs 3–5, white, cream-colored, or pale blue, spotted with brown. Incubation 11–13 days, by female only. Young leave nest 8–10 days after hatching.
Food. Insects and spiders.

Cape May Warbler

Dendroica tigrina

When S. C. Kendeigh examined the breeding birds near Black Sturgeon Lake in Ontario in the 1940's, he found an average of 319 pairs per 100 acres of the surrounding spruce-fir forest. The reason for that high figure was an outbreak of the spruce budworm, a periodic defoliator of conifers in the North. Four species of warblers—Cape May, Blackburnian, bay-breasted, and Tennessee—accounted for 185 of those pairs. When a similar census was taken 20 years later, budworm had virtually disappeared and so had the four warblers.

Local populations of Cape May warblers wax and wane with the budworm cycle. And Cape Mays can wax in a hurry. Most warblers lay four or five eggs. Cape Mays typically lay six or seven, sometimes eight or even nine. If a forest is hit by a budworm infestation, five pairs of Cape Mays may raise 35 young. Assuming a survival and return rate of 75 percent for adults and 50 percent for young, the original 10 birds will have become about 110 by the end of the second summer, more than 270 by the end of the third.

Unhappily for the forestry industry, even that may not be enough. In one area in Maine there were an estimated 1,744,000 budworms per acre, but birds probably consumed no more than 2 percent of the voracious insects.

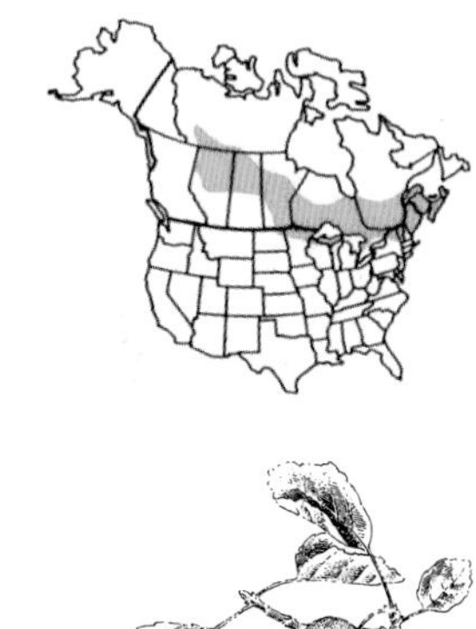

In the fall, Cape May warblers use their sharp bills to probe into fruit and drink the sweet juices.

Recognition. 5–5½ in. long. Yellow-green and streaked above; yellow and streaked below; cheek patch dark rusty; patch behind ear yellow. Female similar but duller; always has yellow patch behind ear.
Habitat. Spruce forests.
Nesting. Nest is a large cup of twigs, stems, and grass, built among short branches near top of conifer 30–60 ft. above ground. Eggs 4–9, cream-colored, heavily spotted with brown and gray. Incubation and nestling periods unknown.
Food. Insects and spiders; also punctures fruit and drinks juice.

Black-throated Blue Warbler

Dendroica caerulescens

Mountain laurel plays so large a part in this bird's life that its name could be "laurel warbler."

Looking just as its name says it should, the male of this species makes identification about as easy as can be. The female, however, is another story, with her olive back and buffy underparts.

Tame and trusting, these warblers can often be observed at close range in forest undergrowths of American yew and such broad-leaved shrubs as mountain laurel and rhododendron—their favored nesting shrubs. One keen observer of the black-throated blue was the naturalist John Burroughs, who first discovered its nest. It was "a thick, firm structure, composed of the finer material of the woods, with a lining of very delicate roots or rootlets." Four frightened fledglings scrambled out of their home, causing the parent birds great alarm. "They threw themselves on the ground at our very feet, and fluttered, and cried, and trailed themselves before us, to draw us away from the place, or distract our attention from the helpless young."

If populations of this species become dense, the males are forced to defend their territories. But as Burroughs observed, their battles seem "indulged in more to satisfy their sense of honor than to hurt each other. . . . In the course of fifteen or twenty minutes they have three or four encounters . . . till finally they withdraw beyond hearing of each other, both, no doubt, claiming victory."

Recognition. 4½–5½ in. long. Male blue-gray on crown and upperparts; face, throat, and flanks black; white wing patch. Female olive-green with small white wing patch; cheeks dark.
Habitat. Woodland thickets.
Nesting. Nest is a cup of bark strips, twigs, grass, fern stems, and leaves, 1–4 ft. above ground in dense thicket. Eggs 3–5, white or cream-colored, spotted with brown and gray. Incubation 12 days, by female. Young leave nest about 10 days after hatching.
Food. Mainly insects; also seeds and berries.

Yellow-rumped Warbler

Dendroica coronata

Easterners know it as the myrtle warbler; westerners call it the Audubon's warbler. But east or west, the yellow-rumped warbler is one of the country's best-known songbirds. "Our most vigorous and successful warbler," one authority has declared it. In the 19th century, Henry David Thoreau noted with delight in his journal that "myrtle birds were fluttering outside as if they wanted in."

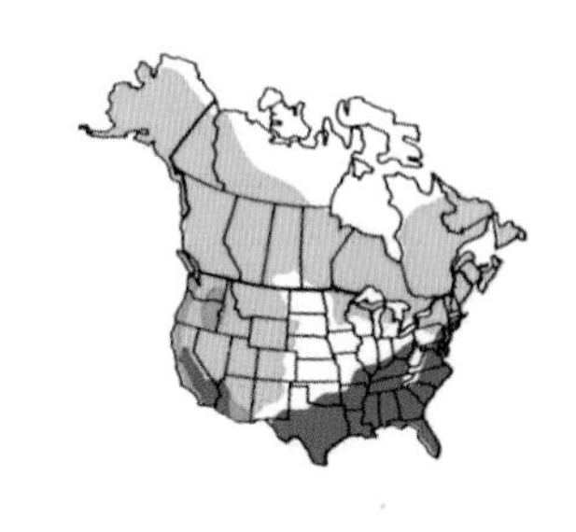

The little bird's four distinct patches of yellow, particularly the yellow splash just above the point where the tail feathers begin, make field recognition, even for a beginning bird watcher, easy enough. Although the Audubon's and myrtle warblers often mate if they have a chance meeting on the right spring morning somewhere in mid-continent, they are still slightly different in appearance, the westerner having a yellow throat, and the easterner being marked with white around the eyes and throat.

One reason for this warbler's popularity is its lifestyle. The yellow-rumped warbler is an early spring migrant in the East, flying joyously northward in great numbers to be ready for the newest crop of wax myrtle and bayberries when they become ripe. It is an easygoing feeder, taking a variety of insects (and the Audubon's delighting all with its sudden upward swoops to feast on an insect swarm), but doing just as well on berries or an occasional snatch of sap from a farmer's maple trees at sugaring time.

Yellow-rumped warblers can be seen from time to time chasing insects over ponds like swallows.

Recognition. 5–6 in. long. Spring males streaked blue-gray above; cap, rump, and sides of breast yellow; breast and sides black; myrtle in East has white eyebrow and throat; Audubon's in West has yellow throat, bold white wing patch. Females similar but duller.
Habitat. Coniferous forests.

Nesting. Nest is a cup of twigs, bark strips, and stems, 3–50 ft. above ground on branch or in foliage of conifer. Eggs 3–5, cream-colored or grayish, spotted with brown and gray. Incubation about 13 days, by female. Young leave nest about 2 weeks after hatching.
Food. Insects, spiders, berries, and seeds.

Black-throated Gray Warbler

Dendroica nigrescens

More like a grouse than a warbler, this species distracts intruders with a skilled "broken wing" act.

Except for a tiny patch of yellow in front of each eye, the black-throated gray is a somewhat drab warbler. But this coloring is to the bird's advantage. In its preferred habitat of young, more open conifer forests or regions of scrub growth such as juniper, pinyon, or manzanita, its streaky plumage of black, white, and grey blends right in with the gray-green foliage.

In keeping with such a somber suit is this bird's businesslike manner of feeding. Particularly in spring, when oak worms are abundant, it forages with well-ordered zeal, methodically checking every leaf and twig for these ubiquitous green caterpillars.

Its nest, too, is carefully crafted and a masterpiece of camouflage. Typical of most warblers' nurseries, it is a neatly woven cup of dried grasses and weathered weed stalks, usually lined with moss and feathers, sometimes with horse or other animal hair, where available. The bird very cleverly conceals the nest in a small clump of leaves, large enough to disguise it but not so obvious that a predator would look for something hidden there. Ever the cautious conservative, the black-throated gray will take all sorts of evasive action to avoid leading you to its nest. But if you should manage to find it, in a flash, the bird becomes a master of the theatrical arts, feigning fits or injury so convincingly, it deserves a standing ovation.

Recognition. 4½–5 in. long. Male gray above; crown, cheek, and bib black; eyebrow, mustache, and underparts white; flanks streaked with black. Female similar, but crown gray, throat white.

Habitat. Dry, open woodlands.

Nesting. Nest is a cup of dry grass, stems, and leaves bound with spider web and lined with moss, feathers, and animal hair, usually 3–10 ft., occasionally to 50 ft., above ground in shrub or tree. Eggs 3–5, white or cream-colored, spotted with brown. Incubation and nestling periods unknown.

Food. Insects, especially small caterpillars.

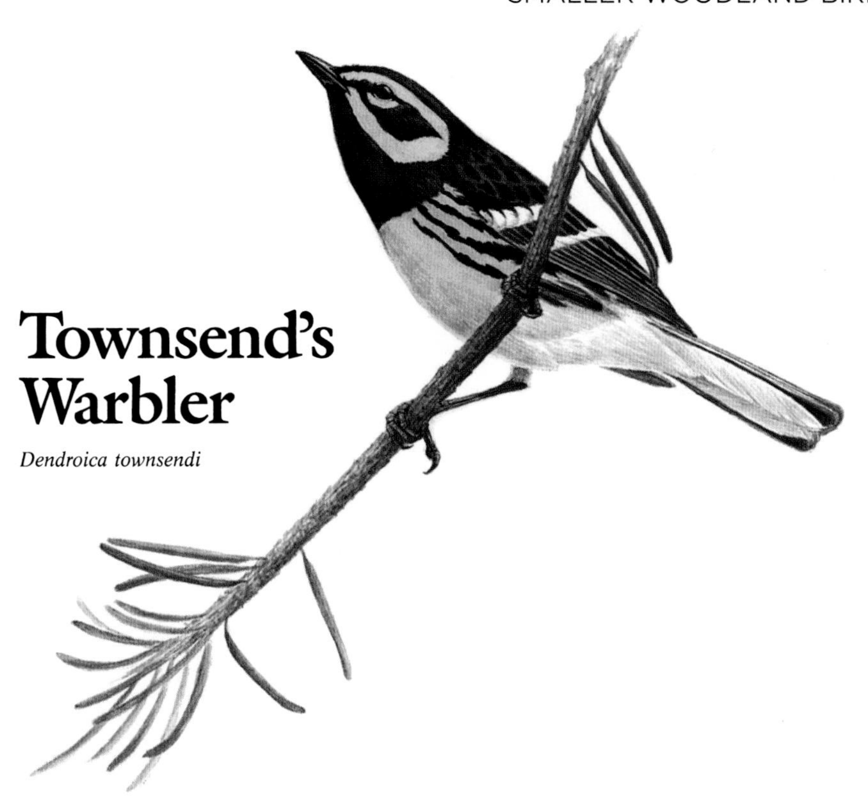

Townsend's Warbler

Dendroica townsendi

It's been said that wood warblers are the butterflies of the bird world. With a flashy, streaked body of black and gold and olive-green and a habit of flitting flirtatiously, the Townsend's warbler especially fits this description.

During nesting season you're not likely to see one. Just like the hermit warbler, with which it shares some of the coniferous forests of the Pacific Northwest, the Townsend's inhabits the tallest trees. It builds its nest at least 8 and perhaps as much as 90 feet above the ground, and sings its wheezy *dzeer, dzeer, dzeer, tseetsee* from the topmost branches. Although it generally lives at lower elevations than the hermit, the Townsend's sometimes mates with its reclusive cousin. And its nests, like the hermit's, have so rarely been spotted that nothing is known about incubation periods or when the young begin to fly.

In the fall Townsend's warblers leave their forest mansions and head for their winter homes in Mexico and Central America. Along the way, many settle for the season on the California coast. It is during their migrations, often in large numbers or flocked with chickadees and other species, that these evasive warblers can be observed at closer range while they forage nearer the ground. Because they feed ravenously on weevils, leafhoppers, and similar pests, the welcome mat is always out for the Townsend's.

Warblers coexist by maintaining separate foraging areas; the top of tall conifers is the Townsend's.

Recognition. 4½–5 in. long. Spring male greenish above; crown, cheek, and bib black; eyebrow, mustache, and underparts yellow; flanks streaked with black. Female similar, but crown greenish, throat yellow.
Habitat. Coniferous forests.
Nesting. Nest, rarely found, is a bulky, shallow cup of grass, weeds, plant fibers, and bark strips, built in conifer 8–15 ft. above ground and probably higher. Eggs 3–5, white, spotted with brown. Incubation and nestling periods unknown.
Food. Insects and spiders.

Hermit Warbler

Dendroica occidentalis

How shy must you be for someone to call you a hermit? In the case of this feathered recluse the name is easy to comprehend. Discovered near the Columbia River by John Townsend in 1835, hermit warblers are prized by listers, those birders who count the number of species they've seen. Some listers count casually. Others seek new species almost obsessively. Sooner or later—after they've seen North America's easily viewed birds—these driven ones will find themselves in the Cascade or Sierra Nevada mountains, staring up toward the steeplelike tops of conifers, hoping that hermits are there.

The birds may be there, but seeing them is a tricky business. Although they forage occasionally at lower levels and sometimes nest there too, hermit warblers spend most of their time in the topmost branches of tall evergreens, in the densest conifer stands they can find. It's a preference that has left more than one lister apoplectic. Some seekers don't even try to see hermit warblers until May or June, when the male's distinctive song gives them a clue to where to look. Others cheat by playing the taped call of a saw-whet owl in prime hermit warbler habitat. If a hermit is in residence, it may well descend from its lofty perch to investigate what it thinks is the intrusion of a predator. If it does, eureka! Another bird goes on the list.

Probing intently for food, the hermit warbler often hangs upside down like a chickadee.

Recognition. 4½–5 in. long. Spring male gray above with 2 white wing bars; most of head yellow; bib black; rest of underparts whitish. Female similar, but throat mostly yellowish.
Habitat. Open coniferous forests.
Nesting. Nest is a neat cup of stems, grass, twigs, and needles, saddled near tip of horizontal branch 2–50 ft. (usually 20–40 ft.) above ground in conifer. Eggs 3–5, whitish, densely spotted with brown and lilac. Incubation and nestling periods unknown.
Food. Insects and spiders.

Black-throated Green Warbler

Dendroica virens

Zree-zree-zree-zoo-zee—a lively, buzzy song emanates from a stand of conifers. *Zroo-zree-zree-zoo-zeeee,* comes the drawn-out, somewhat sleepy reply. So begins a territorial conversation between two black-throated green warblers on a summer morning.

The delightful, lisping, patterned songs of b-t greens, as these birds are affectionately called, are among the easiest to remember. But with so many bird sounds, how can anyone begin to learn them all? For some species, it's easy; their names are built-in memory prompters: bobwhite, chickadee, whip-poor-will, dickcissel, pipit, and so on. For others, a mnemonic, a phrase or word that helps to replay a bird's song or call in your head, is useful.

At the beginning, mnemonics help in remembering songs more easily. "Please, please, pleased to meet you," sings the chestnut-sided warbler. "I am so la-zee!" is the black-throated blue's answer. "Who cooks for you?" roars the barred owl. It's possible to coin a phrase for almost any sound—*Chi-ca-go* for the California quail, *Peter, Peter, Peter* for the tufted titmouse. With the sounds mastered, all sorts of lively dialogue can be imagined: "Jacob, Jacob! Dearie, come here." "José María! Pleased to meet 'ya." "Phoebe! Go back, go back, go backa." "Oh dear me!"

This species usually builds high up in a tree, but some nests are found only a foot above ground.

Recognition. 4½–5½ in. long. Spring male greenish above with 2 white wing bars; face yellow; throat, breast, and sides black. Female similar, but underparts less black.
Habitat. Coniferous and mixed forests.
Nesting. Nest is a tidy cup of grass, twigs, bark strips, plant fibers, and moss, fastened in crotch of tree 1–80 ft. above ground; usually in conifer, occasionally in hardwood. Eggs 4–5, white or cream-colored, coarsely spotted with brown and purple. Incubation 12 days, by female only. Young leave nest 8–10 days after hatching.
Food. Insects and berries.

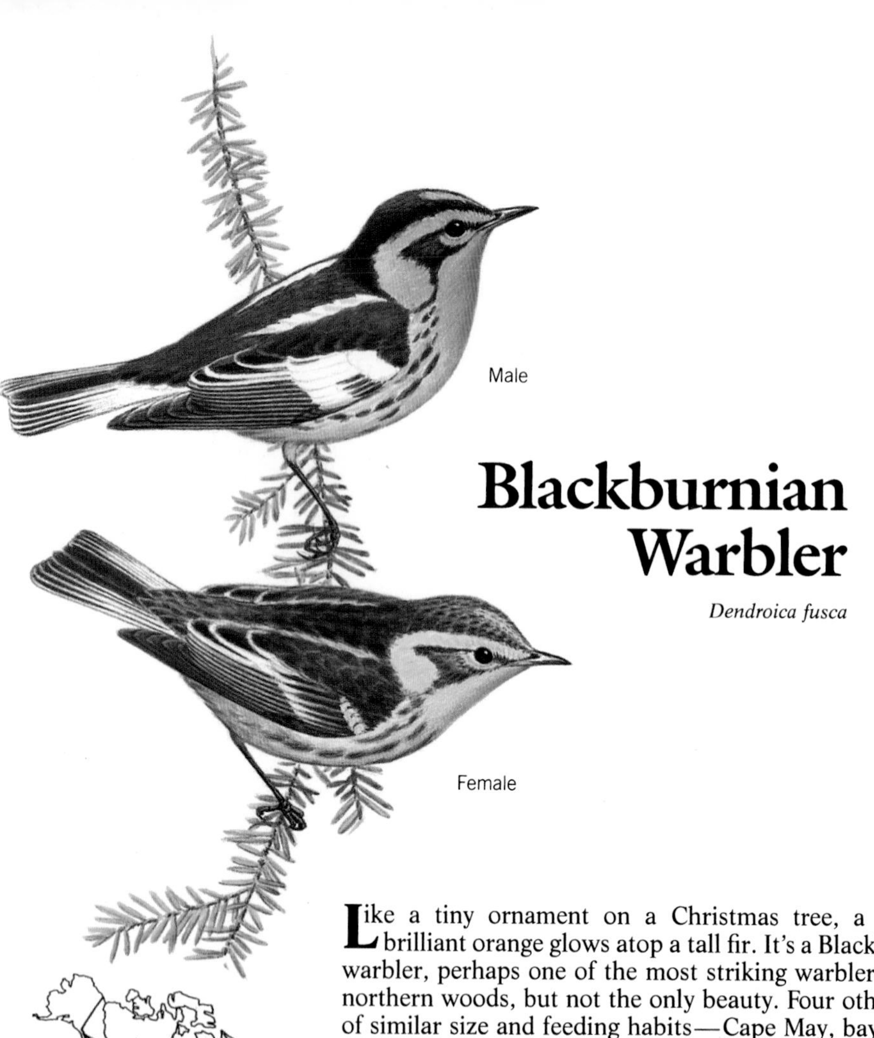

Blackburnian Warbler

Dendroica fusca

Like a tiny ornament on a Christmas tree, a spot of brilliant orange glows atop a tall fir. It's a Blackburnian warbler, perhaps one of the most striking warblers of our northern woods, but not the only beauty. Four other gems of similar size and feeding habits—Cape May, bay-breasted, yellow-rumped, and black-throated green—often share the same habitat, occasionally even the same conifer.

In the 1950's, the late Robert MacArthur, then a graduate student at Yale, set out to answer a question that had puzzled ornithologists: Could all five of these species, which appear to live and feed so similarly, occupy the same ecological niche? If so, they would represent exceptions to the rules of competition.

MacArthur watched these warblers in a Maine forest, mapping each bird's positions and foraging times on a diagram of a tree divided into 16 zones. Then he analyzed his observations, and what had seemed at first a random pattern of activities, started to take a definite shape: Each species fed at particular heights and distances from the trunk and used different foraging techniques. Because of these preferences, they did not occupy the same niche after all, and so were not actually competing with each other. MacArthur's classic work inspired countless students of ecology. His life was short, but his legacy was great.

Blackburnians prefer to nest in tall conifers, but when necessary they settle in deciduous forests.

Recognition. 4½–5½ in. long. Spring male streaked with black and white; throat, eyebrow, and crown patch bright orange; wing patch white. Female similar, but orange areas are yellow, black areas more grayish.

Habitat. Coniferous forests; oak-hickory woodlands in southeastern mountains.

Nesting. Nest is a large, sturdy cup of fine twigs, rootlets, plant down, and lichens, on horizontal branch 5–80 ft. above ground. Eggs 4–5, white or pale greenish, spotted with brown. Incubation about 12 days, by female only. Nestling period unknown.

Food. Insects and berries.

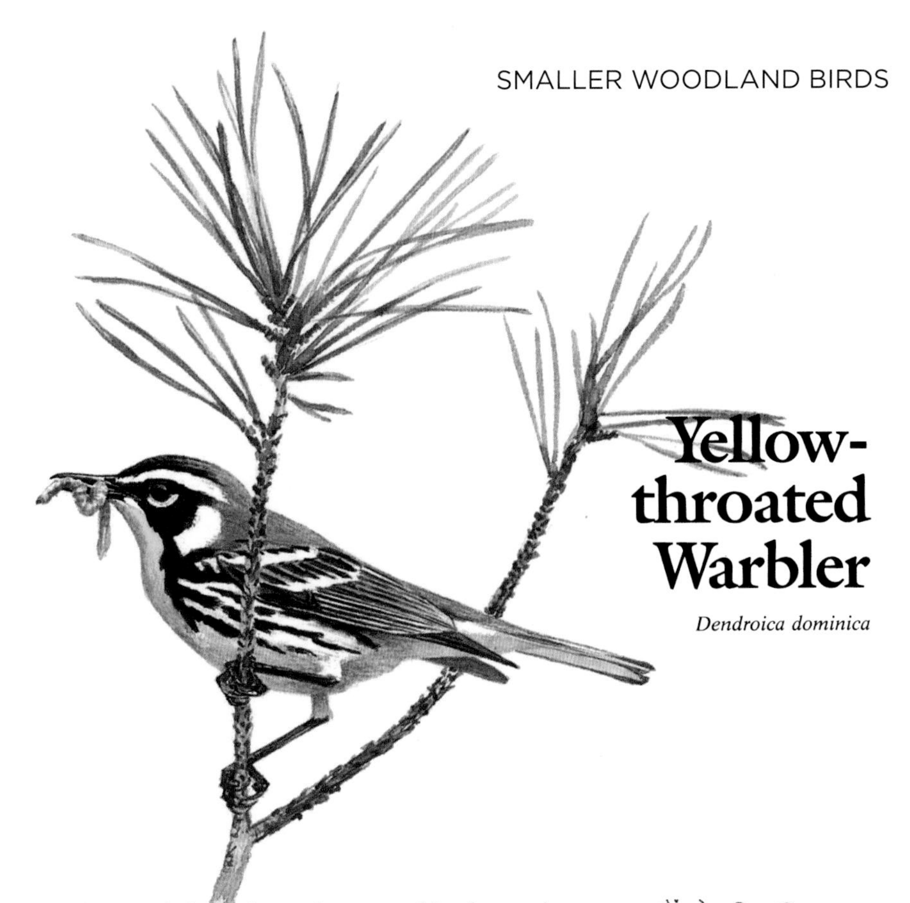

Yellow-throated Warbler

Dendroica dominica

Brightly plumaged throughout the year, this elegant insect-eater nonetheless keeps a low winter profile in the Southeast until the waning days of February, when it becomes a point man for spring. Then its clear song spirals out of lofty heights and shatters the drab silence of winter. In their imaginations, birders who know the song can step into a warm, color-splashed season that is, according to the calendar, still weeks away. Other yellow-throats—those that winter in the balmy climes of the Bahamas, the Virgin Islands, or Puerto Rico—will be along shortly. By early March there's a perceptible migration going on, and it continues well into the real spring.

Southern birds may be solidly committed to nesting by the time those with northern destinations arrive in the United States, only to drag their feet before moving on. One might suppose that those with the longest journeys ahead would be in a hurry to get started, but that would be foolhardy. Zealots who fly too far too fast risk exposure to cold or snow. They may go hungry, and even die. So as a harbinger of spring, the yellow-throated warbler, in common with some other early migrators, has a fine sense of timing, seldom arriving in central Alabama before mid-March, Tennessee before early April, or the northernmost parts of its breeding range before early May.

The yellow-throat's habit of probing for food on tree trunks may account for its long bill.

Recognition. 4½–5½ in. long. Gray on crown and back, with 2 white wing bars; face patterned with black and white; throat bright yellow; rest of underparts white; flanks streaked.
Habitat. Moist deciduous forests; pine groves.
Nesting. Nest is a loose cup of plant down, stems, and grass, usually in hanging mass of Spanish moss, 3–120 ft. above ground in sycamore or pine. Eggs 4–5, grayish or greenish, spotted with reddish and lilac. Incubation about 13 days. Nestling period unknown. Sometimes 2 broods per season.
Food. Insects and spiders.

Grace's Warbler

Dendroica graciae

The identification of this petite warbler reaffirmed a 19th-century taxonomic tradition: naming a new species after the naturalist who discovered it. If the naturalist himself wasn't commemorated, his friends or relations were. Thus, when Dr. Elliott Coues found this pine-loving bird in the Rocky Mountains in 1864, he named it after his 18-year-old sister, Grace Darling Coues, for whom, he said, "my affection and respect keep pace with my appreciation of true loveliness of character." So it had been, too, with the Virginia's warbler, named after the wife of William W. Anderson, a U.S. Army surgeon who first recorded the bird in 1858. A similar honor was accorded the Rev. John Bachman, who had discovered the Bachman's warbler in a South Carolina swamp in the early 1830's.

At one time, identifying a bird meant shooting, trapping, or netting a speciman of one. Squinting through swamp grass and sagebrush, unaided by telephoto lenses or high-powered binoculars, the early explorers could believe only what they held in their hands.

Elliott Coues might never have obtained a good look at the Grace's warbler if he hadn't brought one to hand with a well-aimed load of bird shot. Since this tiny mountain bird spends most of its time in the tops of towering pines, hemlocks, and firs, Coues would have seen only a bright yellow speck from the ground. Today, specimen hunting is no longer generally accepted. In fact, the subject is one of continuing controversy among birders.

The Grace's uses a hovering flight to inspect pine cones for the insect larvae hiding in them.

Recognition. 4½–5 in. long. Gray on crown, cheeks, and back, with black streaks on back and 2 white wing bars; eyebrow and throat bright yellow; rest of underparts white; flanks streaked.
Habitat. Open pine-oak woodlands.
Nesting. Nest, rarely found, is a compact cup of plant fibers, lined with hair and feathers, 20–60 ft. above ground on branch, usually in pine. Eggs 3–5, white or cream-colored, speckled with brown, ringed at large end. Incubation and nestling periods unknown.
Food. Insects.

Pine Warbler

Dendroica pinus

The names of many birds are illogical, if not downright misleading, but in this case a more suitable one could scarcely have been chosen. Pitch pine, jack pine, Norway pine, red pine, white pine, scrub pine—very few types do not figure prominently in the lives of pine warblers. Except during spring and fall migrations, these little birds are rarely spotted too far away from their favorite trees.

Their sturdy nests are found well out on a horizontal pine limb, hidden in tufts of needles at the very tip of a branch or among a cluster of cones. They are built of pine bark, pine needles, and tiny pine twigs, all held together with spider or caterpillar silk (gleaned most likely from around the tree) and lined with plant down, fur, and feathers that also probably come from the immediate vicinity.

The pine warblers' food consists mainly of the insects, spiders, and their kin with whom the birds share the pine trees. Daily they scour the branches, twigs, and needles and creep over the bark searching for their prey—not with the single-minded concentration of brown creepers, but in much the same manner, with their bodies pressed low and feathers brushing the bark. Sometimes their bright plumage becomes streaked with the pitch, or resin, that rubs off on them, and it is clearer than ever that these warblers are a product of the pines.

The pine warbler, like its black-and-white cousin, creeps on the bark of trees.

Recognition. 5–6 in. long. Adult male mainly olive-green; 2 wing bars white; eyebrow and most of underparts yellow. Female duller and darker green.
Habitat. Open pine woodlands, pine barrens, and groves of eastern red cedar.
Nesting. Nest is a compact cup of stems, bark strips, twigs, and pine needles, 15–80 ft. above ground in pine or red cedar, on horizontal branch or in foliage near tip. Eggs 4–5, white or greenish, spotted with brown. Incubation and nestling periods unknown.
Food. Insects, spiders, seeds, and berries.

Kirtland's Warbler

Dendroica kirtlandii

A Kirtland's diet includes not only insects but also the sap of jack pines.

It is a very exacting bird, the Kirtland's warbler. The only place where it nests is in the jack pine plains of central Michigan—an area of about 500 square miles, the smallest breeding range in North America. Even here, these warblers assemble only where Christmas-tree size pines grow densely on sandy soil that also supports a low growth of grass, sweet fern, and blueberry bushes. Tree height is crucial. At less than 5 feet, the pines are usually not dense enough, and when they reach 18 feet, they shade out the undergrowth beneath which the warblers conceal their nests.

Such restricted habitat, of course, keeps the population limited. Adding woe to the Kirtland's struggle for survival is the brown-headed cowbird, which lays its eggs in the warbler's nest. Agencies in Michigan have acted to control the cowbird population and to renew the warblers' habitats through limited burning of old trees (only after a forest fire do jack pines regrow with just the right density). Still, the nesting population remains fairly static, and most are older birds. What happens to the yearlings?

Every fall the Kirtland's warblers leave Michigan to spend the winter in the Bahamas. Ideally, somewhat near the same number should make the return trip, but a large percentage of the yearlings disappear. In the Bahamas? On migration? No one knows, and Kirtland's warblers continue their precarious life on the endangered species list.

Recognition. 6 in. long. Crown and upperparts dark gray; face black with broken white eye-ring; underparts yellow; flanks streaked and spotted with black. Female paler, with no black on face. Bobs tail.
Habitat. Dense stands of young jack pines.
Nesting. Nest is a compact cup of grass and plant fibers, concealed on ground under young jack pine. Eggs 4–5, pinkish or cream-colored, spotted with brown. Incubation about 14 days, by female only. Young leave nest about 9 days after hatching. Rarely 2 broods per season.
Food. Insects.

Bay-breasted Warbler

Dendroica castanea

There's nothing like a plague of spruce budworms to make this warbler's day. Such outbreaks occur cyclically in northern conifer stands where the bay-breasted warblers nest, and can denude vast tracts of spruce and fir. Inflicting the damage are voracious budworm larvae, pests that have been a bane of lumbermen since the early 1900's. Bay-breasted warblers often react to the infestations by creating a population boom of their own to take advantage of the sudden insect bounty.

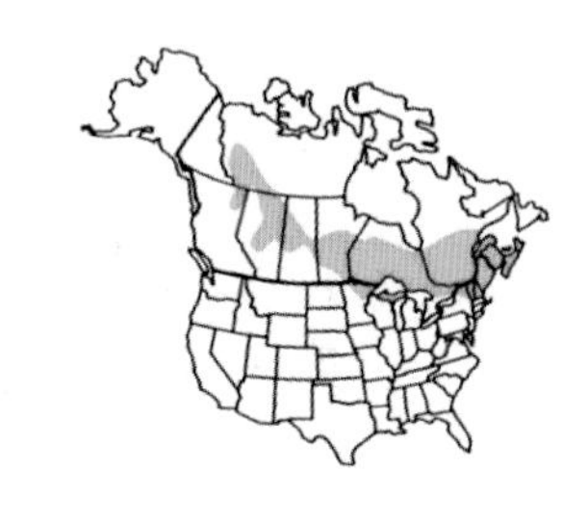

Showing true team spirit, nesting bay-breasts scale down their normal territorial requirements so that more of their kind can squeeze into a budworm-infested area. The prodigious appetites of the additional nestlings are satisfied by an abundant supply of this protein-rich food. The number of bay-breasted warblers produced during such years sometimes exceeds six times the region's annual norm.

Inevitably, the budworms begin eating themselves out of house and home, and the infestation subsides or ends abruptly. Bay-breasted warblers, whose numbers have swollen to peak levels during the budworm explosion, find slim pickings after the decline. As a result, their numbers, too, drop to normal levels, and a more stable bird-insect balance returns to the northern woodlands.

Each warbler species feeds in its own part of a tree; the bay-breast favors level branches.

Recognition. 5–6 in. long. Spring male chestnut on crown, throat, and flanks; face and forehead black; ear patch buff; underparts pale buff. Female similar but less chestnut on flanks. Immatures and fall adults dull green; flanks tinged buff; legs dark.
Habitat. Spruce-fir forests.

Nesting. Nest is a large cup of twigs, grass, bark strips, and rootlets, 5–50 ft. above ground in conifer. Eggs 3–7, white, bluish, or greenish, spotted with brown and lilac. Incubation about 13 days, by female only. Young leave nest about 12 days after hatching.
Food. Insects and berries.

Blackpoll Warbler

Dendroica striata

Imagine a bird that weighs about half an ounce. Picture it migrating more than 10,000 miles a year from Alaska's tree line to the rain forests of South America and back, skimming wooded peaks, skirting urban skyscrapers, and crossing vast stretches of open water in the Atlantic Ocean and Gulf of Mexico, where the fate of weak fliers is to fall in the waves and perish. Now meet this phenomenon—the blackpoll warbler, a dynamo whose traveling prowess is virtually unrivaled among North America's land birds.

The blackpoll's spring journey starts in a somewhat leisurely fashion, covering only about 30 miles a day. Along its spring route through Florida and the eastern United States, it is one of the latest migrants. But then, as it crosses the northern states, the blackpoll picks up speed, finally traveling up to 200 miles a day to reach its nesting area.

These flights are not only long; they are fraught with peril. Resting and feeding during the day and migrating at night, blackpolls and many other warblers successfully battle exhaustion only to crash head-on into tall buildings, television towers, and lighthouses that loom in the dark. Attracted by the lights on these obstructions, hundreds of birds may strike a single structure in one night. But once the blackpolls reach their nesting grounds, conditions are good for them to breed abundantly, and despite the casualties their numbers are sustained year by year.

On the cold northern nesting grounds, blackpolls often line their nests with feathers.

Recognition. 5–6 in. long. Spring male streaked black and white; crown and throat black, cheeks white (pattern similar to chickadee). Spring females, immatures, and fall adults dull greenish; sides lack buff tinge; legs pale.
Habitat. Spruce-fir forests.
Nesting. Nest is a cup of fine twigs, bits of bark, stems, and grass, 1–10 ft. above ground in small spruce or fir. Eggs 3–5, white, pale buff, or greenish, spotted with brown and lilac. Incubation 11 days, by female only. Young leave nest about 12 days after hatching.
Food. Insects and spiders; some berries.

Cerulean Warbler

Dendroica cerulea

Even in the best of circumstances, warblers are a challenging group of birds to learn. Many are treetop-dwellers in deciduous woods, and to all but the keenest observers they are just animated snippets of color and pattern against a constantly changing background. More than most, the cerulean warbler deserves its reputation for bringing bird watchers to their knees. Even when it perches in the open, the male's blue-on-blue combination against an azure sky permits little confidence. Sun, shade, and the motion of leaves stirred by a breeze all conspire against certainty. So when a birder gets even one good look at this heavenly blue warbler, it's a sight to remember.

Like many other birds, adult ceruleans are sexually dimorphic. That is, male and female have markedly different plumages. Though not as glorious as the scene-stealing male, the female is pretty and delicate in shades of buff, gray, and green, with a thatch of blue. Her subtle dress is especially suited to her role as chief incubator of eggs in a nest situated near the end of a high limb, where it pays to be inconspicuous. Among more familiar birds, examples of sexual dimorphism can readily be seen in adult cardinals (he is mostly red, she is mostly brown) and the orioles (bright orange in males, dull orange or dull yellow in females), to name only two.

From dawn to dusk, the male cerulean sings from the highest branches of a tree.

Recognition. 4–5 in. long. Adult male light blue-gray and streaked above with 2 white wing bars; underparts white with dark breast band. Female and immature similar but lack streaks and breast band, often tinged with greenish.
Habitat. Rich deciduous forests.
Nesting. Nest is a tidy cup of bark strips, fibers, moss, and grass, 15–100 ft. above ground, in fork of branch well out from trunk. Eggs 3–5, cream-colored or greenish, spotted with brown. Incubation about 13 days, by female. Nestling period unknown.
Food. Insects.

Black-and-white Warbler

Mniotilta varia

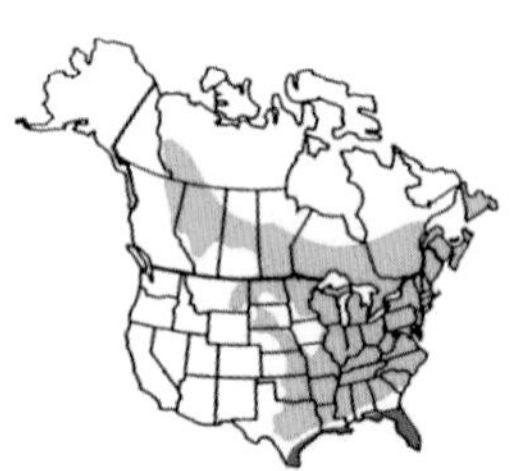

Foraging on bark, black-and-whites can find food in spring before the leaves come out.

These little magicians in sporty racing stripes are abundant birds, especially during spring and fall migration, when they are apt to appear anywhere and everywhere—but seldom far from a tree trunk or sturdy branch.

The search for food, always a high priority, prompts the black-and-white to assume positions that would be out of character for most other warblers. An inspector general of furrows, scales, and warts, a prime consumer of insects and spiders and daddy longlegs, it is a surefooted creeper as it scales the heights of trunks. In fact, it was once called the black-and-white creeper. Since it is equally adept at hitching down tree trunks headfirst or making its rounds of the branches while upside-down, it has also sometimes been called the black-and-white nuthatch.

Though it gives a fair imitation of both creeper and nuthatch, this warbler is more agile than the first and faster (by a hair) than the second, and appears to have a firm toehold on its own identity. Fortunately, it neither hurries nor hides. We can *see* it. It is one of the most cooperative members of a tribe of birds that have heaped misery, frustration, and humiliation upon bird watchers across North America, ever since binoculars replaced shotguns as the primary tools of bird identification. When the black-and-white warbler comes to visit, pleasure awaits in the turn of a focus wheel.

Recognition. 4½–5½ in. long. Streaked with black and white; eyebrow white. Male has black throat, female white throat. Creeps along tree trunks and branches like nuthatch.

Habitat. Deciduous and coniferous forests.

Nesting. Nest is a thick-walled cup of grass, bark strips, twigs, dry leaves, and rootlets, concealed on ground among roots at base of tree, or in shelter of log or overhanging rock; rarely in cavity in stump. Eggs 4–5, white or cream-colored, speckled with brown. Incubation 10–12 days. Young leave nest 8–12 days after hatching.

Food. Insects and spiders.

American Redstart

Setophaga ruticilla

Candelita ("little candle," or "little torch") is the name Latin Americans give these busiest and brightest of little warblers. They seem to dance like candle flames as they scour buds, leaves, twigs, and blossoms for the very tiniest insects, on which they feed. Never still for a moment—at least, so it seems—American redstarts are routinely described as restless, nervous, and fidgety. But their ceaseless ado is not aimless. They are courting, protecting territories, working at some nesting activity, or, most often, searching for food.

They capture their prey in midair as well as on a tree, and may abruptly leave off digging in a crevice to flit swiftly after a flying insect. They are constantly leaping up, dropping down, dancing the length of a limb, or darting suddenly into the air. While so engaged, they incessantly spread their tails and droop their wings. For such showiness, redstarts are accused of being in a state of perpetual display. But it is only by the constant sweeping of their tails and vibrating of their wings that they keep themselves in readiness for those sudden leaps and flutters.

The male does display when courting, of course. He stands on his toes with his body thrown back, his wings fanned to show off their bright orange patches, and his chest puffed out, proclaiming himself the best fellow in the neighborhood.

The male redstart sings and nests in his first spring, before acquiring adult plumage.

Recognition. 4½–6 in. long. Adult male mainly black; sides of breast, wing patches, and tail patches bright orange. Female and young male olive-gray above, with yellow or pale orange patches on wings, tail, and sides of breast. Often fans tail.

Habitat. Deciduous woods and thickets.

Nesting. Nest is a cup of grass, fibers, and lichens, 4–70 ft. above ground in tree or sapling. Eggs 2–5, white, cream-colored, or greenish, spotted with brown and gray. Incubation about 12 days, by female only. Young leave nest about 20 days after hatching.

Food. Insects, spiders, and berries.

Prothonotary Warbler

Protonotaria citrea

The color of the prothonotary warbler—a glowing mixture of yellow and orange—is so like the hoods worn by the prothonotaries, or chief clerks of certain courts, that this bird came to be named for those colorfully clad officials. The prothonotary's radiant hues glow beautifully in the dimly lit, wooded swamps and waterways where it makes its home. Seeking insects and spiders, it explores the rotting and mossy surfaces of floating or partly submerged logs, or creeps for short distances along the rough trunks of waterside trees. Using a thin sharp bill, it tugs and pulls its captives from their little nooks and crannies.

As it climbs, the bird looks for larger recesses to serve its housing needs, for the prothonotary is one of two warblers (the other is the Lucy's warbler) that regularly nest in tree cavities. It prefers a spot right over water but will settle for quarters as close by as possible. An old woodpecker nest is a great find, but any hole will be considered. The males select the sites and seem to have a knack for seeing the potential of any hollow object. A few of the places that prothonotaries have called home include a bridge support, a cigar box, a tin cup, a glass jar—and a coat pocket.

Once the cavity is chosen, the prothonotary warbler stuffs it almost to the brim, mostly with mosses. Thus it forms a soft cup in which it will shortly rear another generation of glowing, orange-yellow, swamp-dwelling prothonotary warblers.

Preferring to reside over water, this species rarely nests outside wetlands.

Recognition. 5–5½ in. long. Male bright orange-yellow below and on head; back olive; wings blue-gray; tail blue-gray with large white patches visible in flight. Female similar but duller. Sometimes creeps on tree trunks like nuthatch.

Habitat. Wooded swamps and streamside forests of the southeastern United States.

Nesting. Nest is a cup of twigs, leaves, and moss, in tree cavity, 3–32 ft. above ground or water. Eggs 3–8, pinkish, spotted with brown and gray. Incubation 12–14 days, by female only. Young leave nest about 11 days after hatching. Sometimes 2 broods per season.

Food. Insects.

Worm-eating Warbler

Helmitheros vermivorus

Despite having a name about as explicit as it can be, the worm-eating warbler rarely, if ever, eats earthworms. It does consume smooth (as opposed to hairy) caterpillars, but most of its diet seems to consist of spiders and such insects as beetles and grasshoppers.

In contrast to the highly active, mainly treetop-dwelling members of its family, the worm-eating warbler lives a less frenetic life, mostly on the ground or within a few feet of it. At times it climbs trees like a creeper, but does not go very high. If this elusive bird is seen at all, it may be spotted walking on the ground, tail cocked high, as if it hadn't a care in the world. But try to approach and it will skitter away, disappearing into the shrubbery.

This warbler's nest, too, is usually found on the ground—under a bush or against a small tree, preferably in a hillside thicket, where concealment is easiest. None too adventurous, the worm-eater makes its home in the same territory year after year, building each new nest within sight of the old ones. And the male stoutly defends his territory, quickly routing any fellow who steps over a boundary.

The worm-eating warbler's song is so like the chipping sparrow's, they can be difficult to tell apart, though the warbler's is a little bit richer and apparently is sung with great gusto. As it sings, the worm-eating songster fluffs out its feathers, lowers its tail, droops its wings, throws back its head, and trills the melody out.

Worm-eating warblers find much of their food in hanging clusters of dead leaves.

Recognition. 5–5½ in. long. Dull olive upperparts; underparts pale buff; head with bold black and buff stripes. Slower-moving than most other warblers; habit of perching motionless for many minutes makes it difficult to spot in foliage.
Habitat. Dry deciduous forests.
Nesting. Nest is a cup of dead leaves, well hidden in leaf litter on forest floor, usually at base of shrub or tree. Eggs 3–6, white, finely spotted with brown. Incubation 13 days, by female only. Young leave nest about 10 days after hatching.
Food. Insects and spiders.

Louisiana Waterthrush

Seiurus motacilla

A pathway of leaves frequently leads to the entrance of a Louisiana waterthrush's nest.

Despite the many characteristics it shares with the northern waterthrush—from being a warbler, to an earthbound lifestyle and teetering, head-and-tail-bobbing stance—the Louisiana waterthrush manages to keep its identity distinct. To begin with, although it does sometimes live in swamplands, it prefers the wooded edges along swift streams, especially in its northerly range. During nesting season it feeds almost entirely in these streams, flipping over leaves in the manner of its more northern relative and eating the aquatic creatures that are dislodged. The rest of the year it also picks into crannies for insects and spiders, and at times darts into the air flycatcher-style to snatch its prey in mid-flight.

Even this southern waterthrush's home is uniquely personalized. Dwelling in a small hole it has gouged into a stream bank and filled with leaves, it often builds a leaf doormat at the front of its nest or a long walkway of leaves, usually leading down toward the stream.

The male does not sing as much as a northern waterthrush; he is usually quiet on migration and he helps his mate with nest building. The Louisiana's song, however, is wilder and perhaps more musical than that of his cousin as it rings with eerie loveliness through the wilderness of his streamside haunts.

Recognition. 6 in. long. Brown above; underparts white and streaked with dark brown; eyebrow broad and white; throat usually unspotted. Walks on ground; bobs tail.
Habitat. Wooded swamps with flowing water and along rushing streams.
Nesting. Nest is a cup of moss, leaves, and grass, hidden among tree roots or in hole in bank, near or over water. Eggs 4–6, white or cream-colored, spotted with brown or gray. Incubation 12–14 days, by female only. Young leave nest about 10 days after hatching.
Food. Insects and other small aquatic animals.

Northern Waterthrush

Seiurus noveboracensis

It looks like a thrush, walks like an ovenbird, and teeters like a spotted sandpiper—but no matter. The northern waterthrush is still a member in good standing of the wood warbler family. The smallest of our waterthrushes, it haunts northern wooded swamplands, silently lifting soggy leaves from the edges of watery pools, and dining on the small crustaceans, mollusks, water beetles, and minnows that dart away from the untimely disturbance. Using all the energy and wariness of a little mouse, the bird also delights in searching for insects over and under debris-tangled logs that lie half-submerged in the mud.

When an incubating female wants to slip out of her mossy nest, hidden among the radiating roots of a fallen tree, again she moves like a little mouse. Stepping out with head down and breast pressed close to the ground, she skulks through the underbrush until she is at least 30 or 40 feet from her nest. Then she straightens up and begins to feed, afterward slipping back home in the same manner.

For all this secrecy, though, the song of the northern waterthrush is a startling, loud, high-pitched whistle that carries for long distances across the watery habitat. The male begins his uninhibited singing during migration and continues through the entire time the female is building a nest and incubating her eggs. He does, however, keep well away from the nesting area, so that he does not call attention to it in the course of advertising his own presence.

Northern waterthrushes lift leaves in their search for food, finding prey other birds miss.

Recognition. 5–6½ in. long. Brown above; underparts tinged with yellow and streaked with dark brown; eyebrow narrow and dull yellow; throat usually spotted. Walks on ground; bobs tail like ovenbird.

Habitat. Northern wooded swamps and bogs.

Nesting. Nest is a cup of leaves, bark strips, and rootlets placed among roots or in decayed stump or log, often over water. Eggs 3–6, cream-colored or buff, spotted with brown and gray. Incubation and nestling periods unknown.

Food. Insects and other small aquatic animals.

Ovenbird

Seiurus aurocapillus

Among North America's many species of warblers, the ovenbird clearly stands out as an individualist. A bird of the forest floor, it is often seen walking among dead leaves or striding along the top of a fallen log, rather than flitting about in warbler fashion or perching on a branch. As it steps along, it places each foot with such precise care and moves with such grace and dignity, that its habit of teetering its tail up and down whenever it pauses gives the bird an incongruous touch of music-hall comedy.

Instead of choosing a treetop, the ovenbird sets its nest on open ground, close beside a road or a trail. Starting with a drift of leaves for camouflage, the bird makes a mound of leaves and weed stems, lines the interior with fine grass and hair, and roofs it over with more brown leaves so carefully placed that the little nest, which resembles a Dutch oven, is practically invisible. Only if an incubating bird flushes from the small opening in the side of the nest is a passerby likely to discover it. And only if an approaching foot threatens to land on the nest will this happen, as the mother leaps out and puts on a crippled-bird act to divert attention from her eggs or nestlings. But these birds sit so tight that a few have actually been trampled by people who stepped on their too-well-concealed nests.

For a warbler, the ovenbird has a fairly noisy and not very musical song. However, flying seems to inspire it. On the wing, the ovenbird pours out a more lyrical strain, especially in the evening.

To divert intruders from the nest ovenbirds dart away at the last moment feigning an injury.

Recognition. 5½–6½ in. long. Brownish above; crown orange with dark brown border on each side; eye-ring white; underparts spotted like thrush's. Walks methodically over ground; song a loud *teacher-teacher-teacher*.
Habitat. Deciduous forests and thickets.
Nesting. Nest is a domed structure of leaves, grass, and stems, with a side entrance, hidden on ground. Eggs 3–6, white, spotted with brown and gray. Incubation 11–14 days, by female. Young leave nest 8–11 days after hatching.
Food. Insects, earthworms, snails, and spiders.

Kentucky Warbler

Oporornis formosus

In April 1976 a spring storm roared out of the north into the Gulf of Mexico, stranding a small ship, like a lost island, some 50 miles at sea. Early the night before, thousands of delicate land birds, most of them insect-eaters such as the Kentucky warbler, had left the Yucatan Peninsula on their annual northward migration. They were fat and primed for the long, nonstop flight across the Gulf. The sky was clear. A light tailwind aided them. The journey seemed promising as the birds started out.

Having flown all night, the migrants were far from land when they collided with the storm front. Chilling rain drenched them, weighting them down. Turbulent headwinds sapped their energy even further. Confused and fatigued, the birds began to drop fatally into the waves.

Aboard the stranded ship, the captain, who was also a birder, made the following entry in his log: "1400 hours: . . . the deck is littered with exhausted birds. We have seen others trying to reach the ship, but they are carried off in the wind, or they fall into the sea. The birds on board are so far gone. As they die the crew brings them to me, and I try to identify them—mostly vireos and warblers. Now and then one surprises us. It struggles up to the railing and flies off into the storm—it has a long way to go before it reaches land . . . ."

Kentucky warblers take much of their food by plucking insects from the undersides of leaves.

Recognition. 5–6 in. long. Olive-green above; underparts bright yellow; yellow spectacles; black on crown and below eye. Female has less black below eye, grayer crown.
Habitat. Moist deciduous forests and thickets.
Nesting. Nest is a large cup of leaves, grass, and stems, hidden under shrub or in patch of weeds. Eggs 3–6, white or cream-colored, spotted with brown. Incubation about 12 days, by female only. Young leave nest about 10 days after hatching.
Food. Insects and spiders, usually snatched from undersides of leaves.

Mourning Warbler

Mourning Warbler *Oporornis philadelphia*
MacGillivray's Warbler *Oporornis tolmiei*

Mourning Warbler

MacGillivray's Warbler

Elusive. Shy. Both are good words to describe this small warbler, whose crepelike markings about its breast account for its name and make it one of the most attractive members of the entire warbler family. The mourning warbler's cautious ways are reflected in its migrating schedule: the bird postpones its springtime flight northward from Central America and northern South America until a good growth of foliage for sure concealment is almost certain to be found at the end of its flight. Then, at last, it comes on in a rush, flying nonstop over the Gulf of Mexico in late May or early June to take up nesting residence in lower Canada and parts of the United States. Only parenthood finally brings the mourning warbler out into the open, the male perching on a branch to serenade his mate and brood concealed somewhere well below.

The western states' counterpart of the mourning warbler is the MacGillivray's warbler. John James Audubon named it in honor of a distinguished Scottish ornithologist, but proper credit for its discovery in 1839 belongs to John Townsend, a naturalist from the same city, Philadelphia, in whose environs the mourning warbler had first been identified 29 years earlier. The MacGillivray's is drawn like a magnet to burned-out spaces where new growth is just beginning to turn fire-blackened hillsides and mountain ranges green again.

Recognition. 5–6 in. long. Olive-green above, dull yellow below. Males have gray hood, black flecks on breast. Mourning has gray face, no eye-ring; MacGillivray's has black face, broken white eye-ring. Females duller.
Habitat. Wooded swamps, moist thickets, and brushy hillsides.

Nesting. Nest is a cup of leaves, stems, and grass, hidden on or near ground in clump of grass, ferns, or weeds. Eggs 3–6, white or cream-colored, spotted with brown. Incubation about 12 days, by female only. Young leave nest 7–9 days after hatching.
Food. Insects and spiders.

Hooded Warbler

Wilsonia citrina

Dauphin Island, Alabama: clouds roll in and the wind quickens. Rain splatters down on the west beach. Birders who have been there know where to look—south, just over the whitecaps of the Gulf of Mexico. Tiny specks appear on the horizon. They grow larger, taking on colors, shapes, and patterns. Insouciant flycatchers. Shy thrushes. Understated vireos. Tanagers, orioles, and buntings in gaudy paint-pot colors. And, not least, cowled hooded warblers—by the hundreds.

To a bird watcher, "fallout" can be a beautiful thing. Not nuclear fallout, of course, but the temporary grounding of migrating birds due to inclement weather. Fallouts are most frequent and most spectacular along the northern Gulf Coast in spring. "Migrant trap" is a catchall term for islands, peninsulas, and wooded mainland areas that offer welcome patches of green where grounded birds concentrate in a small area. To be on hand for a fallout motivates countless birders to visit such famed migrant traps as Dauphin Island; High Island, Texas; and Cameron Parish, Louisiana.

Fallouts do not happen on cue, but the birders' patience is rewarded when the migrants do arrive. They come in waves—tired, hungry, yet very much alive. They forage innocently, splashing this tree with yellow, another with red. They light up the bushes with orange, and pepper the ground with blue. The spectacle ends too soon, in a day, or even hours. But the memory lasts a lifetime.

While male hooded warblers catch flying insects, females hunt closer to the ground.

Recognition. 5–6 in. long. Olive-green above; underparts yellow. Male has black hood, bright yellow face. Female similar, but has only trace of black hood.
Habitat. Moist deciduous forests and woodland thickets of eastern United States.
Nesting. Nest is a tidy cup of leaves, bark strips, and fibers, 1–18 ft. above ground or over water in sapling or shrub. Eggs 3–5, cream-colored, spotted with brown. Incubation about 12 days. Young leave nest about 9 days after hatching. 2 broods per season.
Food. Insects and spiders; often catches insects in flight.

Wilson's Warbler

Wilsonia pusilla

An active bird, the Wilson's warbler restlessly twitches its tail and wings like a kinglet.

Is this colorful songster really a North American warbler? Many Central Americans would say no: the Wilson's warbler is a bird of the rain forest. True, it nests and breeds in northern bogs or alder swales, and it migrates through the United States and Canada. But how much time does it really spend away from its tropical home?

It's a convincing argument when the days are actually counted. Like many other warblers, the Wilson's spends about three months on its breeding grounds, another two or three months migrating north and south, and the rest of the time—six to seven months—on its warm winter range.

Some warblers spend even less time in the North. In the Churchill region of Canada, yellow warblers arrive in spring, build nests, raise their young, and depart within a seven-week span. Admittedly, these restless yellow fliers are not typical. But the fact is that many warbler species have abandoned their nesting grounds by midsummer, and few of them remain after the first days of fall.

When they do depart on their long southern journeys, the Wilson's and other warblers are in fact heading home, back to the land in which scientists say they originated. Were it not for North America's abundant nesting space and explosion of summer insects—both of which are needed to raise young, and neither of which is found in quite the same way in the tropics—birds like the Wilson's warbler might never have developed the urge to fly north at all.

Recognition. 4½–5 in. long. Olive-green above; face and underparts bright yellow; cap black. Many females lack black cap.
Habitat. Forest clearings, brushy areas, and thickets of the North.
Nesting. Nest is a large cup of leaves, bark strips, and stems, 1–10 ft. above ground in shrub or clump of grass, or on ground. Eggs 4–6, white or cream-colored, finely spotted with brown. Incubation about 11 days, by female only. Young leave nest about 10 days after hatching.
Food. Insects, often caught in flight.

Canada Warbler

Wilsonia canadensis

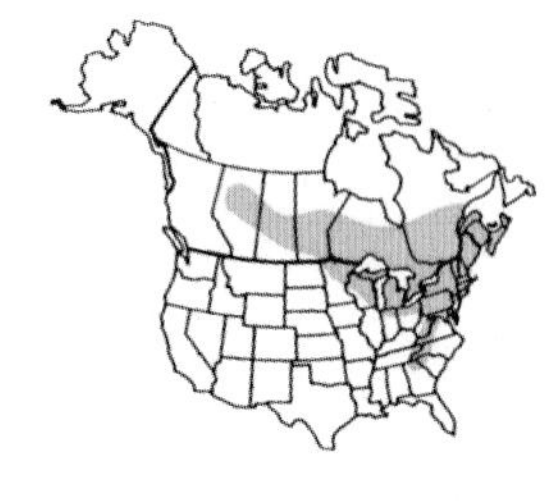

January in the Anchicayá Valley of Colombia finds this wide-awake warbler exercising its right to be different. The difference in this case is a feeding strategy that frees Canada warblers from competing directly with some 300 resident species that crowd this lush wintering ground. While other birds are combing the tops of leaves for insects, Canada warblers pluck food from their undersides. Heavy tropical rains have driven insects and other invertebrates beneath the leaves for shelter—where they afford Canada warblers an ample, untapped food supply.

No musical outbursts will be heard from these birds as they forage through tropical foliage. Canada warblers reserve their singing for the cool, mixed woodland of their northern breeding grounds. There they burst forth with a repertoire of chips and warbles that shames the sibilant whispers of most other warblers. So lusty and vivid is the Canada's cascade of notes that some consider it the most musical song of any North American warbler's. And the male uses this vocal power to defend his territory with singular verve and tenacity. Detecting trespassers at some distance, the feisty male scolds them with an incessant barrage of chipping sounds—one naturalist was subjected to an outburst that reached an impressive 96 chips per minute.

Eager male Canada warblers have been seen trying to feed young even before they hatch.

Recognition. 5–6 in. long. Plain gray above; underparts bright yellow; spectacles yellow. Male has necklace of black spots on breast, lacking or faint in female.
Habitat. Moist, hilly deciduous or mixed forest.
Nesting. Nest is a large cup of leaves, bark strips, and grass, hidden on or near ground in cavity in bank, among rocks or roots, or in clump of vegetation. Eggs 3–5, cream-colored or white, spotted with brown and purple. Incubation and nestling periods unknown.
Food. Insects.

Red-faced Warbler

Cardellina rubrifrons

Searching cones for insects, red-faced warblers dangle upside down like chickadees.

Despite the brilliant face color that makes them standouts in almost any terrain, red-faced warblers are perhaps the least-known members of their family in North America. Living primarily in remote and inaccessible regions, they were once described by an authority as "the United States warblers that inhabit Mexico." But they really are a Mexican species. Their principal breeding grounds are south of the Rio Grande, and those that nest in Arizona and New Mexico are at the most northerly fringe of their range. They arrive in mid-April and make their annual retreat southward in the very early autumn.

Red-faced warblers are birds of the high mountains, nesting on well-drained banks and slopes under towering firs, spruces, or oaks at elevations of 6,000 to 9,000 feet. They hide their nests in the litter of the forest floor, sometimes burying them so deeply in the leaves and needles that they are completely out of sight. And like ovenbirds, these wagtail warblers sit tight on their nests, flushing only at the last moment to avoid being stepped on.

Aside from their nesting habits, however, red-faces are not ground birds. They forage about on conifers, gleaning insects from needles and cones at the tips of limbs, and constantly dart from branches to snatch moths, flies, and beetles in mid-flight. Apparently insects make up the bulk of their diets; but the red-faced warblers' eating habits, like so much else about them, remain largely unknown.

Recognition. 5–5¼ in. long. Gray above; white below; face, sides of neck, and upper breast bright red; crown and sides of head black. White rump visible in flight. Frequently flicks tail while feeding in foliage.
Habitat. Mountain forests of conifers or oaks.
Nesting. Nest is a small cup of leaves, grass, and pine needles, sunk in sloping ground in shelter of shrub, log, or rock. Eggs 3–4, white, spotted with brown. Incubation and nestling periods unknown.
Food. Insects.

Painted Redstart

Myioborus pictus

Their bright red breasts glowing like jewels, pairs of painted redstarts flit unceasingly from twig to bud to branch, from sunlight to shadow, in the canyons and mountainsides of the southwestern United States. They fan their tails, droop their wings, and dart through the air on the trail of gossamer insects. Fluttering to high branches, then flickering to low ones, they meticulously explore tiny chasms in the bark and the wrinkled surfaces of leaves. These painted redstarts behave, in short, exactly like American redstarts—which makes it all the harder to accept the fact that they are actually a Central American species and only distantly related to the North American birds they so closely resemble.

There are, however, significant differences between the species too. American redstarts are known to vary their insect diets with berries and magnolia seeds, while painted redstarts are thought to be wholly insectivorous. Our domestic redstarts build strong and highly decorative nests in trees; painted redstarts locate their more prosaic, utilitarian nests on steeply sloping ground. And while American redstarts have thin, high-pitched voices and lisping songs, painted redstarts have clear voices and their songs are loud and ringing. Moreover—and this is quite unusual for most birds, especially warblers—the female painted redstart sings just as well as the male. Their courtship and bonding fill the springtime air with bold, resounding duets.

Young painted redstarts are so unlike the adults that they seem to be a different species.

Recognition. 5–6 in. long. Mainly black; lower breast scarlet; wing patch, spot below eye, belly, and outer tail feathers white. Often fans tail and darts into air in pursuit of insects.
Habitat. Open oak woodlands.
Nesting. Nest is a large, shallow cup of bark strips, plant fibers, leaves, and grass, concealed on sloping ground or among rocks, in shelter of roots, shrub, or clump of grass. Eggs 3–4, white or cream-colored, finely spotted with brown and reddish. Incubation about 14 days. Nestling period unknown.
Food. Insects.

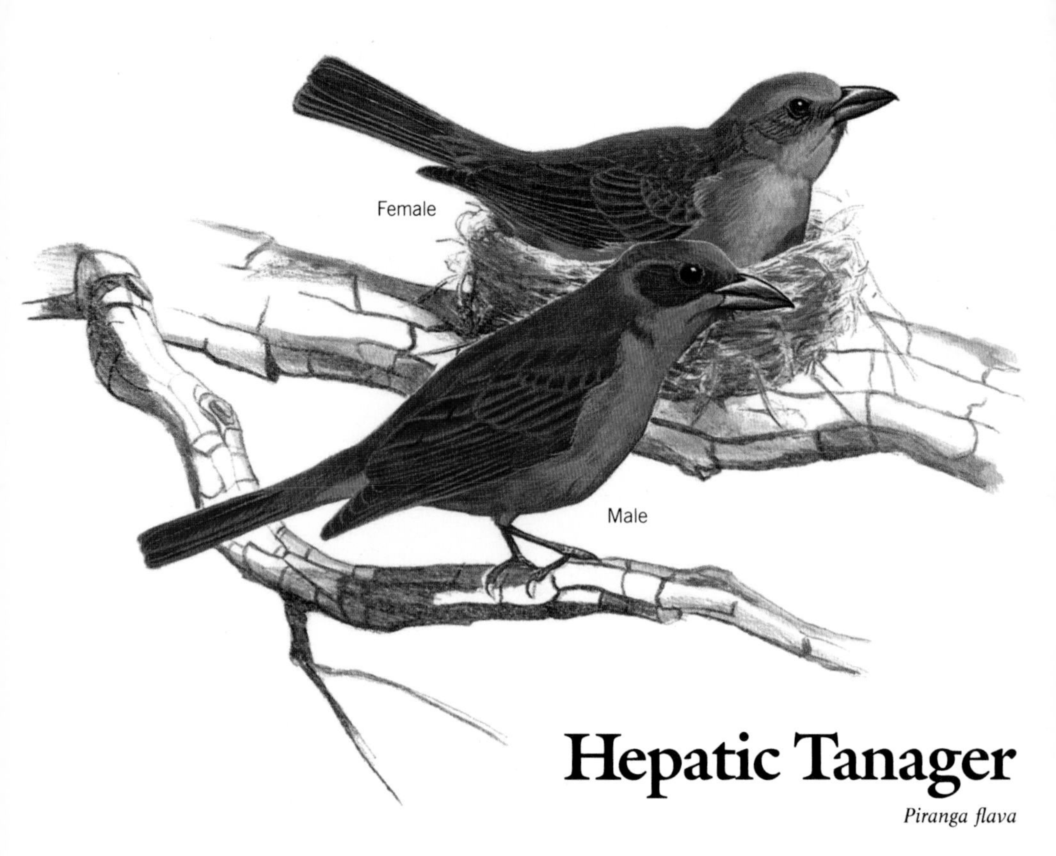

Hepatic Tanager

Piranga flava

Hepatic and summer tanagers look alike, but a hepatic's dark bill is larger, with a toothlike edge.

The United States and Canada get short shrift with respect to the total tanager count in the Western Hemisphere. Most of the 236 known species in this half of the world remain in the tree-canopied tropical forests of Central and South America; only four species undertake the rigors of springtime migration north of Mexico. Yet all four of these are beautiful and interesting birds, and almost every section of the two northern host nations can claim at least one tanager species, all of which rank among the most effective insect-eaters to be found anywhere.

The most thinly distributed and least known of the group is the hepatic tanager, which can sometimes be seen flying over the Mexican border on its way to a summer residence in the American Southwest. Like the other migrant tanagers, hepatics are reliable, effective parents. Egg incubation is the female's chore, but as soon as the young appear the father immediately begins foraging for food. One parent, however, can't do it all, for the mouths of young tanagers—hepatic or otherwise—rarely close during their waking hours, and these gaping circlets keep both parents hopping in their role as caterers. Within six months the young, strong birds will be on their own in their new semitropical quarters hundreds of miles south of where they were born and raised.

Recognition. 7–8 in. long. Bill large and dark; ear patch dark. Male uniform brick-red. Female drab olive above; yellowish below.
Habitat. Pine and pine-oak woodlands in mountain canyons.
Nesting. Nest is a shallow cup of grass, weed stems, and flowers, placed between forked twigs near end of horizontal limb 15–30 ft. above ground. Eggs 3–5, pale blue or greenish blue, spotted with dark red, brown, and lilac. Incubation and nestling periods unknown.
Food. Insects, sometimes caught in flight, and berries.

Summer Tanager

Piranga rubra

This is the bird that gladdens hearts in the South and Southwest all summer long. Anyone who has ever witnessed the happy splashing of a summer tanager in a backyard birdbath, its bright eyes flashing from a swirl of brick-red feathers, can easily understand why.

Most summer tanagers enter the United States through Florida and then fan out in search of pine woods mixed with small oaks. Although they dine on fruits and berries while wintering in the tropics, their eating habits change dramatically once they come north. Southern children, in fact, often first know these visitors by the nickname *red beebird,* which derives from the energetic birds' ability to overtake bees and other insects on the wing. Nor is any wasp nest safe from the arrival of an investigative summer tanager, which apparently considers wasp larvae and pupae among the best of all possible meals.

Where the ranges of the summer tanager and the more northern-ranging scarlet tanager overlap, these peaceful rivals engage in a brilliant singing duel to establish for each other's benefit just where the boundaries between their nesting sites lie. By fall, all the young tanagers, scarlet and summer, are ready for their first migration flight, a journey that may extend thousands of miles south to the Amazon Basin in Brazil.

A summer tanager will brave the stings of bees and wasps to steal their larvae.

Recognition. 7–8 in. long. Bill large and yellowish. Male uniform bright red. Female olive-green above; face and underparts yellowish.
Habitat. Streamside woodlands, willow groves, and oak forests.
Nesting. Nest is a frail, shallow cup of grass, stems, and leaves, saddled on horizontal branch 1–35 ft. above ground in forest tree. Eggs 3–5, pale blue or greenish, spotted with brown and gray. Incubation and nestling periods unknown.
Food. Insects (especially larvae of bees and wasps), spiders, and berries.

Scarlet Tanager

Piranga olivacea

The bird is a delight, this magnificently attired scarlet tanager, a sight just as thrilling to veteran bird watchers as to those catching their first glimpse of it. No bird in summer dress can be recognized faster than the male, as he darts upward in vibrant plumage to broadcast his raspy song from the highest treetop he can find.

And beauty is but one of this bird's distinctions. Scarlet tanagers make the longest flight of the four tanager species that travel north to grace the United States each spring. Once arrived, the male woos a mate by alighting on a low branch and spreading his wings to reveal his rich scarlet back feathers to the females perched above and behind him—who signal their admiration with whistles.

The courtship complete, nest building and incubation are left to the female. The same fiery plumes that made the male a successful suitor also make him a vivid target whose presence near the nest would endanger it. Any predators who do venture too close are lured away by ventriloquism, as the talented male places his song many yards away. And as spring turns to summer, the tanager's appetite for garden pests becomes as astounding as it is useful: one bird ate some 600 tent caterpillars in 15 minutes. Beautiful, distinctive, helpful—all this in one small visitor from the Andes who arrives each year just as the buds are flowering.

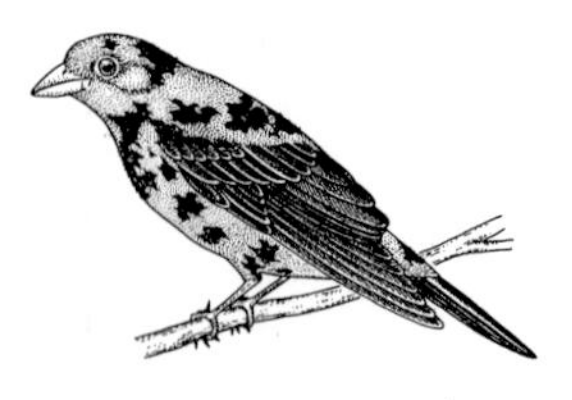

In late summer and fall, a molting male scarlet tanager has an odd, piebald look.

Recognition. 6½–7½ in. long. Breeding male brilliant scarlet, with black wings and tail. Fall male greenish with black wings and tail. Female greenish above; yellowish below; wings dark.
Habitat. Deciduous and pine-oak woodlands; orchards.
Nesting. Nest is a shallow cup of twigs, grass, and stems placed near tip of branch, 4–75 ft. above ground. Eggs 3–5, pale blue or green, finely or boldly spotted with brown. Incubation about 14 days, by female only. Young leave nest 9–11 days after hatching.
Food. Insects, spiders, and berries.

Western Tanager

Piranga ludoviciana

It is not surprising that President Thomas Jefferson was the driving spirit behind Lewis and Clark's famous expedition through the American West in 1804–06. A man of wide-ranging scientific interests, Jefferson specifically charged his 29-year-old former private secretary, Capt. Meriwether Lewis, to be alert for all new forms of plant and animal life on the journey. Somewhere in what is now Idaho the vigilant explorers discovered the western tanager and duly noted the fact in their logbooks.

Like many another westerner, this particular tanager is also prone to wander, and will sometimes turn up as far to the east as New England. In general, however, the beautiful crimson, yellow, and black flier stays true to the high country that runs south from Alaska along the Pacific and inland to the Rockies, where it often delights climbers by flashing into view at elevations as high as 10,000 feet.

Most tanagers are not especially notable singers. Some ornithologists theorize that because these birds are rarely concerned about territorial defense, they have no special need for more than average vocal abilities. Nonetheless, the western tanager does not lack for a broad human audience that admires both its dress and a free wildness in flight that seems a perfect match for the endless expanse of western pine and sky.

Western tanagers build their nests near the very top of western spruces and pines.

Recognition. 6½–7½ in. long. Breeding male mainly bright yellow; face red; back, wings, and tail black, with 1 yellow wing bar, 1 white wing bar. Fall male similar but duller. Female grayish above; yellow below; 2 narrow wing bars.
Habitat. Mountain coniferous forests.
Nesting. Nest is a shallow, loose cup of twigs, moss, and rootlets, placed on horizontal limb 10–65 ft. above ground. Eggs 3–5, pale blue, spotted with brown. Incubation about 13 days, by female only. Nestling period unknown.
Food. Insects; berries and other fruit.

Northern Cardinal

Cardinalis cardinalis

Although some birds are polygamous, most are monogamous at least for a single summer season. In a few species, birds returning to last year's territory remate with the same returning partner, and some birds will reunite for breeding year after year throughout their lives.

Cardinals not only mate for life, they remain together the whole year, generally in unruffled tranquillity. True, at a winter feeding station the male may step aggressively toward his mate, but she usually steps aside or ignores him and goes on feeding. Long before winter is over, they are both singing again, sometimes solo but more often counter singing—one bird trilling several phrases, which the other completes. As spring returns, the male resumes his courtship feeding, bringing his companion tidbits of food, their bills touching briefly as she accepts each offered morsel.

The female builds the nest alone, while the male is close beside her, exuberant in song. The female also incubates the eggs alone, but the male feeds her on the nest, and together they work to exhaustion feeding their first nestlings of each season. Then, as the female starts a second brood, they resume the counter singing and courtship feeding that so beautifully express the harmony of their life.

As a female cardinal begins a second nest, her mate feeds the first brood's fast-growing young.

Recognition. $7\frac{1}{2}$–$8\frac{1}{2}$ in. long. Crested, with stout, conical bill. Male bright red with black face; bill red. Female grayish buff, with red on crest, wings, and tail; bill pink.
Habitat. Brushy woodlands, thickets, garden shrubbery, and parks.
Nesting. Nest is a deep cup of stems, fine twigs, and bark strips, 2–12 ft. above ground in dense thicket or tangle of vines. Eggs 2–5, buff or pale greenish, speckled with brown and lilac. Incubation about 13 days, by female. Young leave nest about 11 days after hatching. Up to 4 broods per season.
Food. Fruits, seeds, and insects.

Rose-breasted Grosbeak

Pheucticus ludovicianus

It is a generally accepted theory in birding circles that brightly feathered male birds avoid the nest-building and incubating duties that go with raising a family because their coloring attracts predators and reveals the exact location of the nest. Yet many rose-breasted grosbeak males, whose shining black is slashed with white that flashes with every move they make, work diligently beside their sparrow-colored mates through the whole nest-building procedure. Some of them actually build the entire nest while the female brings the materials, and even those that don't take part in the work stand beside their toiling mates and sing their lovely songs over and over again.

True, the grosbeaks spend only two or three days putting together their rather flimsy homes, but they make not the slightest attempt to be secretive. Even the females toss away their natural camouflage advantage by singing while they build, and when the brilliantly colored males take their turn at incubating the eggs, they are just as indiscreet, singing joyously for all the world to hear as they sit there warming their future hatchlings. Still, in spite of snakes, squirrels, jays, and other predators—not to mention the cowbirds who so frequently parasitize their nests—these careless, music-loving grosbeaks somehow manage to bring fair numbers of offspring into the world each year.

A male rose-breasted grosbeak sings day and night, even while incubating on the nest.

Recognition. 7–8½ in. long. Bill pale, stout, and conical. Male black above and on head, with white patches on wings; breast bright red; belly white. Female brown above with white patches on wings; breast streaked; eyebrow pale.
Habitat. Open, brushy woods; aspen groves.
Nesting. Nest is a flimsy saucer of twigs, stems, and grass, 5–50 ft. above ground in shrub or tree. Eggs 3–5, pale blue or green, with spots and blotches of brown. Incubation about 13 days, by both sexes. Young leave nest 9–12 days after hatching.
Food. Fruit, seeds, and insects.

Black-headed Grosbeak

Pheucticus melanocephalus

Any novice bird-banding assistant who unsuspectingly reaches into a net to grasp a black-headed grosbeak quickly discovers the power of this large finch's massive bill. One carelessly placed finger pinched in the unrelenting (and painful) grip of a handsome *gros bec*—the French hardly needs to be translated—is undeniable proof of its potent seed-crushing strength.

Bill shape and size are usually a good clue to a bird's diet and lifestyle. Sparrows, for example, have short, strong bills that serve well for crushing small seeds, while the tweezerlike bills of warblers and kinglets are designed for plucking insects and larvae from leaves. The woodpeckers' chisel-shaped bills help them chip into tree trunks in their search for grubs; mergansers' serrated bills aid in grasping the fish they catch; the sharply hooked bills of shrikes, hawks, and owls are suited to tearing the flesh of their prey; and the long, straight bills of American woodcocks have sensitive, flexible tips that allow the birds to search for worms deep in the soil.

The mammoth bills of black-headed grosbeaks enable them to dine on a broad range of foods, from soft berries and insects to large, hard seeds. And despite their huge beaks these birds are graceful eaters, as anyone who has watched a grosbeak carefully shelling one sunflower seed after another at a feeder can easily see.

Like males of other species, female black-headed grosbeaks fight fiercely for territories.

Recognition. 6–8 in. long. Bill pale, stout, and conical. Male black above and on head, with white patches on wings; underparts, collar, and rump bright orange-rusty. Female brown above with white wing patches; breast tinged with buff and streaked; eyebrow pale.
Habitat. Open woodlands, thickets, and parks.

Nesting. Nest is a flimsy saucer of twigs, stems, and rootlets, 4–25 ft. above ground in shrub or tree. Eggs 2–5, pale blue or green, spotted with brown. Incubation about 13 days, by both sexes. Young leave nest about 12 days after hatching.
Food. Seeds, fruit, and insects.

Evening Grosbeak

Coccothraustes vespertinus

Before the beginning of this century, evening grosbeaks lived and nested only in the conifers of northern Canada and the high mountains of the West. Then, early in the 1900's, isolated birds began to wander irregularly south and east for the winter. By 1916 small flocks began appearing erratically throughout the Northeast, and some of the wanderers eventually stayed behind to nest in eastern Canada and the northeastern United States. By the 1960's flocks of evening grosbeaks were patronizing winter feeders as far south as Maryland, and by the 1970's these vagabonds were wintering in Florida and the Gulf states.

No one knows exactly what precipitated the early wanderlust in these once stay-at-home birds, whether population pressure, a food shortage, or some other crisis. Whatever the cause, it affected only small segments of the population in any given winter. But the consequences of this tenuous beginning could be far-reaching. Winter is certainly less arduous in warmer climes, and the increasing prevalence of bird feeders well-stocked with protein-rich sunflower seeds makes survival even easier. It seems possible that evening grosbeaks, aided by carefully tended feeding stations, may evolve into regularly migratory birds—they might even take up nesting in the conifers of the Appalachian Mountains.

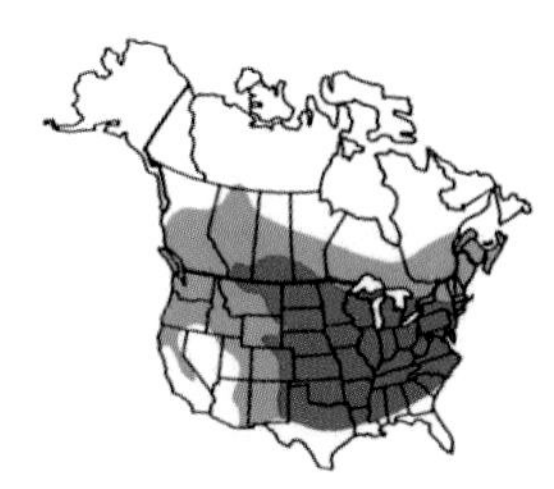

In winter, evening grosbeaks often land on roadways to pick up grit and eat salt crystals.

Recognition. 7–8½ in. long. Bill yellowish, stout, and conical. Male has brownish head, breast, and upper back; eyebrow, rump, and belly yellow; wings black with white patch. Female similar but much grayer.
Habitat. Coniferous forests in summer; wanders widely in winter.

Nesting. Nest is a loose cup of twigs and rootlets, 6–70 ft. above ground in dense foliage near tip of branch. Eggs 2–5, blue or greenish. Incubation 12–14 days, by female only. Young leave nest about 2 weeks after hatching.
Food. Seeds, berries, and insects.

Rufous-sided Towhee

also known as

Eastern Towhee

Pipilo erythrophthalmus

Families of rufous-sided towhees usually stay together throughout the summer.

Scratching about with one foot at a time, like barnyard poultry, rufous-sided towhees gather their food from the very bosom of the earth. It is little wonder, then, that these active birds used to be called ground robins, especially since their well-hidden nests of stems and leaves are nestled so close to the ground.

Today they are named towhees because that is the sound most people hear in their song. But they were once known as chewinks, from the call they make when alarmed. In some parts of the West that call is phonetically transcribed as *shrenk,* while in others it sounds more like the meowing of catbirds. In sections of the Southeast they unquestionably call *Louise!* And although their song is translated into English in most parts of the country as "chip towhee" or "drink your tea," the birds seem to drawl it in the South, and are apt to sing *chip chup chup zeeeeeee,* or just a buzzy *zeeeeeee*, in the West.

Their visual field marks too are almost as varied as their vocal ones. Rufous-sided towhees in Florida have white eyes instead of red ones, and those in the West have two white wing bars plus numerous white spots on their wings and backs. But no matter where they live, all rufous-sided towhees come flying in low to the ground with the side patches in their tails looking like white chiffon streamers trailing dramatically behind them.

Recognition. 7–9 in. long. Eastern male black on head, back, and breast; flanks rusty; belly white; white patches on wings and at corners of tail. Western male similar but spotted with white on back. Females like males but black areas brown or brownish gray.

Habitat. Forest edges, thickets, and shrubbery.

Nesting. Nest is a cup of grass, twigs, and rootlets concealed on ground. Eggs 2–6, cream-colored or greenish, spotted with brown. Incubation about 13 days, by female. Young leave nest 10–12 days after hatching.

Food. Insects, spiders, seeds, and berries.

Bachman's Sparrow

Aimophila aestivalis

Like many another modestly plumed bird, this sprightly sparrow is a greater joy to the ears than to the eyes. Certainly it is among North America's foremost songsters; some would argue that it is in a class by itself. From March to August its mellow song drifts ethereally through the pine and scrub oak woodlands of the southeastern United States in a delightful medley with the caroling of brown-headed nuthatches, eastern bluebirds, pine warblers, and eastern wood-pewees.

Choosing a perch at the top of a snag or at the tip of a broken pine branch, usually at heights of 10 to 30 feet, the Bachman's is a persistent singer, and when undisturbed will render long and intricate performances. Then the advantage is the birder's. Capturing a singing Bachman's in one's binoculars—its head thrown back, its throat muscles working, its chest rising and falling like a great tenor's—is one of the singular joys of bird watching.

At the other extreme is the Bachman's near silence and maddening elusiveness in cold weather. Many a Christmas bird count or friendly attempt to help a winter visitor add to his lifetime list has led to bone-chilling searches of a Bachman's habitat that ended in failure—even as the searchers felt sure that at least one of the thrilling summer choristers was quietly sitting there, watching every move.

Female Bachman's sparrows sometimes build domed nests with entrances on the side.

Recognition. 5¾ in. long. Reddish brown and streaked above and on crown; line through eye reddish; underparts plain grayish buff. Secretive and hard to see.
Habitat. Open pine or oak woodlands, brushy pastures, and palmettos.
Nesting. Nest is a cup of grass and weed stems concealed on ground under tuft of grass or shrub. Eggs 3–5, white. Incubation 12–14 days, by female. Young leave nest about 15 days after hatching. Sometimes 2 broods a season in southern states.
Food. Insects, spiders, and seeds.

Fox Sparrow

Passerella iliaca

The male fox sparrow usually sings from a well concealed perch in a dense thicket.

Without a doubt, fox sparrows are the aristocrats of the sparrow nation. Classified in a genus all their own, they can be seen at a glance to be larger and darker than other sparrows, and with a little closer observation it becomes apparent that anything the others can do, they can do better—even to the matter of scratching.

All sparrows scratch for a living. They scratch down into the leaves, the snow, the earth—scratching with both feet at the same time—to uncover whatever bounty may lie hidden there. But the fox sparrows are better equipped for the job. Their feet are large, with unusually long toes and claws that enable them to dig farther into the moist humus, where all sorts of seeds and minuscule animals may abound. Scattering leaves, snow, and soil with abandon, they scratch longer, deeper, and more profitably than any other sparrows.

It seems only fitting that the songs of these abundantly gifted birds are lovely carols filled with a melody and richness beyond those of many of their fellows. Short and varied, the songs usually rise in pitch at the beginning and drop on their closing lines. And fox sparrows are generous with their talent. They do not wait until they reach their nesting grounds in the thinly populated northern regions of Canada before bursting into song. Rather, their lilting rhapsodies fill the air along their entire migration route from the southern United States, a bouquet of spring music for all whose paths they cross on the way.

Recognition. 6½–7½ in. long. Stocky. Eastern birds mainly rich rusty, streaked on breast and belly, with bright rusty tail. Western birds darker or grayer brown; birds on West Coast have no rusty on tail.
Habitat. Brushy forests and thickets.
Nesting. Nest is a deep, solid cup of twigs, bark strips, rootlets, and moss, concealed at base of thicket or up to 7 ft. above ground. Eggs 3–5, pale blue or greenish, spotted with brown. Incubation 12–14 days, by female. Nestling period unknown.
Food. Seeds, berries, and insects.

White-throated Sparrow

Zonotrichia albicollis

The white-throated sparrow's song may not be the most glorious of sparrow melodies, but it is probably the best known simply because more people hear it. Although white-throats nest chiefly across the rural expanses of Canada, they winter through much of the eastern and southern United States—and they sing all winter long. They sometimes seem to be singing absentmindedly, it's true, but from March into May, when the last migrants leave, white-throats sing with real energy and purpose.

Their song is a simple one, often imitated by birders, and a passable rendition of it will bring the birds flitting through bushes in curious interest, craning their necks and peering about with bright eyes to discover the new arrival with the strange accent. White-throats and other small birds can also be called by "squeaking"—a sound made by a person kissing the back of his hand. Because the sounds are like those of a baby bird in distress, this kind of calling works especially well during the nesting season: parents come flocking, and birders have a chance to see species they might otherwise miss. But, understandably enough, squeaking also upsets the birds and can disrupt the care of their fledglings. Needless to say, therefore, it is a technique to be used with restraint and common sense.

White-throats often sing at night, especially when the moon is full.

Recognition. 6–7 in. long. Brown and streaked above; head striped with black and white or black and tan; throat patch white; patch in front of eyes yellow; breast plain gray. Forages mainly on ground.
Habitat. Brushy conifer or mixed woodlands.
Nesting. Nest is a cup of grass, twigs, and pine needles, concealed under vegetation on ground. Eggs 4–6, cream-colored, bluish, or greenish, spotted with reddish and brown. Incubation 12–14 days, by female. Young leave nest about 12 days after hatching. Rarely 2 broods a season.
Food. Seeds, berries, insects, and buds.

Dark-eyed Junco

Junco hyemalis

How can a single species include birds that vary so wildly in appearance? How can four juncos—formerly called slate-colored, Oregon, white-winged, and gray-headed—now be considered one species, the dark-eyed? Such questions reveal a fundamental dilemma of modern biology: how to classify organisms that live and change.

In the 18th century the famous Swedish botanist Carolus Linnaeus developed a system of Latin names for classifying living things. Birds were categorized primarily on the basis of outward appearance, with distinct plumages indicating different species. A century later Charles Darwin outlined a process of evolution whereby species were no longer fixed permanently but instead changed through time. It is with this concept that the problems of classification began.

In the case of juncos, we know that they have developed different plumages; but in behavior and calls they are similar, and several distinct types mate with one another freely. Will their differences grow more or less pronounced as time passes? It is hard to know. Perhaps, had Darwin lived before Linnaeus, we would have inherited a system that made clearer the subtle changes in a population as it develops from an existing species into a new one.

If danger threatens a junco's nest, the big-footed young can usually escape by running.

Recognition. 5–6½ in. long. Bill pinkish; eyes dark; belly and outer tail feathers white. Eastern (slate-colored) birds gray on head, breast, back, wings, and tail. Western birds similar, but back rusty or brown (gray-headed junco of Rocky Mountains), or back brown and head black (Oregon junco of West Coast).

Habitat. Open woodlands and clearings.
Nesting. Nest is a cup of twigs and grass on ground. Eggs 3–6, white or pale green, spotted with brown. Incubation about 12 days, by female only. Young leave nest 10–13 days after hatching. Often 2 or 3 broods a season.
Food. Seeds, insects, and berries.

Yellow-eyed Junco

Junco phaeonotus

The most conspicuous and active emissaries of nature we encounter day to day, birds occupy a special place in our lives. Beautiful, fascinating, endlessly varied—they are all this and more. But often as interesting as the birds themselves are our reactions to them, especially our impulse to read human qualities into their appearance and behavior. On that basis, we might well label the yellow-eyed junco a fierce bird. The baleful yellow eye of this primarily Mexican species seems to suggest a fiery Latin spirit, and surely the male's courtship antics represent pure avian machismo, as he struts about fanning his white-edged tail and holding it up at a sharp angle.

In fact, the yellow-eye is no fiercer than its dark-eyed cousins, but the effect of such visible traits is undeniable. So it is that we consider the gaze of a forward-looking peregrine falcon intense and fearless, while that of a sideways-looking mourning dove may seem blank or timid. A roosting nighthawk, its eyes half-closed to break up the outline, may look sleepy; a swan, with its long, fine neck, is graceful; and a red-winged blackbird, red epaulets flared, seems unmistakably important.

Our reactions to appearances are part of what makes watching birds so pleasurable—and being human so interesting. We should always bear in mind, of course, that those emotions are ours, not the birds'. Peregrines must have their fears, too.

After fledging, young yellow-eyes gather to feed in flocks of three dozen or more.

Recognition. 5½–6½ in. long. Bill dark above, pale below; eyes yellow. Mainly gray, with brown back; belly and outer tail feathers white. Walks on ground (dark-eyed junco hops).
Habitat. Coniferous mountain forests, usually at elevations of 6,000 feet or more.
Nesting. Nest is a cup of grass and moss, concealed on ground under shrub, log, or rock. Eggs 3–5, grayish or bluish, spotted with brown. Incubation about 15 days, by female only. Young leave nest about 10 days after hatching.
Food. Seeds and insects.

Orchard Oriole

Icterus spurius

Apple blossom time is the right time to watch for orchard orioles. These birds spend just a few weeks in North America to pair, nest, and fledge their young, passing most of the year in Mexico and points south. So quickly do they migrate from their breeding grounds that young birds have been seen arriving in Central America still in juvenile plumage, and adults still in molt, their post-nesting growth of new feathers incomplete.

Though orchard orioles sport the conspicuous plumage typical of their family, the Icteridae, they are not easily seen once courtship and nest building have been completed. Thus the best time for spotting orchard orioles is soon after the male's spring arrival, when he sings and displays territorially, often rising several feet from his perch as if propelled heavenward by the fervor of his song. This impassioned behavior continues until he is paired with a newly arrived female and the two begin to collect grass for their artfully woven nest.

As if to make amends to bird watchers for the brevity of their stay, orchard orioles usually choose nesting sites near settled areas, in belts of trees—especially orchards, as their name implies. There, with luck, a pair may be seen busily at work, weaving the long strands of grass into a nest suspended among the fragrant blossoms.

Besides insects and berries, these orioles eat apple blossoms and other fruit-tree flowers.

Recognition. 7–7½ in. long. Bill slender and straight. Male mainly black, with chestnut rump, wing bars, and underparts. Female olive-green above; underparts yellowish; 2 white wing bars.
Habitat. Streamside woodlands, orchards, and shade trees.
Nesting. Nest is a hanging cup of stems and plant fibers, 6–20 ft. above ground in tree. Eggs 3–7, pale blue or pale gray, spotted with brown, black, and purple. Incubation about 14 days, by female only. Young leave nest about 14 days after hatching.
Food. Insects and fruit.

Hooded Oriole

Icterus cucullatus

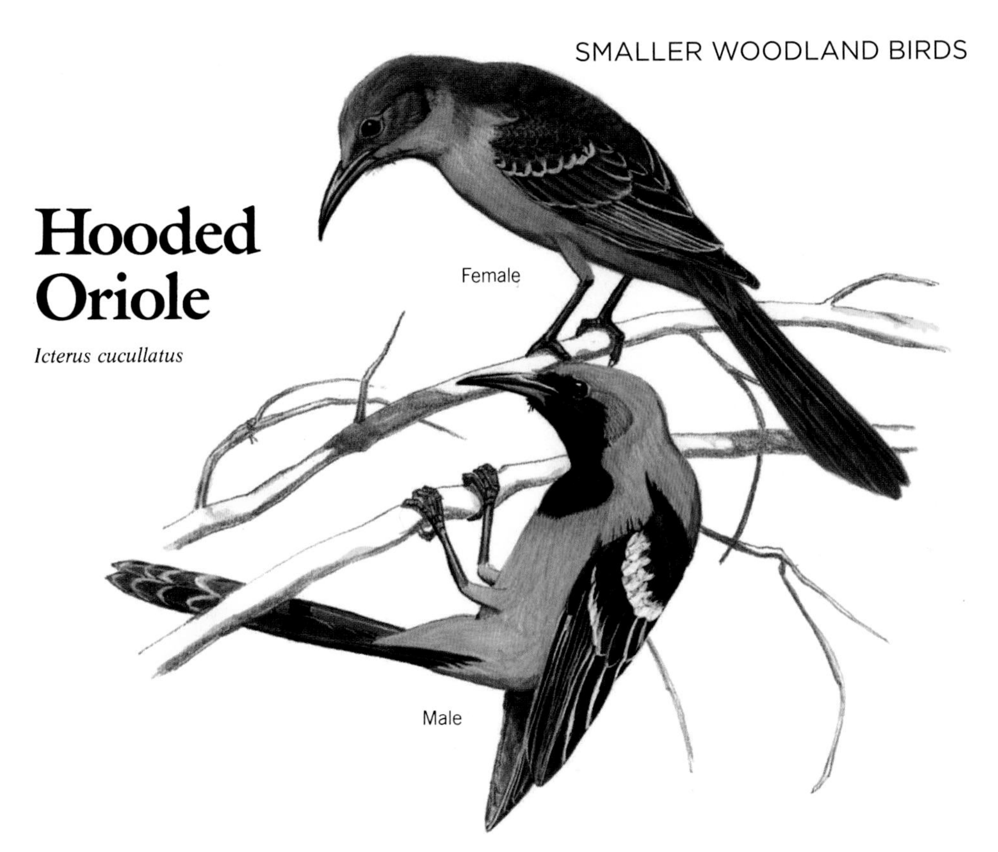

For all living things, the top priority of nature is pure and simple: to perpetuate the species. The flowers of many plants are designed expressly to attract some creature—a bird, a bee, a moth, even a bat—that will pick up pollen and pass it on to another plant of the same kind. Hummingbirds are especially helpful in this process, moving tirelessly among flowers to feed on nectar, all the while transporting pollen grains that stick to their bills and heads.

Hooded orioles are also fond of nectar, but they often claim a flower's reward without providing any transport service in return. With their fine bills they are able to "cheat" plants, piercing the bases of lilies, hibiscus, and other tubular flowers without ever touching the pollen. Not surprisingly, these orioles are frequent visitors at hummingbird feeders, but at least here no one is being shortchanged; there is enough sugar-water to go around.

The growing popularity of hummingbird feeders has been a special boon to the hooded orioles, allowing some to winter well to the north of their normal range. Human help has resulted in summer expansion as well. Some of the orioles' favorite nesting sites are in palms native to the southwestern United States, and extensive plantings of such trees in California have helped hooded orioles expand their natural breeding range northward.

Hooded orioles are often parasitized by a neighbor, the bronzed cowbird.

Recognition. 7–8 in. long. Bill slightly downcurved. Male bright orange-yellow, with black face, back, wings, and tail; 2 white wings bars. Female olive-green above; underparts yellowish; 2 white wing bars.
Habitat. Streamside woods and shade trees.
Nesting. Nest is a thin-walled cup of grass and plant fibers, 6–45 ft. above ground in yucca, palm, or tree. Eggs 3–5, very pale blue or yellowish, spotted with brown, purple, and lilac. Incubation about 14 days, by female only. Young leave nest about 14 days after hatching. 2 or 3 broods a season.
Food. Insects and nectar.

Northern Oriole

also known as

Baltimore Oriole

Icterus galbula

The Baltimore Orioles of major league baseball adopted their name (and their team colors) from a handsome bird dressed in flaming orange and black. That bird, in turn, had received its name in honor of George Calvert, Lord Baltimore, a 17th-century nobleman whose coat of arms bore the same striking colors. But bird names, like those of all flora and fauna, are supposed to reflect scientific fact. Consequently, in the mid-1980's ornithologists concluded that the Baltimore oriole, a bird commonly found east of the Rockies, and its western counterpart, the Bullock's oriole, are actually one species, and united them under a new name: the northern oriole.

The birds themselves, unconcerned with such issues, continue their seasonal cycles of migration, flying to the United States and Canada to nest and fledge their young, then back to Mexico and South America to spend the winter months. It is during bare-branched winter, in fact, that evidence of the northern oriole's summer presence is most readily found. Their signature is a beautifully woven hanging nest, suspended pouch-fashion from the very tips of branches. When the leaves are off the trees, these gently swaying fiber baskets are easily seen, usually high up, and for bird watchers such lovely confirmation of the northern oriole's nesting success is as satisfying as a home run.

The Bullock's (above) and Baltimores look slightly different but interbreed in the Midwest.

Recognition. 7–8½ in. long. Eastern male black with orange underparts, rump, shoulders, and sides of tail; 2 white wing bars. Western male similar but face and eyebrow orange; wing patch white. Females olive above; yellowish below; 2 white wing bars. Western female paler than eastern; belly whitish.

Habitat. Open woodlands and shade trees.
Nesting. Nest is a hanging pouch of plant fibers and string, 6–90 ft. above ground. Eggs 3–6, pale blue, scrawled with black and gray. Incubation 12–14 days, by female. Young leave nest 2 weeks after hatching.
Food. Insects and fruit.

Pine Siskin

Carduelis pinus

A novice bird watcher who puts out a winter bird feeder for the first time, only to have a flock of pine siskins descend upon it, might well be discouraged. So many streaky little brown birds with no outstanding marks—why not some more beautiful ones to start with? But longtime feeder-tenders, especialy those in areas with long, chilly winters, take heart at seeing a swarming flock of siskins alight for a meal. Their chattering and buzzing energy brings life to a cold gray day.

As quickly as siskins appear at a winter feeder, however, they may also disappear. For like the nomadic crossbills, siskins have no fixed migrating patterns, and their presence in any location varies dramatically from year to year. And it is not only in winter that siskins are unpredictable. While most birds show a strong tendency to return to the same breeding area year after year, pine siskins may nest in an area one year and abandon it completely the next, apparently in response to the local food supply.

Pine siskins are a lively, gregarious lot. They also have a feisty side: despite being the smallest of our winter finches, siskins hold their own at feeders against many larger birds. They often flock with their more colorful cousins the American goldfinches, next to whom they may appear a bit drab. But pine siskins have a concealed beauty of their own, a handsome yellow stripe visible only when their wings are open—a bright but fleeting field mark for bird watchers with eagle eyes.

The opportunistic pine siskin sometimes feeds at holes already drilled by sapsuckers.

Recognition. 4½–5½ in. long. Brown and streaked; underparts usually only lightly streaked; yellow patches on wings and tail. Some birds heavily streaked, with less yellow on wings.
Habitat. Coniferous woodlands, thickets of birch and alder trees.
Nesting. Nest is a large cup of twigs, grass, and rootlets, 3–50 ft. above ground, usually in conifer. Eggs 3–5, pale blue-green, spotted with black and lilac. Incubation about 13 days, by female only. Young leave nest about 15 days after hatching.
Food. Seeds and insects.

Purple Finch

Purple Finch *Carpodacus purpureus*
Cassin's Finch *Carpodacus cassinii*

Purple Finch

Cassin's Finch

Perhaps the best known of our "winter finches," purple finches bring a dash of bright color and a rich, bubbling song to parks and woodlands across the eastern United States, most of Canada, and the Pacific states. Much of their range is shared by two close relatives, house finches and Cassin's finches, and the resulting problems of identification have bedeviled bird watchers from coast to coast. In particular, separating a Cassin's from a purple finch requires a closer look at bird "topography."

For experienced birders, topography involves creating a mental map of each bird by dividing its body into separate areas. To help beginners, field guides usually contain a diagram labeling the different parts of a bird's body. By consulting the diagram and the field guide's text on purple and Cassin's finches, for instance, a bird watcher will learn that the back is reddish in a male purple finch but brownish in a male Cassin's, and that the head pattern of eyebrow, cheek, and mustache is more prominent in the female purple than in the female Cassin's.

Such are the keys to sorting out similar species—clearly not a skill easily mastered. But for anyone who has sampled the pleasures of bird watching, becoming familiar with their topography is an essential step toward "seeing" the birds more knowledgeably, and thus enjoying them more.

Recognition. 5½–6½ in. long. Streaked, with notched tail. Male with red head, back, rump, and breast. Female brown, heavily streaked below; eyebrow, cheeks, and whisker mark whitish. Male Cassin's similar, but browner back contrasts with red crown; female lacks distinct face markings.

Habitat. Woodlands, gardens with trees, parks.
Nesting. Nest is a cup of twigs and grass 6–50 ft. above ground, usually in conifer. Eggs 3–6, pale greenish, spotted with brown and black. Incubation about 13 days, by female. Young leave nest about 14 days after hatching.
Food. Seeds and fruits; insects in summer.

Pine Grosbeak

Pinicola enucleator

For most of us, pine grosbeaks are only occasional cold-weather visitors, birds that from time to time appear at feeders far south of their normal range. But for those who know their regular haunts, "mopes" (as Newfoundlanders often call these frequently motionless birds) are the embodiment of the northern woodlands. To encounter a small group of these beautiful creatures quietly feeding in a spruce tree, their soft calls further muffled by a blanket of snow, is one of the special pleasures of winter.

In plumage, pine grosbeaks resemble white-winged crossbills, but the two differ in size, proportions, and, most importantly, bill shape. Pine grosbeaks do not have the distinctive bills that enable crossbills to pry open cones and extract their seeds. Instead, these cosmopolitan birds enjoy more varied fare, eating the seeds, buds, and berries of many northern trees and shrubs.

Pine grosbeaks delight viewers not only on our continent but across much of the Northern Hemisphere. Their distribution, like that of many far-northern birds, is circumpolar. Willow ptarmigans, great gray owls, northern hawk-owls, three-toed woodpeckers, redpolls, crossbills, and Lapland longspurs are but a few of "our" species that make themselves equally at home in the northern reaches of Europe and Asia.

In winter, mountain ash berries are one of the pine grosbeak's favorite meals.

Recognition. 8–10 in. long. Bill stubby and conical. Male pink, with 2 white wing bars. Female gray with 2 white wing bars; head and rump tinged with olive. Tame; usually seen in small flocks that feed quietly.
Habitat. Coniferous forests.
Nesting. Nest is a loose cup of twigs and rootlets, 2–30 ft. above ground in birch or conifer. Eggs 2–6, pale greenish, spotted with brown and gray. Incubation about 14 days, by female only. Young leave nest about 20 days after hatching.
Food. Seeds, berries, nuts, and insects.

White-winged Crossbill

Loxia leucoptera

To appreciate the special abilities of the red and white-winged crossbills, pick up an unopened pine cone and try pulling out a seed or two. It's no easy trick—even when both hands, all ten fingers, and a liberal dose of elbow grease are applied. The tough scales protecting the seeds press tightly against one another, and the seeds are hidden deep inside the cone. But to crossbills, pulling them out is child's play. Those seeds are their primary food, and both red and white-winged crossbills can open the cones with such speed that anyone standing below a feeding flock will be showered by falling scales and empty stems.

Though the overlapping tips make the crossbills' mandibles look like gardening shears, the bill is actually used more as a probe and lever than as a cutting device. The birds hang head-down over the end of a limb to work at the cone from above, or else yank it loose, carry it to a solid perch, and pin it down with their feet. Then they insert their bills between the scales and use the mandibles to hold the scales open while their flexible tongues lift out the seeds.

Despite these specialized tools, crossbills see no reason to limit the variety of their diet. In addition to the seeds of conifers, white-wings regularly consume berries, sunflower seeds, aphids, caterpillars, and other delicacies.

Unlike a red crossbill, the white-wing will readily dig into cones fallen to the ground.

Recognition. 6–6¾ in. long. Male pink; wings and tail black; 2 white wing bars. Female grayish-olive, streaked; 2 white wing bars. Crossed bill hard to see in field. Feeds in small flocks; hangs upside down like parrot at pine cones.
Habitat. Northern coniferous forests, especially spruce and fir.

Nesting. Nest is a cup of twigs and grass 3–70 ft. above ground in spruce or fir. Eggs 2–5, pale blue or greenish, spotted with brown and black. Incubation about 14 days. Nestling period unknown.
Food. Seeds, berries, and insects.

Red Crossbill

Loxia curvirostra

The crossbills are the most unpredictable members of what is probably the least predictable group of birds in North America. All the northern finches—a group that also includes evening and pine grosbeaks, common and hoary redpolls, pine siskins, and purple finches—are notoriously errant migrants. A majority of these species spend most winters within their breeding ranges in the coniferous forests of Canada or the northern and western United States. But every three or four years—or sometimes six or seven—various finch species come south in large numbers, an irregular population shift known as an irruption. And those capricious creatures seem as reliable as Swiss clocks compared to the fickle crossbills, who irrupt southward only once every 15 or 20 years.

Like none of our other finches—and very few birds anywhere in the world—the crossbills are true nomads. They set up temporary colonies wherever they find enough food, and seem unconcerned with normal seasonal rhythms, frequently mating and raising their young in the dead of winter. Clearly, a nest one year is no promise that these habitual vagrants will return the next. The first nest of red crossbills ever discovered in the United States was found in New York City in April 1875. New York's bird watchers are still waiting for the second one.

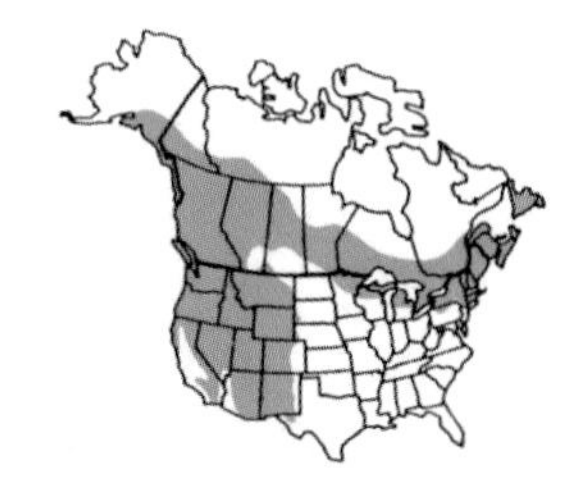

A bumper crop of cones often spurs red crossbills to start nesting in midwinter.

Recognition. 5½–6½ in. long. Male dull red; wings and tail blackish; no wing bars. Female grayish-olive; wings plain and dark. Crossed bill hard to see in field. Feeds in flocks; hangs upside down like parrot at pine cones.
Habitat. Pine forests.
Nesting. Nest is a cup of twigs and bark strips, on horizontal branch of conifer 5–80 ft. above ground. Eggs 3–5, bluish with brown and black spots. Incubation about 14 days, by female only. Young leave nest about 17 days after hatching. Sometimes nests in January.
Food. Mainly conifer seeds; a few insects.

Ducklike Birds

As a group, the larger water birds of North America—the ducks geese, and other wide-ranging families—may be our best-known birds. From multicolored wood ducks to majestic swans, from freshwater anhingas to seagoing puffins, they are unsurpassed in their diversity of appearance, behavior, and habitat. And they surely provide some of our richest, most evocative images of nature—whether great flocks of Canada geese passing high overhead, a solitary mallard mother gliding across a pond with her brood close behind, or the cry of a loon echoing in the wilderness.

362	Redhead	**377**	Common Merganser
363	Canvasback	**378**	Red-breasted Merganser
364	Greater Scaup	**379**	Clapper Rail
365	Lesser Scaup	**380**	Virginia Rail
366	Common Eider	**381**	Sora
367	King Eider	**382**	Purple Gallinule
368	Harlequin Duck	**383**	Common Moorhen
369	Oldsquaw	**384**	American Coot
370	Black Scoter	**385**	Dovekie
371	Surf Scoter	**386**	Common Murre
372	White-winged Scoter	**387**	Black Guillemot
373	Common Goldeneye	**388**	Marbled Murrelet
374	Bufflehead	**389**	Rhinoceros Auklet
375	Ruddy Duck	**390**	Tufted Puffin
376	Hooded Merganser	**391**	Atlantic Puffin

Red-throated Loon

Gavia stellata

Few outside the Arctic ever see the red throat of a red-throated loon. This handsome emblem appears only during the mating season and has disappeared by the time migrating birds reach their coastal wintering areas in October. It may still be absent when the birds head north in April. Most people know the red-throat in winter dress only, slim and gray with a star-studded back.

This smallest of the loons is also the most agile, at least in an aquatic environment. A loon's feet are positioned well to the rear of its body—perfect for swimming, but ill-suited for walking. Walking, in fact, is so difficult that for short distances loons may simply push themselves along on their bellies. Getting airborne is no easy task for most of them, either. Common and Arctic loons need open water with lots of room for a running start in order to take off. If either of these lands on a wet parking lot, mistaking it for a dark body of water, the bird will be grounded, unable to take flight. Only the red-throat can launch itself from land and escape this trap.

In spring, even before the shallow tundra pools are freed from their sheath of winter ice, red-throated loons reach the Arctic. Arctic foxes and jaegers take a toll in eggs and young birds, but the red-throat is a dogged defender, and its daggerlike bill a weapon that commands respect.

The lightweight red-throat is the only loon able to take off directly from the ground.

Recognition. 24–27 in. long. Bill slender and uptilted. Breeding adult has plain back; head gray, throat dark rusty. Winter adult pale, with less contrast between upperparts and underparts than in other loons.
Habitat. Oceans, bays, and inlets; nests on tundra lakes.

Nesting. Nest is a muddy depression or large mound of wet plants or mud at edge of pond or lake. Eggs 1–3, olive or brown. Incubation 24–29 days, mainly by female. Young downy, leave nest soon after hatching; fly in 6 weeks.
Food. Fish, crustaceans, and insects.

Pacific Loon

Gavia pacifica

Until recently this sturdy diving bird was known as the Arctic loon *(Gavia arctica)* and was thought to be biologically indistinguishable from the Arctic loons of Europe and Asia. But ornithologists now believe that the Arctic and Pacific loons are two distinct species. Still, only a fine line separates them. In truth, this bird is both an "Arctic" and a "Pacific" loon, each name reflecting an important part of its double-sided life.

From May to September, Pacific loons reside in the Arctic, west of Hudson Bay. Shortly after the ice retreats, adult birds establish themselves in one of the tundra lakes, where they court, nest, and raise their young in double time—as quick as an Arctic summer. By October, when winter returns, the Pacific loons have taken up residence in coastal waters from Alaska to Baja California, only a handful making it to the East Coast each winter.

Small fish are the mainstay of any loon's diet, and the Pacifics are marvelously adept at catching them. Beneath the water, loons' body feathers compress, forcing out air and reducing drag. With powerful thrusts of their rear-set legs, these sleek projectiles fairly fly underwater. When they surface, droplets running in rivulets off their waterproof backs, the Pacific loons triumphantly throw back their heads and swallow their fish headfirst and whole.

Pacific loons hatch a day or more apart, so the older bird is larger than its nestmate.

Recognition. 23–29 in. long. Bill slender and horizontal. Breeding adult pale gray on crown and back of neck; back black with large white spots; throat black with white stripes. Winter adult has white underparts, dark upperparts.
Habitat. Open ocean, bays, tidal channels; nests on Arctic lakes.

Nesting. Nest is a muddy depression or mound of wet plants or mud at shore of lake. Eggs 1–2, olive or brown with dark spots. Incubation 28–30 days, by female only. Young downy, leave nest soon after hatching; fly in 6 weeks.
Food. Fish, crustaceans, and insects.

Common Loon

Gavia immer

From the foggy surface of some nameless northern lake, the common loon calls. Some liken it to a cry, others to a yodel or a laugh. However it is perceived, no other sound in nature so typifies the great American wilderness. The shivering cry cuts easily through fog—and through the polished veneer of civilization, awakening feelings in the human soul that go deeper than memory.

It's a large, heavy bird, but in its aquatic element the loon swims and dives with consummate grace. Disappearing beneath the surface with hardly a ripple, a common loon may plunge to depths exceeding 150 feet and remain below for a minute or more in search of small fish. Cold, clear water, an ample food supply, and a measure of solitude are the qualities this bird seeks in a lake. Its bulky nest is placed on or near the water, where newly hatched young begin to swim and dive in just a day or two, and by October cold winds have propelled adults and their fully fledged young to the coasts. Indeed, many young loons appear to spend their entire first summer at sea.

Over much of North America the common loon population is declining. The shrinking of America's wilderness is partly to blame, but the contamination of northern lakes by acid rain is also thought to be a factor. In coastal waters loons encounter floating oil, which coats feathers, destroying their buoyancy and insulating properties. Oil-fouled birds drown, die of exposure, or are poisoned by the oil they swallow trying to clean their blackened feathers.

During its first few weeks, a young common loon usually travels on its parents' backs.

Recognition. 28–36 in. long. Bill stout and horizontal. Breeding adult has black back checkered with many white spots; head and neck black with white necklace. Winter adult larger than other loons; bill pale; dark crown and hind neck shade into white throat.
Habitat. Oceans, bays, tidal channels, and lakes; nests on northern lakes; rarely winters inland.
Nesting. Nest is a muddy depression or mound of plants or mud at shore of lake. Eggs 1–3, olive or brown with dark spots. Incubation about 29 days, by both sexes. Young downy, leave nest soon after hatching; fly at 12 weeks.
Food. Fish, crustaceans, frogs, and insects.

Pied-billed Grebe

Podilymbus podiceps

"Look there! Is that some kind of . . . ?" But it's too late. The bird is gone. Where a snaky head and neck seemed to protrude amid the lily pads just a moment ago, there is nothing now, not even a ripple of evidence. An illusion? Imagination? No, just the "water witch"—the common and endearingly elusive pied-billed grebe.

In less than a minute, the periscope head of this puckish bird will reappear—somewhere. Convinced that no danger is imminent, the ringed-billed wraith of the reeds may even bob to the surface, water running off breast feathers as silky as otter fur. But it takes only an incautious movement, just a pointing finger or raised binoculars, to send the pied-bill under again. Commercial hunters, who a century ago shot the bird for feathers to adorn women's hats, swore it could dive at the flash of the muzzle and be safely submerged by the time the shot struck the water.

Flight seems something of an afterthought to the pied-billed grebe, a last resort. Takeoffs require both an aquatic runway and a flapping, foot-splashing start. And once airborne, the bird labors mightily to stay there. Fortunately, it doesn't have to fly too often. Over much of its range, the pied-billed grebe is a permanent resident of freshwater ponds and marshes. In winter, northern pied-bills retreat only far enough to find open water. Occasionally they misjudge the reach of winter and get caught on a freezing lake. When this happens, they are grounded by the ice—and will be lucky to see the next thaw.

Downy young pied-bills sporting their bold stripes present a sharp contrast to the adults.

Recognition. 12–15 in. long. Stocky; dull brown; feathers under tail white. Bill short and pale, with dark ring. Dives frequently; often hides in aquatic vegetation with only bill above surface.
Habitat. Ponds, marshes, and quiet streams.
Nesting. Nest is a floating mass of marsh plants and mud anchored to reeds or bushes. Eggs 2–10, pale bluish or greenish, but soon stained with mud. Incubation about 23 days, by both sexes. Young downy; leave nest soon after hatching. Sometimes 2 broods a season.
Food. Insects, crustaceans, and fish.

Horned Grebe

Podiceps auritus

From November to April this grebe wears the colors of a winter sea—storm-gray and tarnished silver. But as the sun grows stronger, the chemistry of spring works a wondrous transformation. The bird's pale throat and sides turn to chestnut. Gaunt white cheeks fill out, becoming sleek and black. Plumes that sprout from its temples make it look like some golden-winged Mercury of the marsh. But whether the horned grebe is clad in winter drab or spring finery, its eye is always scarlet. In all the world, no eye of any other species of bird seems quite so red.

Like those of other grebes, the nest of this marsh dweller is a masterpiece of function and design. It floats! Tucked among the reeds at the edge of a prairie pond, anchored by the rushes to keep it from drifting, the bulky nest bobs like a little dinghy. And maintenance is easy: as the nest looses buoyancy, more plant matter is simply added to the pile.

Soon after hatching, the young grebes take to the water. Though capable swimmers, they will be happy to accept a ride on a parent's back. And if danger threatens, the chicks instinctively know enough to dive. Propelled by powerful kicks from their oversized feet, they plunge headfirst into a bed of algae or secure themselves below by grasping a submerged stalk. Presently, the danger passed, they bob to the surface again, like a handful of downy corks.

The grebe protects its stomach from fish bones with a lining made of its own feathers.

Recognition. 12½–15 in. long. Bill short and thin. Breeding adult has black head, with bushy golden ear tufts; neck and sides rusty; back black. Winter adult has black crown and back; white cheeks, throat, and breast.
Habitat. Marshes, ponds, and slow streams.
Nesting. Nest is a floating mass of marsh plants and mud anchored to reeds or bushes. Eggs 3–6, pale bluish or greenish, but soon stained with mud. Incubation about 25 days, by both sexes. Young downy; leave nest soon after hatching. Occasionally 2 broods a season.
Food. Fish, crustaceans, and insects.

Red-necked Grebe

Podiceps grisegena

Up and down the Atlantic seaboard, birders and baymen can tell when the Great Lakes freeze up without referring to the daily weather map. Suddenly, red-necked grebes reach the coast in numbers. Winter ice has finally squeezed the last lingering birds out of their freshwater hideouts.

The red-neck is a large, stocky street-brawler of a grebe. It likes its breeding marshes deep and its wintering areas salty and rough. In summer, with its rufous throat and broad white cheek patches, this grebe is a handsome bird, the dandy of the freshwater marsh. In winter, when birds don the cold gray colors of the sea, the "dirty-necked" grebe may be distinguished from the similar horned grebe by its size and shape, as well as by its dusky throat.

During migration or in winter, if there is excess food red-necked grebes may occasionally be found in small groups. But for the most part they are solitary birds. Scattered among the eiders, oldsquaws, and scoters that enliven the rocky shores of New England, red-necks can be found diving for fish. If there are several birds in the same area, they maintain a respectful distance from one another and show even less tolerance for would-be observers. Taken by surprise, one of these burly birds will dive quickly and speed off, a shimmering white streak cleaving a path through gray-green water—then in less than a minute pop safely to the surface, many yards away.

Mainly fish-eaters, red-necks also take crayfish and insects in lakes where fish are scarce.

Reoognition. 17–22 in. long. Bill long and pointed. Breeding adult has rusty neck; cheeks whitish; bill yellowish. Winter adult dull gray with grayish cheeks.
Habitat. Marshes, lakes, and ponds; coastal bays in winter.
Nesting. Nest is a floating mass of marsh plants and mud anchored to reeds or bushes. Eggs 3–6, pale bluish, but soon stained with mud. Incubation about 23 days, by both sexes. Young downy; leave nest soon after hatching.
Food. Fish, insects, and crustaceans.

Eared Grebe

Podiceps nigricollis

They seem a bit crazed, these eared grebes. Their ear tufts are eccentric and unkempt. The red eyes appear unfocused, ablaze with a mad inner light. Courting birds fairly churn the surface of a marsh into foam with their frenzied dancing, and whole nesting colonies may suddenly break into a chorus whose volume and persistence reminded one observer of "a first-class frog pond in March."

Even by grebe standards, the nesting habits of this species are unorthodox. Their floating nests are thoughtless affairs—piles of stalks and stems, plotted with little apparent concern for concealment. Sitting amid the flotsam, a nesting bird simply molds the material around itself until the semblance of a nest is formed. Then, as if overcome by second thoughts, it abandons the nest and begins work on a new one. This may happen several times, and in a large enough colony the marsh may be filled with so many active and abandoned nests that it resembles an enormous salad bowl.

Colonial nesting and unwary habits made eared grebes a prime target for commercial hunters during the late 19th century. The bird's plumage was in high demand in the fashion industry, and the feathered skin of one eared grebe's breast fetched 20 cents in New York—a substantial sum in those days. As a result, many thousands of nesting birds were destroyed. Happily, under protection, this wonderfully eccentric water bird has made a strong recovery.

When it leaves its nest, an adult eared grebe covers the eggs with wet vegetation.

Recognition. 12½–13½ in. long. Bill short and thin. Breeding adult mainly black, with black crest; wispy golden ear tufts. Winter adult mainly blackish, with black crown and white patch behind ears. Often seen in flocks.
Habitat. Marshes, ponds, and slow streams.
Nesting. Nest is a floating mass of marsh plants and mud anchored to reeds or bushes. Eggs 3–9, whitish, but soon stained with mud. Incubation about 22 days, by both sexes. Young downy; leave nest with parents soon after hatching. Usually nests in colonies. Sometimes 2 broods a season.
Food. Insects, fish, and crustaceans.

Western Grebe

Western Grebe *Aechmophorus occidentalis*
Clark's Grebe *Aechmophorus clarkii*

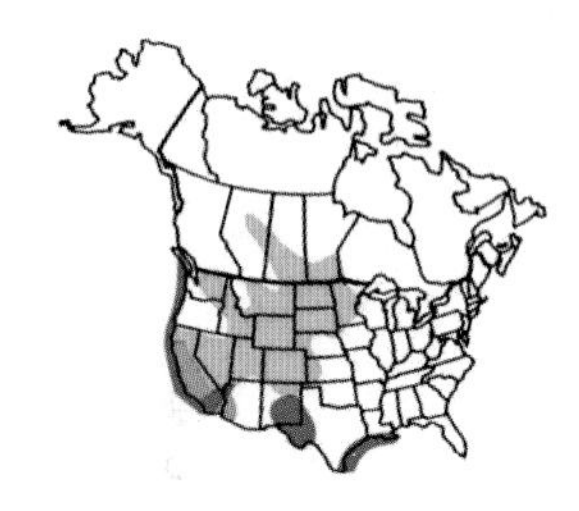

It's May, and the pageant of spring is in full swing across the prairies. Over lakes and marshes mallard and pintail drakes race to win a female's favor. Black terns glide in tandem, and the territorial chorus of yellow-headed blackbirds makes the bulrushes ring. But among all those vying for recognition, the western grebe is clearly best in show. In all categories—grace, poise, innovation, and skill—the courtship routine of this bird rates a perfect 10.

As courtship begins, western grebes swim side by side, arching their long, graceful necks in a backward bow. Suddenly, the birds spring upright, lobed feet pattering across the lake, their necks bowed in demure S-shaped curves. At the conclusion of this reckless dash, the pair drops lightly to the surface and glides. But this is merely a warm-up. Swimming toward each other now, the birds dive . . . then emerge, rearing high above the surface, breast touching breast, a sprig of moss clenched in each upturned bill. After pirouetting in tandem—one, two, three times—the birds finally part and settle to the surface.

A close relative of the western grebe (and only recently made a separate species) is the Clark's grebe. The two birds are as similar in range as they are in appearance, nesting in colonies among the reeds of freshwater marshes and wintering on coastlines, bays, and inland lakes.

Except for the white cheek feathers encircling its eye, the Clark's is a near twin of the western grebe.

Recognition. 22–29 in. long. Bill long and thin; neck long, slender, and black and white. On western, bill is dull yellow-green; black crown extends below eye. On Clark's, bill is bright yellow; black crown does not extend below eye.
Habitat. Prairie marshes and open bays.
Nesting. Nest is a floating mass of marsh plants and mud anchored to reeds or bushes. Eggs 3–4, pale bluish or buff, but soon stained with mud. Incubation about 23 days, by both sexes. Young downy; leave nest with parents soon after hatching. Nests in large, noisy colonies.
Food. Fish, crustaceans, and insects.

Double-crested Cormorant

Phalacrocorax auritus

It is almost as though nature had played a practical joke in designing the double-crested cormorant. Here is a bird unlikely ever to win any beauty contests—and a bird whose habits are scarcely more appealing than its looks. It feeds its young regurgitated food. Its slovenly nest may include such items as pocket combs, bobby pins, and plastic cutlery. The only sound it makes is a vaguely piglike grunt. Hardworking fishermen have long resented the double-crest as a much too successful rival: it can achieve spectacular speeds on (or under) the water once it spots its next meal, accelerating with tandem thrusts of its large webbed feet and powerful wings. Finally, the odor that arises from an established cormorant colony roosting on a hot afternoon is decidedly not for sensitive noses.

Nevertheless, at nesting time this same double-crested cormorant performs a ceremonial changing of the guard that is one of the most elegant, appealing rituals in the avian kingdom. When a sitting bird is ready for a break, the other member of the mated pair flies to the nest, walks regally around the sitting partner, nudges it tenderly, and finally tucks its head under the other's wing. The sitting parent then flies off, sometimes joining its fellow birds in a line of rising and dipping low-altitude flight—a scene captivating enough to persuade the poetically inclined observer that a writhing sea monster is passing by.

A cormorant's large feet, with all four toes joined by webbing, give it extra speed underwater.

Recognition. 29–36 in. long. All black, with small orange-yellow throat patch. Long tail. Swims with bill pointed upward. Flies in long lines and wedges.
Habitat. Seacoasts, harbors, lakes, and wooded swamps where fish are plentiful.
Nesting. Nest is a mass of sticks or seaweed on rock ledge or up to 50 ft. above ground in tree. Eggs 2–9, chalky blue. Incubation about 25 days, by both sexes. Young leave nest 5–6 weeks after hatching; become independent at 10 weeks. Nests in colonies.
Food. Fish and some crustaceans.

Anhinga

Anhinga anhinga

A silent denizen of silent places, the prehistoric-looking anhinga belongs to hidden swamps and seemingly primeval ooze. There are only four anhinga species worldwide, and only one belongs solely to the Western Hemisphere. It is actually a tropical bird: those found in Florida and other southern states are living at the northerly limit of their range. The body plumage of an anhinga is so thick and compact that it appears to be clad in fur rather than feathers, and it clambers clumsily in trees and bushes on the large, sharply clawed, fully webbed feet of a water bird.

The anhinga is a heavy bird with solid bones, but internal air sacs allow it to float high on the surface of the water or let itself sink into the depths—almost without a ripple. Usually rising only far enough above the water for its head and long neck to be exposed, it pursues fish or frogs, water snakes, or young alligators with surprising agility. Submerging, it impales a victim on its sharp beak, then returns to the surface and tosses the unlucky creature into the air to be caught and swallowed headfirst.

As supple in water as it is clumsy on land, the anhinga is most serenely at home in the air. When not otherwise engaged, it spirals upward to breathtaking heights. There, with others of its kind, it rides the wind for hours on end—for no apparent reason other than the simple joy of soaring.

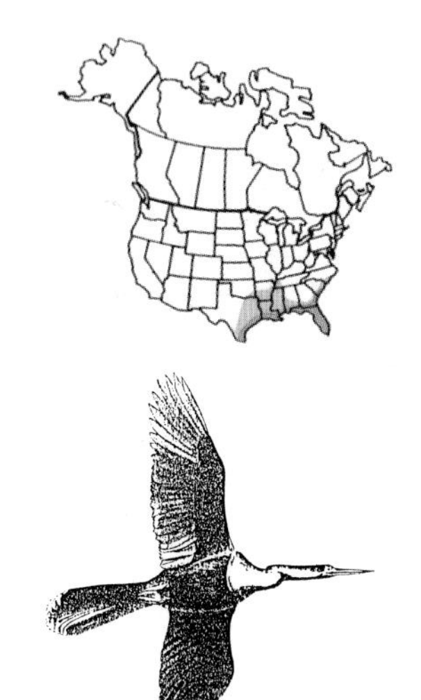

Unlike the cormorants, anhingas use their long wings to soar on high-altitude wind currents.

Recognition. 32–36 in. long. Mainly black, with long slender neck, long pointed bill, and long fan-shaped tail. Male has white plumes on back; female has buff head, neck, and breast.
Habitat. Swamps, marshes, ponds, and rivers with surrounding woodlands.
Nesting. Nest is a small mass of twigs, lined with green foliage, up to 40 ft. above water in bush or tree; sometimes uses old heron's nest. Eggs 1–5, chalky blue. Incubation 25–28 days, by both sexes. Nestling period unknown. Sometimes nests in colonies, often with herons.
Food. Fish, frogs, and crustaceans.

Tundra Swan

Cygnus columbianus

Yes, the tundra swan does indeed sing a beautiful and haunting death song. In 1898, Daniel G. Elliot, a noted authority on ducks, swans, and geese who knew every sound a tundra swan ordinarily uttered, wrote of having been with a hunting party on Currituck Sound, North Carolina, when a member of his group shot and mortally wounded a swan flying overhead. The swan set its wings and, Elliot wrote, "sailing slowly down, began its death song, continuing it until it reached the water nearly half a mile away." The song was not like any other swan note he had ever heard. Elliot inquired among local hunters and found that they too had heard that sad and beautiful song as a dying swan fell through the air.

Then, in 1955, H. A. Hochbaum, a scientist who specialized in the study of waterfowl, observed that before they take off into the air, tundra swans always sing what he chose to call a "departure song." The noted authority John K. Terres later described this departure song as "one of the most beautiful utterances of waterfowl—a melodious, soft, muted series of notes . . . ." Hochbaum himself believed that the departure song was "probably the swan song of legend, for when a swan is shot and falls crippled to the water, it utters this call as it tries in vain to rejoin its fellows in the sky."

Tundra swans nest early in spring, when snow and ice still hamper human travel.

Recognition. 47–58 in. long. Large, white, and long-necked. Neck held straight up, not curved. Bill black, with small yellow spot in front of eye. Young tinged with dusky; bill pinkish. Usually travels in flocks; call notes musical.

Habitat. Lakes, bays, and estuaries; nests on tundra or in marshes.

Nesting. Nest is a mound of leaves and grass, near edge of water. Eggs 2–7, dull white. Incubation about 32 days, by female only. Young downy; leave nest soon after hatching; stay with parents until following spring.

Food. Aquatic plants and mollusks.

Trumpeter Swan

Cygnus buccinator

Trumpeter swans had all but vanished by 1912, when the ornithologist Edward Howe Forbush wrote sadly, "The trumpetings that were once heard over the breadth of a great continent will soon be heard no more." Naturalists had long railed against the killing. But hunters found these large and beautiful birds irresistible, slaughtering them for their down and for their skin, which was made into powder puffs. To make matters worse, gourmets had developed a fondness for the swan's eggs, and there were no game laws to limit the massacre.

Luckily, just as the species seemed lost, with fewer than a hundred trumpeters known to exist, a national conservation policy was beginning to emerge. A few pairs of the birds had survived in Yellowstone National Park, protected by the Lacey Act of 1894, which outlawed hunting inside park boundaries. Swans nesting outside the park received protection in 1916 under the Migratory Bird Treaty. Then, in 1935, the federal government designated some 22,000 acres of prime trumpeter nesting habitat in southwestern Montana as the Red Rock Lakes Migratory Waterfowl Refuge. Slowly, the great birds came back. Several hundred now nest at Red Rock Lakes and several thousand throughout North America, their deep, resonant calls trumpeting a major conservation victory.

Trumpeter swans use their long necks to reach food on the bottom of marshes and lakes.

Recognition. 59–72 in. long. Large, white, and long-necked. Neck held straight up, not curved. Bill black, without yellow spot in front of eye. Young tinged with dusky; bill pinkish. Usually seen in small parties; call deep and trumpetlike.
Habitat. Mountain lakes and rivers; some winter on coastal bays.

Nesting. Nest is a mound of plant material, near edge of water. Eggs 2–13, cream-colored. Incubation about 33 days, by female only. Young downy; leave nest soon after hatching; stay with parents until following spring.
Food. Aquatic plants and insects.

Mute Swan

Cygnus olor

The loud hisses and upraised wings of a male mute swan warn intruders away from its nest.

Music and literature have helped to make this magnificent bird famous throughout the world. The Danish writer Hans Christian Andersen immortalized its transition from ugly duckling—which, of course, it is not—into a flier of dazzling white beauty. And the French composer Camille Saint-Saëns captured the bird's grace with the soaring tones of a solo cello in his composition *The Swan.*

Originally an inhabitant of Europe and Asia, the mute swan was introduced into the United States in the mid-1800's to grace the ponds of estates and parks. Today, thousands of descendants of those original pairs are living wild, and their numbers continue to multiply.

Mute swans are widely noted as mates for life, but even in the orderly world of swans this is not quite the case, at least not all the time. About 5 percent of their matings appear to break up, although the reasons for that remain unclear. Mute swans are also known to be powerful birds—perhaps even stronger than the bald or golden eagles. The orange-colored bill of a mute, used to dislodge aquatic plants for food, can lash out with terrifying speed as the swan launches a neck blow strong enough to break the arm of a full-grown man.

Recognition. 56–62 in. long. Large, white, and long-necked. Neck held in graceful curve. Bill orange, with black knob. Young tinged with dusky; bill grayish, without knob. Often swims with wings held saillike over back.
Habitat. Ponds and lakes in residential areas.
Nesting. Nest is a large mass of plant material near edge of water. Eggs 4–6, blue-gray. Incubation 34–38 days, mostly by female. Young downy; leave nest soon after hatching; stay with parents 4 months. Usually nests in isolated pairs; sometimes in loose colonies.
Food. Aquatic plants.

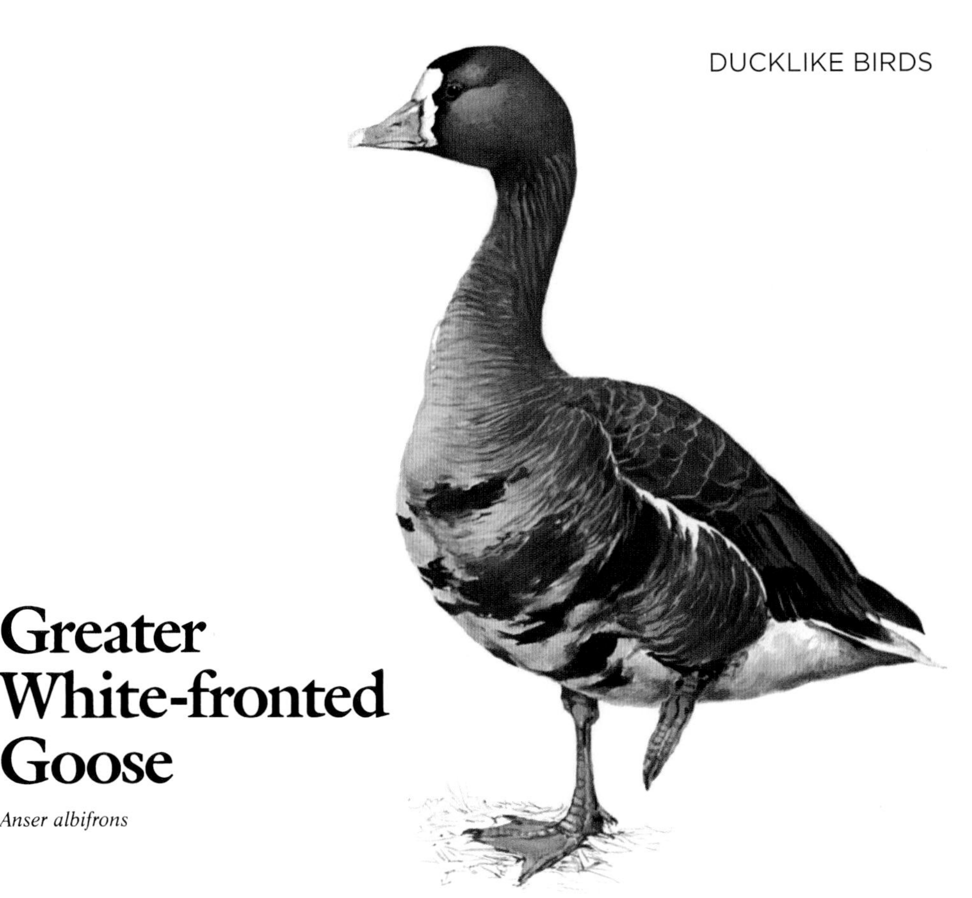

Greater White-fronted Goose

Anser albifrons

Geese are monogamous—they take one mate and form a uniquely enduring pair-bond. In some birds, pair-bonding involves no long-term commitment. It may be a perfunctory "Charmed, I'm sure" followed by mating, a familiar pattern among grouse and other species that hold no territory. And a female hummingbird is deserted by the male before the blush has left her cheeks. The ties that bind most other birds are made of stronger stuff: they may last the entire breeding season, or even through succeeding years. In such cases the males often help in various ways—although some get by doing little more than troubadoring in the trees.

Geese, however, seem to approach the human ideal in terms of faithfulness. The male stays with his family, protecting the nest and the young. Male and female usually remain together during migration and throughout winter's lean times, year after year, until death finally parts them. Some ornithologists hold that the long-lasting pair-bond is really just a convenience—that mated geese are both attached to their nesting area rather than to each other. Others, however, can cite cases of fidelity and sacrifice that have few parallels, avian or human.

Where their own numbers are few, white-fronts will tag along with flocks of Canada geese.

Recognition. 26–34 in. long. Brown, with stubby pink or orange bill; face, rump, and feathers under tail white; belly with black bars; feet orange. Usually seen in flocks.
Habitat. Fields, prairies, and marshes; nests on Arctic tundra.
Nesting. Nest is a shallow depression, lined with down and feathers, near edge of water. Eggs 4–7, cream-colored. Incubation about 28 days, by female only. Young downy; leave nest soon after hatching; stay with parents until following spring. Usually nests in loose colonies.
Food. Aquatic plants, grain, and insects.

Snow Goose

Chen caerulescens

Flocks of snow geese often fly in a U formation, not the familiar wedge of the Canada geese.

The name snow goose is applied today to all the lovely birds that used to be considered three separate species: the blue goose, the lesser snow goose, and the greater snow goose. The blue goose was discovered to be a dark form of the lesser snow goose, which winters in California and the Gulf states. And the greater snow goose is simply a larger form of the same bird, one that winters along parts of the East Coast. All the forms breed in the Arctic regions of Siberia and North America.

In migration, flocks of snow geese fly very high—at altitudes of perhaps a thousand feet—not in a V formation as the Canada goose does, but in a long, curved, often U-shaped line. Usually, when these migrating flocks come to a watery resting place, they suddenly lift their bodies upward, dangle their legs, and braking, with their wings down-bent, float softly earthward until, with a few quick wingbeats, they touch the water. But there are times when, for reasons unknown, the geese come in helter-skelter, zigzagging down, almost falling the thousand feet into the water, while the whole flock honks its high-pitched yet somehow musical clamor—and comes unceremoniously to rest.

Recognition. 25–31 in. long. White, with black wing tips; bill pink with black "lips." Dark form (blue goose) has dark gray body, white head and neck. Usually seen in flocks, often of hundreds. Call note a high-pitched yelp.

Habitat. Marshes, fields, and lagoons; nests on Arctic tundra.

Nesting. Nest is a depression lined with grass, stems, and down. Eggs 3–8, white. Incubation 22–25 days, by female only. Young downy; leave nest soon after hatching; stay with parents until following spring.

Food. Aquatic plants, grasses, and grain.

Brant

Branta bernicla

Flying far and fast in its spring migration—as far north as the windswept bays and boulder-strewn coastlines of Greenland—the brant is an avid consumer of grasses, grains, and aquatic vegetation. The bird is an especially good harvester of eelgrass, its preferred food, often pulling it up by the roots at low tide, then feasting at leisure as goose and grass float together when the tide comes in. In 1931, however, a virulent blight nearly wiped out the Atlantic Ocean's supply of this nutritious food. Abruptly, the brant population was at the brink of extinction, except for a relatively adventurous few who switched over to sea lettuce. This good breeding stock survived, and eventually the eelgrass also returned. Today the brant eats both of these aquatic plants, prudently maintaining the feeding habit formed in dire necessity, even though the crisis itself has long since passed.

Every bit as sociable as the larger Canada goose, brant feed and fly in flocks of from 20 to 50. With the approach of cold weather they migrate south, often wintering along the Atlantic coastline from Cape Cod to North Carolina; particularly large groups of them mass in the Brigantine and Barnegat protected areas of New Jersey.

When a hard freeze bars access to marine plants, brant often graze on seaside golf courses.

Recognition. 22–26 in. long. Head, neck, and breast black, with white patch on neck; feathers under tail white. East Coast birds white below; West Coast birds (black brant) dark brown below. Call note a low croak.

Habitat. Rocky or marshy coasts; fond of sandy points; nests on Arctic tundra.

Nesting. Nest is a hollow lined with plant matter, down, and feathers. Eggs 1–7, dull white. Incubation 22–26 days, by female only. Young downy; leave nest soon after hatching; stay with parents until following spring.

Food. Aquatic plants, including marine algae; insects, crustaceans, and mollusks.

Canada Goose

Branta canadensis

Unlike males of most duck species, male geese stay with each brood for nearly a year.

Few events in nature are more thrilling to behold than a V formation of Canada geese, flying at speeds of up to 45 miles per hour, males and females honking greetings and signals to each other. The elaborate courtship displays and mating rituals of these stately birds have enchanted observers from John James Audubon onward. The gander—his wingspread is as much as five feet and his weight approaches 14 pounds—is also a fierce defender of his mate and offspring. After first giving fair warning, the male begins the vigorous head-pumping that signals attack—and an aroused gander will charge any suspected enemy, even one as large as an elk.

As such fierce protectiveness suggests, Canada geese are exemplary parents, staying with their young for almost three-quarters of a year. But the young also learn survival early. In their first day in the water, goslings can swim four or five yards underwater; if danger threatens on land, they can flatten themselves out to resemble rocks or sand mounds. The fortunate ones survive to become adults, and some pairs may have as long as 20 years together before succumbing to disease, a hunter's bullet, or a predator who senses that their fighting days are finally over.

Recognition. 22–48 in. long. Races vary greatly in size. Head and neck black, with bold white cheek patch. Body gray-brown; feathers under tail white. Flocks usually travel in V formation; call note a loud honking.

Habitat. Ponds, lakes, rivers, freshwater and salt marshes, and grainfields.

Nesting. Nest is a large hollow lined with plant matter and down. Eggs 2–12, white. Incubation 25–30 days, by female only. Young downy; leave nest soon after hatching; stay with parents until following spring.

Food. Aquatic plants, grass, grain, and small aquatic animals.

Fulvous Whistling-Duck

Dendrocygna bicolor

If a bird doesn't look like a duck, or act like a duck, how come it ends up being one? Observers of the fulvous whistling-duck can't help asking this question. Trailing long legs behind it in flight, even gliding on occasion, this Gulf Coast wanderer looks disconcertingly like a heron. When it does beat its wings—flapping them slowly and unevenly—it resembles an ibis. Even its flock formation in the air is a haphazard, cloudlike arrangement that bears little resemblance to the trim ranks of most ducks in flight.

Their unorthodox ways continue as they drop to the ground. Just before landing, these birds hang their necks down like geese. Once down, they crowd together, crane their necks, and stand uncommonly straight before walking off without one trace of a ducklike waddle. If suddenly threatened, they stand totally still with their long necks upstretched, as a bittern might. When the danger passes, they walk along like so many foraging geese, thrusting their bills deep in the mud.

Fulvous whistling-ducks don't quack like black ducks; they squeal out slurred whistles. And they don't feed during the day like mallards, but instead descend upon rice fields at night. They do, however, swim with webbed feet, eat with duck bills, and stay warm with puffy down. That makes them ducks. They just try to keep it a secret.

Several whistling-ducks may lay in one nest; up to 60 eggs have been found in a single clutch.

Recognition. 18–21 in. long. Long-necked and long-legged. Mainly bright buff; rump and stripes on side white; back and wings dark.
Habitat. Freshwater marshes, weedy ponds, wooded swamps, and croplands.
Nesting. Nest is a plant-lined hollow hidden in tall grass or, rarely, in tree cavity 4–30 ft. above ground. Eggs 10–20, cream-colored or buff. Incubation 24–26 days, by both sexes. Young downy; leave nest soon after hatching; first fly at 8–9 weeks.
Food. Mainly seeds and grain.

Wood Duck

Aix sponsa

Male

Equally at home in woodland or water, the gentle wood duck is widely regarded as the loveliest of American waterfowl—indeed as one of the most beautiful birds in the world. A pair of wood ducks floating contentedly downstream together, sunlight dappling the multicolored crest of the male and the delicately patterned body of his mate, appear to be as lightly and magically suspended as two drifting leaves.

Placid though it appears, the wood duck has surprising strength, coming off the water like a rocket when disturbed. Nevertheless, by the early 20th century hunters had almost wiped out this splendid bird. Protective laws ended the slaughter in time, fortunately, and the wood duck is now making a healthy comeback.

Mated wood ducks are highly compatible, and the female's affection extends even to her own birthplace, to which she may return with her mate year after year to raise successive broods. The babies—anywhere from 8 to 14 of them—are as engaging a sight as their handsome parents. Although born as high as 50 feet above ground in a tree cavity, they soon scramble out on tiny clawed feet and take their first anxious leap into thin air—floating downward like puffballs to where their mother waits to take them for the first swim of their lives.

Just a day after hatching, young wood ducks drop to the water and swim off with their mother.

Recognition. 17–20 in. long. Crested and long-tailed. Male has bold white face pattern; bill red; breast chestnut; flanks buff. Female dark gray with white eye patch. Usually seen in pairs.
Habitat. Wooded swamps, ponds, and marshes.
Nesting. Nest is a shallow cup of white down in a cavity 5–50 ft. above ground. Eggs 8–14, dull white or tan. Incubation 28–32 days, by female only. Young downy; leave nest soon after hatching; first fly at about 7 weeks. Sometimes 2 broods a season.
Food. Aquatic plants, nuts, and fruit; also insects, small fish, and crustaceans.

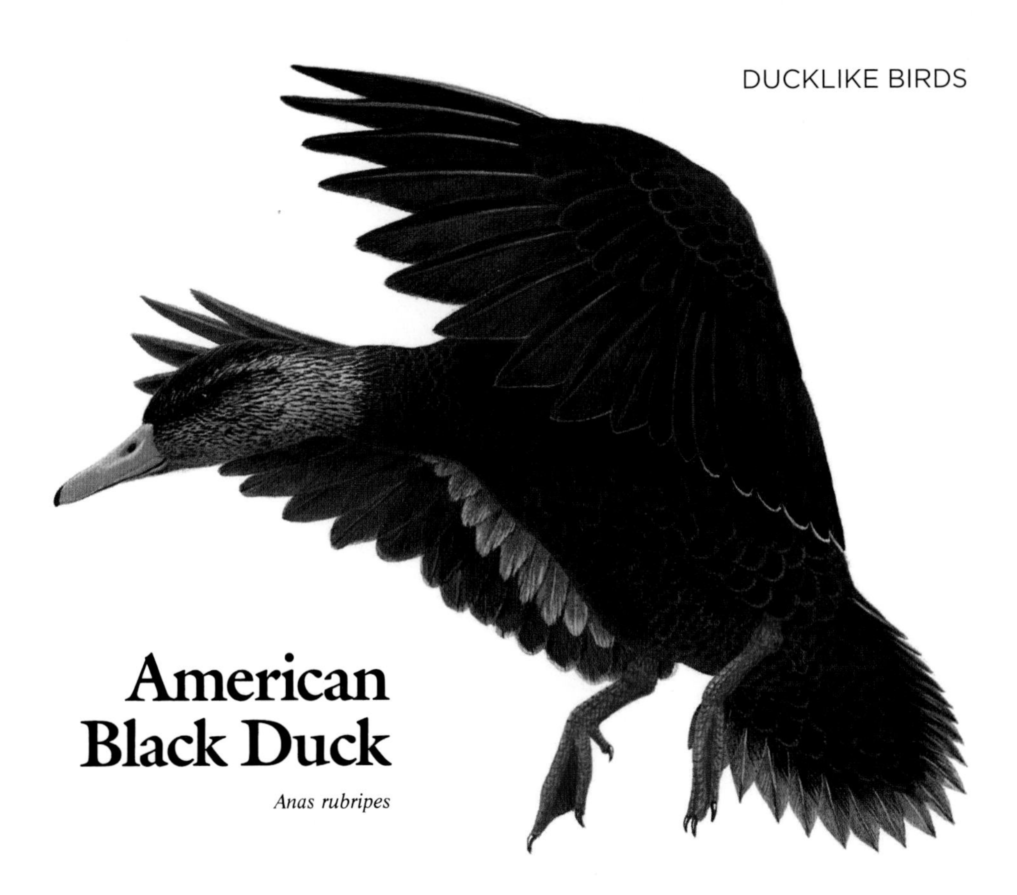

American Black Duck

Anas rubripes

Hunters may be the ones who originally misnamed it, for this fast flier does seem black against the sky. Actually, the male black duck's true color is sooty brown, with underwing patches of white that he shows off to good effect during courtship flights around an eligible female.

Whether living on rivers, marshes, or ponds, black ducks are known for their ability to stay well fed year-round. In spring these dabbling ducks go tails-up, often in unison like an aquatic drill team, to root about for submerged plant life. In summer they add frogs, toads, and snails to their diets. In fall they visit farmlands to shovel up grains, and in winter they will settle down in unfrozen salt marshes where various grasses are still available.

They also have a well-founded reputation for wariness. Black ducks can take to the air instantly and fly out of danger at a speed of 25 miles an hour or more. That keen sensitivity to potential danger develops early in life. Mothers may hatch a brood a mile or more from water, and the offspring must then proceed at their faltering waddle toward water soon after piercing their shells. Once swimming, they are fair game for snapping turtles or bullfrogs in the shallows, and for large fish lurking farther out. But by August they are flying, and almost ready to embark on their own lives as adults.

When ducklings are foraging, the mother often keeps her distance to avoid competing for food.

Recognition. 21–25 in. long. Mainly dark, dusky brown; head and neck paler, yellower brown; undersides of wings white; wing patch violet; feet orange or red. Male has yellow-green bill; female has mottled bill.
Habitat. Marshes, lakes, and ponds.
Nesting. Nest is a down-lined hollow of grass and stems hidden in vegetation near water. Eggs 5–17, cream-colored or greenish. Incubation 26–28 days, by female only. Young downy; leave nest soon after hatching; fly at about 9 weeks.
Food. Aquatic plants, worms, snails, and seeds.

Mottled Duck

Anas fulvigula

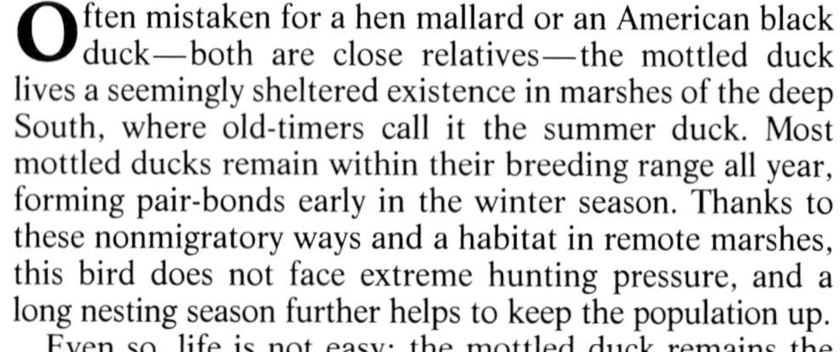

Often mistaken for a hen mallard or an American black duck—both are close relatives—the mottled duck lives a seemingly sheltered existence in marshes of the deep South, where old-timers call it the summer duck. Most mottled ducks remain within their breeding range all year, forming pair-bonds early in the winter season. Thanks to these nonmigratory ways and a habitat in remote marshes, this bird does not face extreme hunting pressure, and a long nesting season further helps to keep the population up.

Even so, life is not easy: the mottled duck remains the least numerous duck in North America. Nesting success is poor, despite a long season, and the female may abandon her eggs when disturbed. The nest itself—placed at or slightly above ground level—is ever at risk of inundation by heavy rains or floods, and of raiding by raccoons, possums, skunks, snakes, and grackles. In one case, a bird made five nesting attempts totaling 34 eggs before success was met. Once the young have hatched, they are prey to large fishes, turtles, alligators—and, in coastal marshes, hordes of blue crabs. Only one brood, however depleted it may be, is raised in a season. But despite all the hazards of nesting, the greatest threat to the mottled duck comes from man's continuing appropriation of marshland for farming and development.

The mottled duck takes more fish and other animal food than its relative the mallard.

Recognition. 20 in. long. Mainly sandy brown; cheeks and throat pale sandy buff; wing patch greenish with white rear border; bill yellow, lacks dark spots found on mallard's bill.
Habitat. Marshes and ponds.
Nesting. Nest is a down-lined hollow of grass, reeds, and stems hidden in vegetation near water. Eggs 8–11, cream-colored or greenish. Incubation 26–28 days, by female only. Young downy; leave nest soon after hatching; age at first flight unknown.
Food. Aquatic insects, snails, fish, seeds, grain, and aquatic plants.

Mallard

Anas platyrhynchos

Probably the best known and most abundant wild duck in the Northern Hemisphere, the mallard is the ancestor of almost every breed of domestic duck. Indeed the mallard itself is readily domesticated, and has been a familiar dooryard duck for centuries. In the wild it is an opportunistic feeder, eating snails, aquatic insects, grasshoppers, fish eggs—practically anything it happens to find. But given a choice, the mallard will feast on seeds day after day, and as a result its flesh is usually quite palatable. Little wonder, then, that it has been a valued addition to poultry yards on three continents—above all in China, where it has provided generations of people with eggs, meat, down, and feathers in almost limitless abundance.

Completely circumpolar, the wild mallard breeds across much of Europe and Asia, and some flocks in those regions winter as far south as Africa and India. In North America it summers across most of Canada and the United States, and its winter range extends well into Central America. Still, no matter how great its numbers in the wild or how widespread its distribution, the mallard continues to be known as a domestic duck almost everywhere. And the loud, resonating, down-the-scale *Quack!* of the female mallard has become, perhaps more than any other, the call that means *duck* to people the world over.

Awkward but effective, tipping up is a standard feeding method for mallards and other dabblers.

Recognition. 20–28 in. long. Male has glossy green head; bill yellow; breast chestnut; wing patch blue, bordered with 2 white stripes. Female sandy brown; bill mottled with orange; wing patch blue, bordered with 2 white stripes; tail feathers pale.
Habitat. Lakes, marshes, swamps, and parks.

Nesting. Nest is a down-lined hollow of grass and stems hidden in vegetation near water. Eggs 5–14, white or pale green. Incubation 26–29 days, by female only. Young downy; leave nest soon after hatching; first fly at about 8 weeks.
Food. Seeds, snails, insects, and small fish.

Northern Pintail

Anas acuta

The Greyhounds of Waterfowl—even casual observation will explain why that nickname could easily be applied to these birds, the slender, swift-flying, graceful pintails. They sit alert, high upon the water, and their long, needlelike tails distinguish them in a rather jaunty way from their fellows.

An abundant circumpolar species, pintails prefer the shallow ponds and potholes that are the natural haunts of dabbling ducks everywhere. The water needs to be shallow because a dabbler—in addition to "dabbling," or skimming bits of food from the surface with its bill—feeds by the method known as tipping up. Unlike a diving duck, a dabbler tips its tail straight up above the water, paddling its feet for balance, and stretches out its neck to dig into the bottom mud for food. The pintail prefers the seeds of grasses, sedges, and pondweeds, which it strains out of the mud, but it will also eat snails, insects, crayfish, and other aquatic animals small enough to swallow.

Pintails and their fellow dabblers have at least one great advantage over diving ducks and most other waterfowl. They do not need to run over the water's surface to get into the air. Because a dabbling duck's wings are larger in proportion to its weight, it can simply leap from the water and instantly launch itself into flight—a life-or-death talent in a world full of hidden dangers.

Although she lacks bright colors, a female pintail is set apart by the slenderness of her neck.

Recognition. Male 25–29 in. long; female 20–22 in. long. Male's head brown; neck slender with white stripe; body mainly gray, with long, thin black feathers in center of tail. Female slender, long-necked, gray-brown; bill gray.

Habitat. Marshes and ponds.

Nesting. Nest is a down-lined hollow of plant material. Eggs 6–12, cream-colored or greenish. Incubation about 26 days, by female only. Young downy; leave nest soon after hatching; first fly at about 7 weeks.

Food. Seeds, snails, insects, crustaceans, and small fish.

Blue-winged Teal

Anas discors

It is undeniably a dabbling duck, the blue-winged teal, but a dabbler with a difference. Instead of tipping up to feed—tail pointed skyward above the water, head down in the mud below—a blue-wing often just skims off floating vegetable and animal matter with its bill. An especially ambitious one may reach down with its head and neck to dig up seeds and aquatic life in the shallowest part of whatever body of water it happens to be in.

And that body of water can be very small indeed. Not only can teals and other dabblers spring into the air without a run on an open surface, but they can also fly slowly enough to drop with precision onto a puddle ringed closely with cattails, weeds, and sedges. Those plants are doubly valuable, dropping seeds into the water and also providing hiding places, both for small ducklings and later for adults during their flightless molting period.

In the privacy of a pond, large or small, dabbling ducks tend to be very noisy. All the females call out their descending sequences of quacks, which range from the teals' crisp but rather quiet delivery to the deeper, hoarser tones of the pintails. The males, in these secure situations, often join in with a great variety of whistles and peeps (and sometimes croaks), adding their thinner voices to the general chorus of dabbling duck well-being.

Unlike most ducks, the male blue-wing wears his drab eclipse plumage all through autumn.

Recognition. 14–16 in. long. Small, with chalky blue patches on wings. Spring male has blue-gray head with white crescent on face. Female and winter male grayish brown, with pale blue wing patches.
Habitat. Marshes and shallow ponds.
Nesting. Nest is a down-lined hollow of grass, hidden in vegetation near water. Eggs 6–15, white or cream-colored. Incubation about 24 days, by female only. Young downy; leave nest soon after hatching; first fly at about 6 weeks.
Food. Seeds, aquatic plants, snails, and insects.

Cinnamon Teal

Anas cyanoptera

A scent on the wind can mean death for a young cinnamon teal. That scent is the smell of its own cracked shells—the smell of a duckling beginning its life. As enticing to marshland predators as barbecue smells are to humans, this eggshell aroma serves as a dinner bell for raccoons and other carnivores that plunder duck nests for eggs and hatching young. Their success depends on timing. Should they zero in on a nest in which only one duckling has hatched, while a dozen others remain encased in their shells, all but the free bird will perish. But if all the eggs have hatched and the ducklings have been whisked to safety, the only loss will be a hungry raccoon's lunch.

The trick is getting all the youngsters to hatch simultaneously, so that the mother can clear them out in a group. Cinnamon teals and other waterfowl manage this through delayed incubation. Instead of beginning her incubation duties with the laying of her first pinkish-white egg, a teal mother-to-be waits for up to two weeks, until all her eggs have been laid. Then she begins warming all of them, thus insuring that they will develop at the same rate and from the same starting point. The result: three weeks later, one egg will show telltale cracks. Within a few hours all the eggs will have hatched, and a dozen ducklings will have followed their mother toward safety.

The cinnamon teal (top) has a slightly longer bill than its close relative the blue-wing.

Recognition. 14–17 in. long. Small, with chalky blue patches on wings. Breeding male rich cinnamon. Female gray-brown, with pale blue patches on wings.
Habitat. Marshes and shallow ponds.
Nesting. Nest is a shallow cup of grass, lined with down, hidden in vegetation near water. Eggs 9–12, white or pinkish buff. Incubation about 25 days, by female only. Young downy; leave nest soon after hatching; begin to fly at about 7 weeks.
Food. Seeds, aquatic plants, snails, and insects.

Green-winged Teal

Anas crecca

Pint-sized bantams among the dabbling ducks, green-winged teals are compensated by nature with the grace and agility to fly in tight formation at high speed in flocks of considerable size. Even when only commuting from one pond to another, these flocks bank and turn, twist and straighten, rise and descend in faultless precision and obvious play. The females typically voice their contentment in a decrescendo of four faint quacks. In contrast, the males utter high-pitched, rather abrupt whistles, along with lower-pitched trills that resemble those of courting spring peepers. But these are not courting calls. Green-winged teals have a very special note—*KRICK-et*—which they use only during their courtship displays.

Such displays vary among dabbling ducks: some of the best effects are used by them all, but different species never use them in the same sequence. In one performance, the courting male swims rapidly about, lifts his head and tail, unfolds his wings, and poses before the female in all his glory. In another, the male rises upright on the water, arches his neck, whistles, then tosses a billful of water in a rainbow of tiny droplets toward his intended mate. The females all spend a great deal of time ignoring these antics—but they are also quick to provoke their suitors into attacking any other males who intrude on the scene.

The Eurasian green-wing, with its horizontal white stripe, makes occasional visits to North America.

Recognition. 13–16 in. long. Small, with green wing patch. Male gray; head rusty; ear patch green. Female gray-brown; belly whitish. American males have vertical white stripe on side; Eurasian birds (sometimes seen in North America) have horizontal stripe on side.
Habitat. Marshes, ponds, lakes, and mud flats.

Nesting. Nest is a hollow of grass, lined with down and hidden in vegetation near water. Eggs 7–15, dull white, greenish, or buff. Incubation about 24 days, by female only. Young downy; leave nest soon after hatching; begin to fly at about 6 weeks.
Food. Insects, seeds, and aquatic plants.

Northern Shoveler

Anas clypeata

In 1840, John James Audubon, a gourmet of sorts who enjoyed cooking birds as well as painting them, wrote, "No sportsman who is a judge will ever pass a shoveler to shoot a canvasback." Since the lordly canvasback was considered the ultimate in good eating—its pound of delicate flesh sweetly flavored by the bird's fondness for wild celery—a reader might wonder just how long Audubon-the-epicure had been in the woods without a square meal.

Handsome as it is, the northern shoveler wins little esteem as a table trophy; many a hunter would rather give it away in bag-limit quantities than eat it himself. It may be smoked, marinated, broiled, baked, or barbecued, but its true identity will still be revealed to a discriminating palate.

Taking its food from both the bottom ooze and the surface of a lake or pond, the shoveler has a bill with comblike "teeth" along the upper and lower mandibles. These enable it to strain the best that a pond has to offer not only of plant seeds, but also of fingernail clams, beetles, midge larvae, and other creatures that taste better to birds than to people. Like many dabbling ducks, shovelers also gather at wastewater treatment ponds, where food is plentiful but of dubious origin. So, recalling the ancient story of the Trojan horse, modern gourmets would be well advised to beware of neighborly hunters bearing gifts.

Despite her modest brown plumage, a female shoveler is easily identified by her bill.

Recognition. 17–20 in. long. Long, shovel-shaped bill. Male's head green; breast white; flanks rusty. Female sandy brown with chalky blue wing patches.
Habitat. Marshes, shallow ponds, and lakes.
Nesting. Nest is a hollow of grass, lined with down and concealed in vegetation, usually near water. Eggs 6–14, buff or greenish. Incubation about 26 days, by female only. Young downy; leave nest soon after hatching; begin to fly at about 7 weeks.
Food. Crustaceans, insects, mollusks, seeds, and aquatic plants.

Gadwall

Anas strepera

When Arthur Cleveland Bent, one of American ornithology's great figures, set eyes on gadwalls for the first time, the sight made a profound impression. The year was 1905, the place Saskatchewan. On a cold, rainy day in June, Bent found himself sloshing through a network of vast marshy sloughs near Crane Lake. His mission was to study the waterfowl there as part of a research project sponsored by the Smithsonian Institution. Kicking through shallows, Bent triggered a waterfowl spectacle. Hundreds, perhaps thousands, of ducks arose from wet meadows in clouds that looked like mosquito swarms. These haphazard flocks circled overhead, darkening the sky. Wings beat madly. Calls filled the air. All was confusion and ducks.

A good many of the birds were gadwalls, and for days Bent observed their mating rituals—the wild calling and whistling, the furtive construction of nests in tall grass, and the flights in which males dashed past their mates so close that their wings touched with a clatter. These were North American waterfowl in their heyday, their pristine marshes not yet filled in or drained for human use. There was only abundance. "I shall never forget the sights that I saw," Bent later wrote of that trip, as he compiled the series of more than 20 volumes that remains the most comprehensive, widely used resource on North American birdlife.

In the fall gadwalls often leave the water to forage for acorns under trees.

Recognition. 19–23 in. long. Small white wing patch shows in flight. Male mainly gray; head pale sandy brown; feathers under tail black. Female sandy brown; best distinguished by white wing patches.

Habitat. Ponds, lakes, and marshes.

Nesting. Nest is a hollow lined with plant matter and some down, usually near water, often on an island. Eggs 7–15, cream-colored. Incubation 25–28 days, by female only. Young downy; leave nest soon after hatching; begin to fly at 7–9 weeks.

Food. Mainly aquatic plants and seeds; a few insects and mollusks.

American Wigeon

Anas americana

What some birds won't go through for a taste of wild celery. Take the American wigeon, a duck with diving limitations that would seem to rule out its enjoyment of a plant that grows deep in the water. But the wigeon has a simple solution: let other ducks get the celery, then grab some of it. The unwitting donors are redheads, scaups, and canvasbacks—all diving ducks adept at rooting up celery from the bottom. Tailoring its migration routes and wintering locations to theirs, the crafty wigeon shadows them like pilot fish trailing a shark.

As a flock of redheads feeds over a celery bed, a redhead dives—and a nearby wigeon springs into action. Anticipating where its victim will surface, the wigeon swims over and waits. When the redhead pops up with a billful of celery stems, the opportunist brazenly snatches a strand.

Is the redhead outraged? Does it drive the wigeon away? On the contrary, diving ducks put upon by wigeons accept their role as providers with evident tolerance. Perhaps they can sense that wigeons help them in return—by warning them of danger. More restless than many other ducks, wigeons are quick to sense trouble and impart their alarm to the divers around them, leaping from the water in a flurry of alarm quacks and rattling wings. Surely insurance like that is worth a celery strand or two.

Wigeons sometimes wander among rows on newly planted fields, pulling up shoots.

Recognition. 18–23 in. long. Large white wing patch shows in flight. Male has white forehead; green ear patch. Female sandy brown; bill pale blue-gray; most reliably distinguished by white wing patches.
Habitat. Lakes, ponds, and marshes.
Nesting. Nest is a hollow lined with grass and much down, in tall weeds, often some distance from water. Eggs 6–12, cream-colored. Incubation about 25 days, by female only. Young downy; leave nest soon after hatching; begin to fly at 7–8 weeks.
Food. Seeds and foliage of aquatic plants; some insects and mollusks.

Ring-necked Duck

Aythya collaris

One of the hazards of being a duck (or a goose, or a swan) lies at the bottom of the waters in which it feeds. It is accumulated lead shot—shot that missed the mark on the day it was fired from a gun and now poses a far more insidious threat, in the form of lead poisoning.

All waterfowl are susceptible, but the ring-necked duck and other popular game birds are at especially high risk because they probe the bottom for seeds and tubers of water plants. In so doing they are apt to ingest large numbers of lead pellets, which go to the gizzard and are broken down there by abrasion. Some of the lead compounds are then absorbed into the bloodstream, damaging kidneys and liver and leading inexorably to death. It is a slow and agonizing process; the end, when it comes, is a merciful act. The annual death toll of waterfowl alone is estimated at between 1,500,000 and 3,000,000. Even the bald eagle, which preys on waterfowl, is not exempt from lethal, second-hand ingestion of lead.

An estimated 6,000 tons of lead shot had rained down on ponds and marshes yearly before the U.S. Fish and Wildlife Service introduced a gradual ban on its use, intended to be in full force by 1991. Even if the ban does become law, though, countless pellets already lie in the earth, each with the potential to kill birds not yet born.

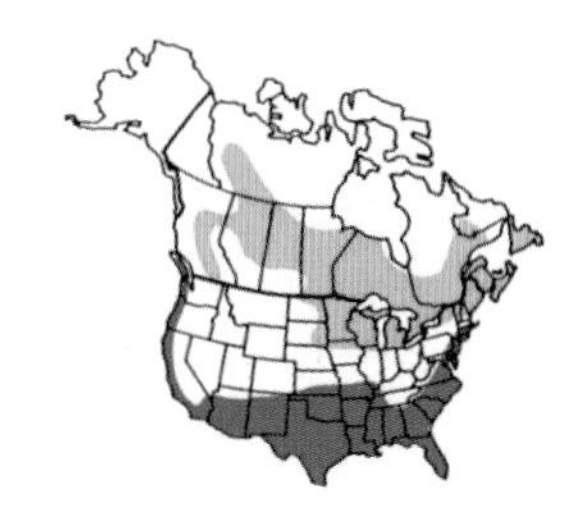

The ring-neck is one of the few ducks that nest on boggy ponds in the boreal forest.

Recognition. 15–19 in. long. Stocky. Male mainly black; bill gray with white band near tip; flanks white; wings black, with gray border visible in flight. Female gray-brown; eye-ring whitish; bill gray with white band near tip.
Habitat. Ponds and lakes in wooded country; marshes and estuaries.

Nesting. Nest is a hollow of grass and down, in vegetation near water. Eggs 6–14, greenish. Incubation about 26 days, by female only. Young downy; leave nest soon after hatching; first fly at 7–8 weeks.
Food. Seeds and foliage of aquatic plants; insects, snails, and crustaceans.

Redhead

Aythya americana

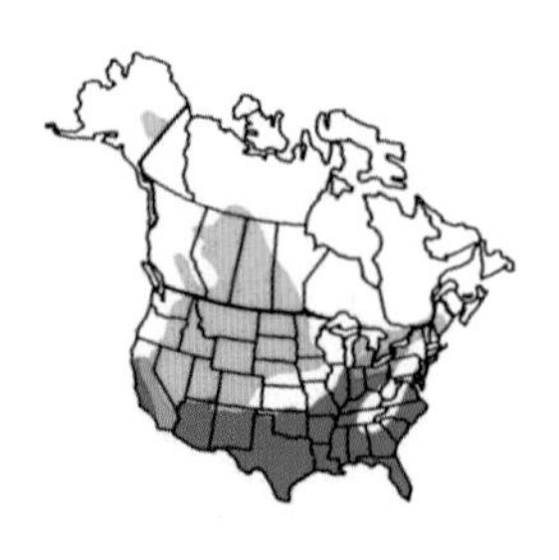

Some birds produce only one egg a year; this one has eggs to spare. After filling her own nest with a dozen or so, the female redhead often goes on a veritable binge of egg-laying. She might begin by visiting the nest of another duck species to make a deposit of eggs. Then she may lay in the nest of another redhead—and be joined there by still other redhead females, each with eggs to contribute. The result of this convergence is a "dump nest," often containing more eggs than any bird could hope to incubate. One researcher watched 13 different redheads lay their eggs in a single dump nest; 87 eggs were found in another.

While the purpose of dump nests remains unclear, some biologists think they may be a step on the road to complete parasitism of the kind that cowbirds practice. Redheads still incubate their own eggs and raise their own young, but they show less attachment to their nests than other ducks, and dump nesting may be teaching them how to propagate the species without all the burdens of maternity.

In any case, the redhead might not survive long enough to play out whatever role nature planned for it. Industrial development and recurring drought have reduced its numbers so sharply that hunting of it has been banned in key areas. Action like this helped bring the wood duck back; with luck, the redhead may star in its own success story.

The deserted nest has become a redhead symbol, so often do the females lay their eggs elsewhere.

Recognition. 18–22 in. long. Stocky. Male has rusty head; bill blue-gray with black tip; breast black; back gray. Female has brown head and body; bill blue-gray with black tip.
Habitat. Marshes, ponds, lakes, and bays.
Nesting. Nest is a shallow cup of plant material, sparsely lined with down, in tall vegetation near water. Eggs 10–16, buff. Incubation about 24 days, by female only. Young downy; leave nest soon after hatching; first fly at 8–10 weeks. Sometimes lays in nests of other water birds.
Food. Foliage and seeds of aquatic plants; some insects and mollusks.

Canvasback

Aythya valisineria

Looking as though each body had literally been wrapped in a small square of canvas, the wary canvasbacks gather in huge rafts—close groupings of birds afloat on the water—on their coastal and inland wintering grounds. There, far from shore, they feed, rest, and sleep. Clumsy on solid ground, canvasbacks rarely come ashore except when nesting in the prairie sloughs and marshes of the Northwest. Even there, they build their semi-floating nests among plants growing in water nearly two feet deep.

During the two to three months of incubating and rearing their broods, the females are on their own. Once the eggs are laid, the males gather in small groups and head off to deeper waters for a season of quiet diving and fishing. Toward summer's end, mothers and their young rejoin the fathers, and in October the main migration flights take place. Canvasbacks make these high-altitude journeys in enormous V-shaped flocks, some of them heading almost due east to winter on the mid-Atlantic coast, in the waters off Virginia and North Carolina.

In recent years the canvasback population has fallen sharply. Hunters have long prized them as the best-tasting of all ducks; but far more damaging has been the drainage of northern marshes—for agricultural needs—at a heavy cost to the canvasbacks' prime nesting habitat.

In a downy young canvasback there is little hint yet of the long, sloping bill of an adult.

Recognition. 19–24 in. long. Stocky, with sloping bill profile. Male has rusty head and neck; bill black; body white; breast black. Female has brown head, grayish back. Birds are usually seen in flocks.
Habitat. Lakes, ponds, marshes, and bays.
Nesting. Nest is a solid cup of grass and stems, lined with down, concealed in tall weeds or grass near water. Eggs 7–12, greenish. Incubation 24–27 days, by female only. Young downy; leave nest soon after hatching; begin to fly at 10–12 weeks.
Food. Mainly roots and tubers of aquatic plants.

Greater Scaup

Aythya marila

Parental desertion is widespread among ducks, especially among diving ducks like the greater scaup. Although they may have been abandoned, the unlucky ducklings are not necessarily alone, for there are many others in the same circumstances. They may join flocks composed of ducklings of different age groups, and from different broods. Young ducklings follow older ones as they would follow their own mothers. Unfortunately, such motherless babes, and the older youngsters they latch onto, are unwise in the ways of the world. Without parental guidance, all are particularly vulnerable to predation.

Some orphaned ducklings, while still flightless, become part of a crèche, increasing their chances for survival. In one of its general uses, the word *crèche* means a nursery or foundling home. Ornithologically, its meaning is much the same. Duckling crèches may be composed of a few broods or many, accompanied by a few adult females. The adults may be the natural parents of at least some of the young, or they may be without families of their own. A waterfowl crèche seldom numbers more than a hundred ducklings and a few faithful guardians, and it may benefit parents as well as hatchlings. After weeks of heavy-duty incubation, adults are often emaciated and weak. In forfeiting their roles to surrogates, they are free to rest and forage at leisure while the young, under a sort of blanket protection, move forward to their own independence.

Two or more female scaups sometimes pool their broods and tend them together.

Recognition. 16–20 in. long. Stocky; long white wing stripe shows in flight. Male has rounded head with greenish gloss; body pale gray; breast and rump black; bill pale blue. Female dark brown; face patch white; bill pale blue. Often seen in very large flocks.
Habitat. Lakes, ponds, bays, and estuaries.

Nesting. Nest is a hollow lined with plant matter and down, often in open site. Eggs 8–11, olive-buff. Incubation 24–28 days, by female only. Young downy; leave nest soon after hatching; first fly at 5–6 weeks.
Food. Aquatic plants and mollusks.

Lesser Scaup

Aythya affinis

Ducks are among the best known of what biologists call precocial birds. They come into the world with eyes wide open, wearing a birthday suit of natal down for insulation. Precocial birds have the edge on altricial birds—which hatch naked, blind, and helpless—in being ready to embark on an active, hazardous life from day one.

Ducklings are instinctive swimmers and fast learners. And they know how to follow orders. When a mother duck says "jump" to her brood of day-old nestlings, each one in turn scrambles out of its nest, wherever and of whatever sort it may be. It then follows her on what may be a perilous course to safety before being allowed to look for food.

The coming days are busy, as the youngsters learn how to recognize danger, plot escape routes, find food, and do everything else needed to survive. Adults offer protection and guidance, but the young must fend for themselves. Precocity helps their cause. At the age of three days, a lesser scaup is remarkably skillful at feeding itself. In a matter of seconds it can dive, catch a minnow, and return to the surface. And survival demands no less. Before long, even before learning to fly, a duckling may be on its own, abandoned by its mother at the onset of her post-nuptial molt. And it had better have learned its lessons well.

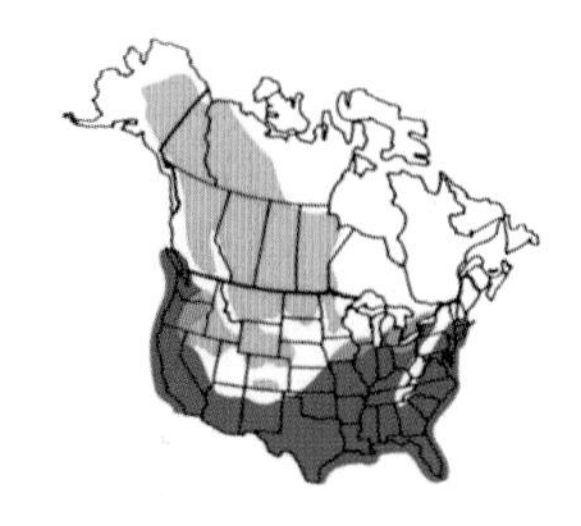

The lesser scaup (right) closely resembles the greater, but has a much steeper forehead.

Recognition. 15–19 in. long. Stocky; short white wing stripe shows in flight. Male's head has purple gloss, peaked crown; body pale gray; breast and rump black; bill pale blue. Female dark brown; face patch white; bill pale blue. Often seen in large flocks.
Habitat. Lakes, ponds, bays, and estuaries.

Nesting. Nest is a hollow lined with plant matter and down, in vegetation near water. Eggs 6–15, olive-buff. Incubation 26–27 days, by female only. Young downy; leave nest soon after hatching; first fly at about 7 weeks.
Food. Aquatic plants and mollusks.

Common Eider

Somateria mollissima

Surf crashes on the ice-rimmed shore of an isolated bay. Snow swirls in a winter gale; the air temperature is far below zero. Barely visible in the Arctic dusk, scores of common eiders swim and dive in the icy water. How do they survive long after other birds have been driven south?

Inside and out, eiders are designed expressly for life in the frigid waters of the Arctic. They are the biggest of North America's ducks, and large body size means less surface area per pound—and so less heat loss. Beneath their outer feathers is a dense coat of down, and under their skin is a thick layer of fat. Body parts sticking out beyond the protective feathers are also specialized. An eider's feet, constantly immersed in frigid water, have arteries and veins passing close to each other. This allows blood flowing into the feet to warm the blood going back, limiting heat loss and keeping the rest of the body warmer. Its beak has feathers farther out than on most birds, and its wings, not needed for long migrations, are shorter.

The warmest quilts and sleeping bags are filled with eiderdown, for good reason: it is the best natural insulation known. So as a winter storm howls through the night, someone wrapped in a down quilt can enjoy the same warmth an eider feels in its own home-grown comforter.

Eiders often fly low in single file over the water, sheltered in the troughs between waves.

Recognition. 23–27 in. long. Large and stocky, with long, sloping bill profile, narrow shield on forehead. Male mainly white; crown, forehead, and flanks black. Female warm brown with fine black bars on sides.
Habitat. Rocky seacoasts.
Nesting. Nest is a hollow lined with plant matter and large amounts of down, usually not concealed. Eggs 3–5, brown or greenish. Incubation 24–27 days, by female only. Young downy; leave nest soon after hatching; first fly at about 8 weeks. Often nests in colonies.
Food. Mollusks, starfish, crustaceans, and fish.

King Eider

Somateria spectabilis

In the Arctic summer, millions of birds struggle to feed themselves and their young. For eiders, as for most species, the key to survival is to find food not used by other birds. Common eiders seek it offshore, diving into the ocean depths for shellfish. Meanwhile, king eiders, which normally also use the sea, have moved inland to nest near freshwater lakes and ponds. There they feed on insect larvae and aquatic plants. Once the eggs are laid, though, the drakes return to the Arctic Ocean, leaving hens and nestlings to follow as soon as the young can fly.

Along the coast many of the king eiders mix with the large flocks of common eiders diving for mussels beyond the surf. But others draw a little farther offshore, into deeper waters. Searching for mussels, crabs, and other animals, some descend as far as 180 feet, well below the range of their rivals. In those cold, dark waters the king eiders find food that is theirs alone.

Swimming eiders throw themselves forward to dive. Their paddling feet drive them through the icy sea, but to get to the really deep areas they must literally fly under water. They use their stubby wings to propel themselves downward and then back up. At times they don't even stop at the surface—and instead can be seen bursting from the water in full flight.

A king eider drake coming out of his summer eclipse plumage is drab but still recognizable.

Recognition. 19–25 in. long. Male mainly black and white; wing patch white; bill and shield over bill bright orange. Female warm brown with fine crescent-shaped markings on sides.
Habitat. Rocky seacoasts.
Nesting. Nest is a flattened hollow of grass, lined with much down, usually near water. Eggs 4–7, olive-buff. Incubation about 23 days, by female only. Young downy; leave nest soon after hatching; age at first flight unknown.
Food. Mollusks, starfish, insects, crustaceans, and fish.

Harlequin Duck

Histrionicus histrionicus

When we think of ducks, most of us call to mind scenes of mallards feeding on a quiet pond, of buffleheads resting on a sheltered bay, or of scoters riding the swells behind the breakers at a beach. We seldom think of rock-pounding surf and turbulent streams, for few of us have met harlequin ducks in their element.

Harlequins are lovely creatures. The rounded head, steep forehead, and short bill give both the male and female a delicate profile, and the male's subtle but rich colors, combined with his bizarre pattern of white spots and stripes, create the unlikely look of a whimsical porcelain bird. But harlequins are far from delicate; indeed, they seem to thrive on rough conditions.

Lucky birders usually find them in winter along cold rocky coasts, where the birds' swimming talents keep them out of harm's way as they dive in the choppy surf. Come spring, however, they become torrent ducks, working their way up rushing mountain streams in much the same way as salmon do. Propelled by their feet and by partially spread wings, the harlequins dive to the gravelly bottom for aquatic insects and small crustaceans. Even more remarkably, harlequin ducks have been seen, like American dippers, actually walking on stream bottoms, angling their bodies down into the current to keep their feet on the gravel.

With her pale face spots, the female harlequin looks like a miniature scoter.

Recognition. 15–21 in. long. Male slate-gray, with chestnut flanks; white spots on head; white stripes on back and sides of breast. Female dark brown with small white spots on head.
Habitat. Rocky shores and mountain streams.
Nesting. Nest is a hollow lined with grass and down, concealed in thick vegetation near water. Eggs 5–10, buff or cream-colored. Incubation 27–33 days, by female only. Young downy; leave nest soon after hatching; begin to fly at about 6 weeks.
Food. Aquatic insects, mollusks, crustaceans, and small fish.

Oldsquaw

also known as

Long-tailed Duck

Clangula hyemalis

On cold winter mornings shrouded in mist, only the tops of fishing boats remain visible as they move out to sea. Ducks are nowhere to be seen. But a flock of oldsquaws, baying softly like a distant pack of hounds, can almost always be heard. *Ha-ha-way,* the Indians named them for their distinctive calls. Their English moniker is perhaps not a kind one, and their Latin name is even more direct: "winter noise." Oldsquaws are indeed the most garrulous of our ducks, but few returning fishermen have not welcomed their chatty sounds as a sign of land nearby.

Long-tailed and shining white, male oldsquaws are certainly among the most elegant ducks in winter. But on a tundra pond in the Arctic during summer they scarcely seem recognizable—the males are mostly dark brown with a white cheek patch, a reversal of their winter dress. Typically, male ducks have only one adult plumage, masked briefly by an "eclipse" plumage in late summer, and most female ducks do not change at all. But oldsquaws sport a total of five adult plumages: winter, summer, and eclipse for the male, winter and summer for the female.

The diving ability of oldsquaws is as exceptional as their voice and plumage: they've been known to reach depths of more than 200 feet. Mollusks and crustaceans are the real attraction—but perhaps it's quieter down there, too.

The winter oldsquaw is our only sea duck with a mostly white body and solid black wings.

Recognition. 15–23 in. long. Wings all dark in flight. Winter male mainly white, with dark patch on face; bold black and white pattern on body; central tail feathers long, pointed, and black. Female has white head with dark patches; dark back. Summer male has black head and neck with white face patch.

Habitat. Bays and estuaries; nests on tundra.
Nesting. Nest is a hollow of grass and down, in low vegetation or among rocks. Eggs 5–11, buff or cream-colored. Incubation 23–25 days, by female only. Young downy; leave nest soon after hatching; begin to fly at about 5 weeks.
Food. Aquatic plants, shrimp, and insects.

Black Scoter

Melanitta nigra

It has been conjectured that the name "scoter" comes from *scoot* and refers to the speedy way these black ducks depart from the water's surface, lifting themselves into the air far more easily than other diving ducks. But they are equally at home below the surface, and just as likely to dive as to take flight when startled. Using both legs and wings for underwater propulsion, they can descend 100 feet or more—and scoters are built to withstand the pressures of such depths, with powerful muscles on a sturdy frame and tough, elastic skin with feathers firmly attached.

Curiously, although these ducks nest far up in Arctic regions, small flocks of what appear to be adult birds can be seen year-round on their southern wintering grounds. As it turns out, black scoters do not breed until their third summer, and all the full-size but not ready for breeding birds stay behind when the older ones make their flights to the nesting grounds. It is on these spring migrations that most pairing takes place. Many of the males winter apart from the rest of their kind and are not available for pairing until the flocks merge on the way north. Then, more than at any other time, the melodious, almost bell-like whistle of the male black scoter is heard. And the female replies with a similar song—though her voice, at least to human ears, transforms it into a harsh and grating croak.

One of the black scoter's favorite foods is the blue mussel—fortunately found on both coasts.

Recognition. 17–21 in. long. Stocky with silvery flash under wings in flight. Male black with bright orange knob on bill. Female dark brown with pale cheeks.
Habitat. Open bays and ocean; nests on northern lakes.
Nesting. Nest is a hollow of grass and down concealed near water. Eggs 5–8, buff to pinkish. Incubation 27–31 days, by female only. Young downy; leave nest soon after hatching; first fly at about 6 weeks.
Food. Mollusks, crustaceans, and aquatic insects and plants.

Surf Scoter

Melanitta perspicillata

Like all the other sea ducks, surf scoters dive and swim beneath the water, float in rafts on its surface, and engage in well-defined courtship rituals. They just happen to do these things more spectacularly than their relatives.

Surf scoters are daytime feeders, and each morning they gather into immense, nearly silent rafts on the surface of the sea. Then all at once they dive, as though on cue—and of the entire horde not one bird is left in sight. In half a minute or so they pop back to the surface, not as a unit this time but singly or in small groups. True to their name, surf scoters spend a great deal of time in the white-capped waves of the shoreline. There, within the glassy curl of a wave about to break, they often dash along parallel to the shore, perfectly at home in the pounding water.

Pairings for next summer's nesting usually take place on the wintering grounds, so that much of the scoters' courtship behavior is aimed at keeping rivals away from their chosen mates. But the actual choosing remains mysterious, since it takes place underwater. When several males are courting a female, they swim wildly about, pursuing her and chasing each other for hours. Then, suddenly, the whole courting group dives. When they come up, the female swims away with the male of her choice—and the also-rans go off together, petulantly chasing and pecking at one another.

A female surf scoter has two white spots on her face, but no white in her wings.

Recognition. 17–21 in. long. Stocky, with no white on wings. Male black with white patches on head; bill brightly patterned with orange, black, and white. Female dark brown with light spots on sides of head.

Habitat. Open bays and ocean; nests on northern lakes and tundra.

Nesting. Nest is a hollow of grass and down, often some distance from water. Eggs 5–8, buff or pinkish. Incubation by female only; period unknown. Young downy; leave nest soon after hatching; age at first flight unknown.

Food. Crustaceans and aquatic plants.

White-winged Scoter

Melanitta fusca

The white-wing adds no down to cushion its nest until all the eggs have been laid.

Fish are adapted for life in water, terrestrial animals for life on land, and birds for life in the air. If living things were really that simple, our studies of them would certainly be neater. But they would also be duller, for some of the most interesting aspects of many organisms are the adaptations they make to live outside "their" element.

White-winged scoters are sea ducks, and along with many other water birds they have adapted to life in the water. Some of the adjustments seem fairly simple—added insulation, waterproofing of feathers, webbed feet for paddling in the water. But the more striking changes are those that allow a bird both to fly and to swim—two somewhat incompatible tasks.

Most diving birds use one of two styles beneath the surface. The "submarines" close their wings upon diving and propel themselves with powerful webbed feet, set so far back on the body that walking on land is awkward. In contrast, the "fliers" rely on their wings underwater, but because wing size has been reduced, many are clumsy in the air. Then there are compromisers like the white-winged scoters, which dive using feet *and* partly spread wings—not a "neat" solution, but an effective one.

Recognition. 19–24 in. long. Stocky, with white patch on wings visible in flight. Male black; small white mark behind eye. Female dark brown with pale spots on sides of head.
Habitat. Open bays and ocean; nests on northern lakes and rivers.
Nesting. Nest is a hollow lined with plant matter and down, concealed or exposed on ground. Eggs 6–14, buff or pink. Incubation about 28 days, by female only. Young downy; leave nest soon after hatching; first fly at 9–11 weeks.
Food. Mollusks, crustaceans, sand dollars, and aquatic plants.

Common Goldeneye

Bucephala clangula

Any mother coaxing her child into swimming for the first time can sympathize with a common goldeneye mother. Before she can coax her newborn chicks into the water she must persuade them to make a frightening leap out of their nest high in a tree.

As soon as a pair of common goldeneyes reach their breeding grounds, the female goes off in search of an old woodpecker nest or tree cavity. She lines it with down from her breast, then deposits one egg each morning until, on average, 10 or 12 lie hidden there. Only when the last egg is laid does she begin incubating, and about 30 days later all her ducklings hatch within a few hours of each other. She broods them for another day or two, until they are strong enough for their long drop to the ground.

After determining that no enemies are lurking nearby, she goes excitedly in and out of the nest until all her babes are scrambling up the sides of the cavity. As they near the entrance, she drops to the base of the tree and calls to the hesitant little ones until, with tiny wings beating wildly, they drop . . . down through the empty air to make their powder-puff landings on the earth below. In scarcely more than a minute all are down and ready to go. The mother quickly gathers her little brood and hurries them away to begin their new lives upon the waters.

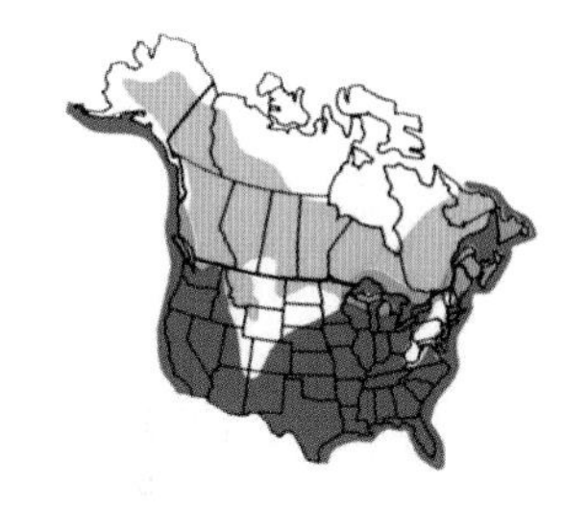

Goldeneyes are sometimes called "whistlers" for the sound their wings make in flight.

Recognition. 16–20 in. long. Stocky. Male mainly white; head round, glossy green with round white spot at base of bill; back black. Female gray, with chocolate-brown head; bill dark with yellowish tip. Wings whistle in flight.
Habitat. Lakes, large rivers, and bays.
Nesting. Nest is a mass of down in tree cavity or bird box, 6–60 ft. above ground. Eggs 6–15, pale green. Incubation 27–32 days, by female only. Young downy; leave nest soon after hatching; begin to fly at 8–9 weeks.
Food. Aquatic insects, aquatic plants, mollusks, and crustaceans.

Bufflehead

Bucephala albeola

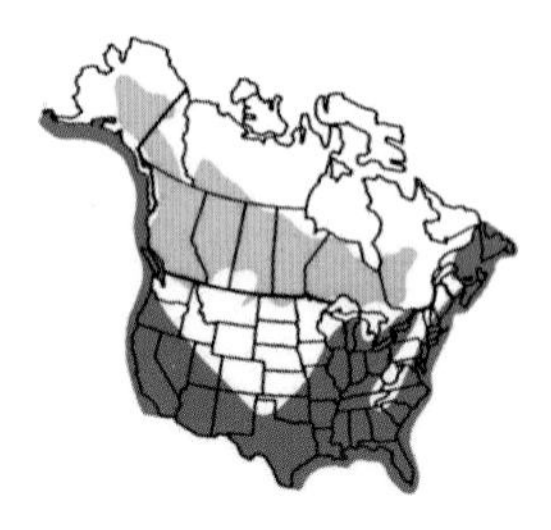

It is the very smallest of our sea ducks, but its big shaggy head looks just enough like a buffalo's head to account for its name. There the resemblance ends. The bufflehead migrates from far-flung wintering grounds to summer in the North Woods, where each breeding pair claims an entire small lake on its arrival. The female is clearly in charge: returning each year to the area of her birth, she locates an old woodpecker hole for her nest, lines it with feathers, and lays one egg each morning for 6 to 11 days. Once the last egg has been laid, she begins to incubate—and her mate soon departs to spend the rest of the summer on a molting ground with other unemployed males.

The ducklings hatch in about a month, and the mother broods them for a day or two, until, like wood duck and goldeneye mothers, she coaxes them to make their long leap to the ground. But then she leads her young to a new territory, where she alone must guard them until they learn to fly and can go with her to join the males.

All this activity, from claiming a lake to reuniting at the molting area, takes almost exactly 120 days—the same period most bufflehead nesting sites are ice-free each year. Thus has experience taught the buffleheads to synchronize their breeding cycle with the precise time slot in which there will be enough food for them and their young.

The black and white pattern on a young bufflehead is a sharp contrast to the adult male's.

Recognition. 13–16 in. long. Small and chunky. Male black and white; head black with large white patch behind eye. Female dark brown above, paler below, with bold white spot on either side of head.
Habitat. Lakes, large rivers, and bays.
Nesting. Nest is a mass of down and feathers, in a woodpecker hole or other tree cavity, 5–20 ft. above ground. Eggs 6–11, pale buff. Incubation about 29 days, by female only. Young downy; leave nest soon after hatching; begin to fly at 7–8 weeks.
Food. Aquatic insects, snails, crustaceans, and aquatic plants.

Ruddy Duck

Oxyura jamaicensis

Anyone who takes the time to watch nesting ruddy ducks will surely be impressed by the labors of the females. Not only do they lay large numbers of eggs—up to 17 in a clutch—but their eggs are larger than those of mallards, birds that weigh twice as much. It would be natural to marvel at a bird whose nest soon contains 20 or 25 eggs. Should the number reach 40, 50, even 60, outright disbelief would be well justified: not even the most zealous female could incubate so many. In fact, that hardworking mother-to-be has had some unrequested help.

Brood parasitism—the laying of one bird's eggs in another's nest—provides a successful lifestyle for a small number of the world's birds. Cowbirds and some cuckoos, for example, have numerous "host" species whose nests they appropriate. Such interspecies intrusions are usually easy to detect, but other cases are far less obvious. Ostriches, some game birds, and many ducks—ruddies among them—lay eggs in the nests of others of their own kind, and it's hard to spot the intruders. This same-species parasitism seems to be motivated either by the destruction of a female's nest or by a shortage of nest sites, and it often works. Occasionally, however, several parasitic females converge for unknown reasons to lay eggs in the same nest—a "dump nest," as it's called. It may be that each assumes one of the others will take up the nest duties; clearly, none of them has any intention of doing so herself.

Like the adults, young ruddies are expert divers but can only shuffle about on land.

Recognition. 14–16 in. long. Small and stocky, with long tail. Breeding male rusty; crown black; cheeks white; bill bright blue. Female and winter male dull brown with dark crown; cheeks pale, crossed by dark line on female.
Habitat. Marshes and weedy ponds.
Nesting. Nest is a floating mass of stems and leaves, anchored to marsh vegetation. Eggs 5–17, white or cream-colored. Incubation about 24 days, by female only. Young downy; leave nest soon after hatching; first fly at 6–7 weeks. Sometimes 2 broods a season.
Food. Seeds and foliage of aquatic plants; insects and snails.

Hooded Merganser

Lophodytes cucullatus

Flashing his handsome black and white crest, the little hoodie swims excitedly around the indifferent female. The more she ignores him, the more entranced he becomes; the faster he swims, the more his elegant body quivers and trembles. Finally, at fever pitch, he leaps vertically from the water and executes a perfect somersault before coming to rest briefly amidst a shower of shining droplets. Unimpressed, the female continues to feed.

And so begins another breeding season for the hooded merganser. But if the strenuous work of courtship is the male's burden, all other domestic responsibilities fall to the female. Once mated, it is she who seeks a nest site, looking for a suitable tree cavity or competing with wood ducks for a nest box in the wet woodlands both species require for raising their young. Lining the nest hole with down from her breast, she then lays her glossy white eggs and incubates them, alone, for some four to five weeks.

Her solitude ends abruptly once the ducklings hatch. Quickly rousted out of the nest and into the water, the fledglings stick burr-tight to their mother's side, their swimming a marvel of group coordination. Despite the mother's watchful care, some of the young are lost to birds of prey, snapping turtles, or even hungry pike. But for those who survive, the breeding season draws to a close with a new generation of hoodies well-fledged, and the promise of another vibrant spring.

The white on a hoodie's crest signals agitation; most of the time it is discreetly folded down.

Recognition. 16–19 in. long. Slender, with crest and thin bill. Male mainly black; large white patch on crest bordered with black; flanks rusty. Female dull brown, paler below, with buff or orange tinge on crest.

Habitat. Wooded lakes, ponds, and streams; also on estuaries and marshes in winter.

Nesting. Nest is a mass of down and feathers in a tree cavity or hollow log. Eggs 6–18, white. Incubation 29–37 days, by female only. Young downy; leave nest soon after hatching; begin to fly at about 9 weeks.

Food. Mainly crustaceans, mollusks, aquatic insects, and frogs.

Common Merganser

Mergus merganser

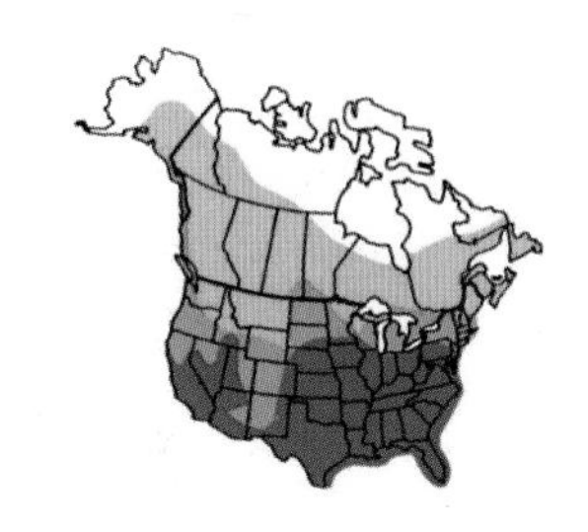

Found in almost any place north of the equator with a suitable habitat, this merganser owes the name "common" to the sheer size of its distribution, which exceeds that of any other mergansers. But it might be called common anyway, simply because—being as large as a goose—it is so easy to see. (The name *merganser* in fact means "diving goose," and the preferred British name for the species is *goosander,* meaning "goose-duck.") The bulky drake with his white highlights stands out against the dark waters of the rivers and lakes he favors. A breeding male is especially conspicuous: his large dark head, with feathered mane and coral-red beak, together with his black back and broad white breast, all add up to a very eye-catching bird.

Common mergs are sociable birds, typically banding together in small groups that make them even easier to find. Like hooded mergansers, the males abandon their mates shortly after incubation begins and gather in large flocks to molt. Then as the breeding season ends, the largest flocks of the year assemble in autumn and early winter. But not all fly south. These hardy mergs like to stay as close to their breeding grounds as possible, braving the encroaching ice and diving for fish in frigid waters. In short, they keep themselves visible to the bitter end, simply by hanging around long after most other ducks have departed.

Like other diving ducks, mergansers must taxi across the water to take flight.

Recognition. 21–27 in. long. Large and streamlined. Male has green head; breast and flanks white; back black; bill red. Female has rusty head with ragged crest; breast white.
Habitat. Wooded lakes and rivers; winters on large rivers.
Nesting. Nest is a mass of down in tree cavity, hollow log, or crevice among rocks. Eggs 6–17, buff or yellowish. Incubation 28–32 days, by female only. Young downy; leave nest soon after hatching; begin to fly at 9–10 weeks.
Food. Fish, crustaceans, mollusks, frogs, and salamanders.

Red-breasted Merganser

Mergus serrator

A young red-breasted merganser reveals no hint of the adult's dramatically colored plumage.

Sea robin: does the name suggest a red-breasted duck riding the waves of coastal waters? If so, then this old nickname for the red-breasted merganser can be an aid to identification, especially where its range overlaps that of its close relative the common merganser. A few visual keys usually help birders tell the difference between them. The most obvious is the red-breasted merganser's spiky, swept-back double crest, a sharp contrast to the smooth head and mane of the common merg. Then, too, red-breasts rarely nest in the tree holes or nest boxes common mergs prefer, but almost always in sheltered hollows on the ground. And unlike his male cousins, a red-breasted drake may be seen with his mate attending to their young brood.

Of course, all mergansers have common traits as well. Usually seen flying low and single-file over the water, they are strong swimmers and divers. And few fish wriggle free from a merganser's bill—long and tapering, with serrations that account for the nickname "sawbill duck." A red-breast rising from the deep with a silvery fish in its beak creates an unforgettable image. In an instant it maneuvers the slippery creature into position and gulps it down head-first—a sight keenly reminiscent of that other red-breast, the robin, disposing of a worm in a country garden.

Recognition. 20–26 in. long. Large and streamlined. Male has green head with ragged crest; breast streaked with buff; flanks gray; back black; bill red. Female has rusty head, ragged crest; breast and flanks gray-brown.
Habitat. Wooded lakes, ponds, and rivers; winters mainly on salt water.

Nesting. Nest is a hollow lined with grass and down, hidden in dense vegetation, in burrow, or among rocks. Eggs 6–16, greenish buff. Incubation 28–35 days, by female only. Young downy; leave nest soon after hatching; begin to fly at about 8 weeks.
Food. Mainly small fish.

Clapper Rail

Rallus longirostris

Noisy, noisy, noisy birds. The trademark clattering of the clapper rails permeates the reeds of salt marshes from coast to coast—yet who ever sees one of them? Feather colors and patterns blend with their surroundings so well that the birds cannot be seen until they move. And when they move, they run on sprinter's legs, their thin, wedge-shaped bodies melting through the close-grown reeds. Only their cackling, wooden-clapper voices disclose that these elusive birds are swarming through the marshes.

But the clapper rails' natural enemies are legion. At every stage of their lives, rails are the constant victims of bobcats, opossums, skunks, minks, and raccoons. Hawks and fish crows prey on them; so do snakes, large fish, and sea turtles. Storms wash the broken eggs, broken nests, and broken bodies of nestlings and adults into windrows of disaster along the high-tide line. Even the moon collaborates with human beings to wreak havoc: hunters waiting in their boats gun the rails down as they climb the reeds to escape the full moon tides.

One last ruse, however, remains in the rails' repertoire. They can swim when necessary (though not for long distances) and they can dive (though not very deep), so when one is pursued it races into the water and sinks beneath the surface. Then, holding tight to a reedy base with its spreading toes, it pokes the tip of its bill into the air like a snorkeling tube. And there it remains, immobile, until the enemy of the moment goes off in search of easier prey.

In a seldom-seen ritual of courtship, a male clapper rail offers food to the female.

Recognition. 14–16½ in. long. Bill long; tail very short. Breast gray-brown (rusty brown on West and Gulf Coast birds), flanks barred; feathers under tail white. Secretive; loud, chattering call heard more often than bird is seen.
Habitat. Salt marshes.
Nesting. Nest is a cup of marsh grass, hidden in dense vegetation in drier part of marsh. Eggs 6–14, buff or greenish, spotted with brown. Incubation about 3 weeks, by both sexes. Young downy; leave nest soon after hatching; become independent at 5–6 weeks.
Food. Crabs, mollusks, and insects.

Virginia Rail

Rallus limicola

In spring across much of North America, when resident birds have begun to sing and the early migrants have returned from their winter haunts farther south, a cattail marsh still seems mostly brown and lifeless. The last season's stalks have been twisted and bent by wind or snow, and the new shoots have not yet emerged. It seems an inauspicious time for a visit, but in fact life has reappeared—the rails are back, setting up territories and seeking food, and spring is the best time to see them.

Years ago, only hunters or naturalists would have waded through a marsh and glimpsed a rail as it flew feebly away. Now, with the aid of recordings to lure them in, it is easier to see rails—but no less magical. At dawn along the edge of the cattails, it is a delight to watch a Virginia rail slip into view through the reeds—its slender body adding another sense to the English idiom "as thin as a rail."

Once in the open, a Virginia rail seems hesitant, taking several quick strides, a slow one, then perhaps another quick one. With each step it bobs its head and flicks its tail in a syncopated gait that makes it seem always slightly off balance. Weaving along the edge of the marsh, it strides over floating vegetation on exaggerated toes, probes for food with its long bill, then finally disappears again into the dense cattails, like a ghost through a wall.

A day-old Virginia rail spends most of its time walking, but it can also swim if it has to.

Recognition. 8½–10½ in. long. Bill long; tail very short. Breast pale rusty; cheeks gray; flanks barred with black. Secretive; more often heard than seen.

Habitat. Freshwater and salt marshes.

Nesting. Nest is a loose cup of grasses and reeds, concealed in marsh vegetation. Eggs 5–12, buff, spotted with brown. Incubation about 3 weeks, by both sexes. Young downy; leave nest soon after hatching; period of parental care unknown.

Food. Mainly insects, mollusks, and small fish; some seeds and aquatic vegetation.

Sora

Porzana carolina

"The sora, or Carolina rail, is unquestionably *the* rail of North America. . . .Throughout its wide breeding range its cries are among the most characteristic voices of the marshes." So wrote Arthur Cleveland Bent in his *Life Histories of North American Birds* over 60 years ago. A businessman with a lifelong interest in birds, Bent was approached in 1910 by the Smithsonian Institution to write a projected six-volume series. He agreed—though in the end Bent's monumental work filled 26 books, the last three compiled after his death in 1954 at the age of 88.

Habitat, nesting and courtship behavior, food, plumages, and migration were among the topics Bent covered. He recorded his own observations and those of his collaborators, and in a distinctly old-school style combined information with unabashed sentiment. Of soras in fall, he noted that "a sudden noise, such as the report of a gun or the splash of a paddle or a stone thrown into the grass, will start a chorus of cries ringing from one end of the marsh to the other. In such places they remain until driven farther south by the first frosts. . . . in late September or early October, a marsh, which was teeming with rails the day before, may be found entirely deserted, every bird having departed during the night. They have started on their autumn wanderings, their fall migration."

For at least the first few days after they hatch, young soras are dutifully fed by their parents.

Recognition. 8–10 in. long. Bill short and yellow; tail very short. Breast gray; face black; flanks barred with black. Young birds similar, but browner, without black on face.
Habitat. Freshwater and salt marshes; weedy pond margins.
Nesting. Nest is a loose cup of grasses and reeds, concealed in marsh vegetation. Eggs 6–18, buff, spotted with brown. Incubation about 3 weeks, by both sexes. Young downy; leave nest soon after hatching; period of parental care unknown.
Food. Mollusks, insects, seeds, and duckweed.

Purple Gallinule

Porphyrula martinica

A museum visitor spying a purple gallinule on display might think that a mischievous taxidermist had replaced its feet with those of a much larger bird. But to watch a live member of the species stride across water lilies in a southeastern marsh is to understand the purpose of these outsized features, which allow this dazzling bird to walk over—if not on—water in its aquatic world.

What a contrast it is to watch the same gallinule attempt a short flight to another part of the marsh, long feet dangling behind as the bird flops above the reeds. Clearly, flight is not the gallinule's forte. Even so, year after year individuals appear far from their normal range, some reaching California, Minnesota, and even Newfoundland in their wanderings. And therein lies a curious characteristic of gallinules and other members of the rail family.

By all accounts, rails and their relatives are poor and reluctant day-to-day fliers. Yet they have traversed vast expanses of water to colonize remote islands around the world. So visitors to Tristan da Cunha, 1,800 miles off the coast of Brazil, should not be too surprised to see a vagrant purple gallinule, though they might wonder why it's there. One expert offers the intriguing theory that birds so reluctant to fly may be equally reluctant to land—and thus, once off course, wander far into the world's outer reaches.

The gallinule's long toes make it one of the few birds able to walk on lily pads.

Recognition. 12–14 in. long. Unmistakable. Green above; head and underparts blue; shield on forehead pale blue; bill red with yellow tip; legs long and yellow. Young mainly greenish buff; bill dark; legs yellowish.
Habitat. Freshwater marshes and weedy ponds.
Nesting. Nest is a shallow cup of grass, reeds, and leaves, woven into marsh vegetation. Eggs 5–10, buff, spotted with brown. Incubation period unknown. Young downy; leave nest soon after hatching; period of parental care unknown.
Food. Seeds, grain, foliage, insects, frogs, and birds' eggs.

Common Moorhen

Gallinula chloropus

For years, these chickenlike birds with the yellow-green feet were called moorhens everywhere in the world—except America. Here they were officially known as common gallinules (from a Latin word for a small hen) and locally referred to as water chickens, mud hens, and chicken-foot coots, among other monikers. But American ornithology finally bowed to international usage, and *Gallinula chloropus* is now recognized as the common moorhen.

Unlike its relatives the rails, the common moorhen does not need extensive spreads of cattails and reeds; a small patch of cattails beside a pond or river will suffice. Not particularly shy, it will peck away at seeds and berries on dry land, prowl lightly over lily pads, swim, dive, or simply tip up like any puddle duck, paying little heed to onlookers.

The nest is wedged among rushes and reeds no more than a foot or two above water level, and a ramp leading down to the water is used by the parents in all their comings and goings. The 10 or so eggs are placed in the center of the nest, and incubation begins when the first egg is laid. This means that the first hatchling is nearly two weeks old when the last breaks out of its shell. But the parents have built several nestlike platforms nearby for brooding the earliest chicks, and with both adults sharing the chores, all the little moorhens are well cared for.

A pair of female moorhens sometimes lay in the same nest, filling it to overflowing.

Recognition. 12–15 in. long. Mainly slate-gray; bill and shield on forehead red; back brown; stripe along side white. Young similar but duller, with dark bill.
Habitat. Marshes, ponds, and lakeshores.
Nesting. Nest is a cup of marsh vegetation, often with "runway" of reeds and stems, hidden in marsh. Eggs 4–17, buff or pale rusty, spotted with brown. Incubation about 3 weeks, by both sexes. Young downy; leave nest soon after hatching; period of parental care unknown.
Food. Seeds, roots, and leaves of aquatic plants; insects and snails.

American Coot

Fulica americana

Coots are a funny sort. Ducklike in their aquatic habits and outline but really closer to rails, moorhens, and gallinules, they seem to have adapted a gallinule's body to paddling about in the water. Their odd, lobed toes—much like a grebe's—serve well in place of a duck's webbed feet.

Coots are also a gregarious lot. Find them in winter on a large lake and they'll often be clustered into a tight flock, sometimes hundreds of them. From a distance the dark mass appears to be a creature moving across the lake's surface, slowly stretching out, then pulling back together as the laggards patter and splash to catch up.

On lakes shared with ducks or swans, coots sometimes act in curious ways. One or two birds may swim over to some canvasbacks or mallards, then pirouette while pecking at the water. At first it appears to be a courtship ritual aimed at the ducks. In fact, the coots are feeding on aquatic plants and animals stirred up by the feet of the waterfowl. Such an encounter with birds of another kind, called commensal feeding, is beneficial to the coot at no cost whatever to the ducks. There are also cases of commensalism in which both sides profit. Great egrets often mingle with white ibises, for example, the ibises stirring up fish for the egrets, while the latter keep a sharp eye out for danger, protecting the ibises as well as themselves.

The coot's lobed feet are useful in swimming, and its sharp claws make formidable weapons.

Recognition. 13–16 in. long. Mainly dark gray; head blackish; bill and small shield on forehead white; feathers under tail white; legs and toes long; toes lobed. Young similar but paler.
Habitat. Ponds, marshes, and lagoons.
Nesting. Nest is a partly floating platform of reeds and marsh vegetation. Eggs 2–22, buff or pinkish, spotted with brown. Incubation about 24 days, by both sexes. Young downy; leave nest soon after hatching; become independent 7–8 weeks after hatching.
Food. Aquatic vegetation, fish, mollusks, and insects; occasionally eggs of other birds.

Dovekie

Alle alle

The Arctic fox knows as little of food chains as the gyrfalcon does of population dynamics. What these two creatures do know is that spring means the dovekies are coming. And the dovekies help keep them alive.

When winter darkness ends and the high Arctic brightens toward a time of midnight sun, the first flocks of dovekies leave the open water and approach their cliffside nesting grounds. Millions more soon follow, in fluttering swarms that form thick black lines sometimes stretching over the waves for miles. Foxes, falcons—and Arctic people—await. Each group forms a strand in a food web dovekies weave through sheer abundance.

Gyrfalcons, glaucous gulls, and ravens haunt crowded dovekie colonies for chicks that will fatten their own young. Arctic foxes also den near dovekie rookeries, using dovekie meat to layer on fat for winter. Eskimos, too, depend on these quail-size birds—by capturing them for food, and by earning an income from the sale of fox pelts, which would not be available if the foxes had no dovekies to pounce on. So it is that strands of life intertwine.

At the center of that web, the dovekies continue to thrive. Arctic weather discourages human intrusion, and the seas assure a bounty of marine crustaceans. Allowed to fish clean waters and return to undisturbed nesting grounds, millions of dovekies will continue to fuel the Arctic summer, and the sky will still darken with birds.

Even in winter, few dovekies stray from the food-rich waters near the pack ice.

Recognition. 8–9 in. long. Small, chunky, and short-winged, with stubby bill. Breeding adult mainly black, with white belly. Winter adult similar, but white on cheeks and breast.
Habitat. Open ocean, often close to shore.
Nesting. Builds no nest; single white egg laid in rocky crevice. Incubation about 24 days, by both sexes. Young downy; leave nest about 4 weeks after hatching and swim out to sea with parents. Often nests in huge colonies on Arctic islands and seacoasts.
Food. Small marine crustaceans; some fish.

Common Murre

Uria aalge

To observe a murre colony is to wonder how so much order can hide behind so much chaos. The confusion is born of numbers and proximity. Shoulder to shoulder, beak to beak, thousands of these sleek black seabirds cram together on the rock ledges of their cliffside nesting grounds. Here, on platforms so narrow they can hardly turn around, adult murres lay single eggs that are pear-shaped for good reason. If moved, they will turn in a tight circle; round eggs would roll right off the cliff.

Once the eggs hatch, true pandemonium erupts. Squadrons of parent birds take off for the sea, diving for small cod and herring that will feed their young. Meanwhile, others are returning from similar missions, creating a blizzard of arrivals and departures. An incoming bird's main challenge is simply to figure out whose chick is whose. Looking utterly identical (at least to human eyes), 10,000 yammering youngsters await; yet each parent bird banks, sets its wings, and alights by just the right chick.

This room service will end in about two weeks, as parents call from the sea far below. Chicks will teeter on their cliff-edge platforms, then tumble softly downward. Buoyed by updrafts, cushioned by soft down feathers, they will land with a faint splash, find their parents beside them, and swim off toward life in the sea.

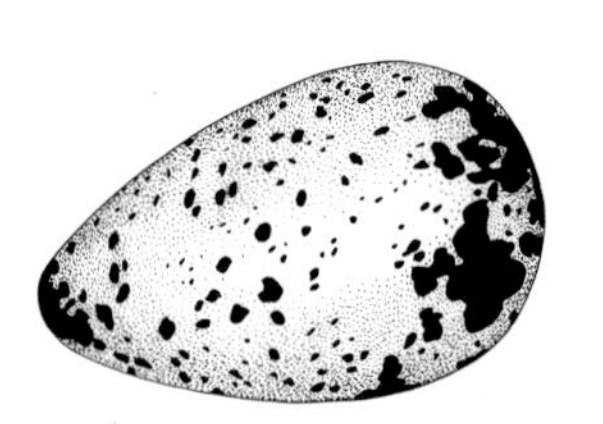

The pear shape of a murre's egg lets it roll harmlessly in a circle instead of tumbling off the ledge.

Recognition. 16–17 in. long. Penguinlike, with long, slender bill. Breeding adult blackish brown above and on head and neck; underparts white. Winter adult has much white on face; black line behind eye.
Habitat. Open ocean; nests on sea cliffs.
Nesting. Builds no nest; single egg, white, greenish, or brown with dark spots, laid on narrow ledge. Incubation 28–34 days, by both sexes. Young downy; leave nest about 25 days after hatching and swim out to sea with parents. Nests in crowded colonies on islands and protected seacoasts.
Food. Mainly fish, shrimp, and squid.

Black Guillemot

Black Guillemot *Cepphus grylle*
Pigeon Guillemot *Cepphus columba*

Dangling 200 feet above the water, dodging frightened birds that rocketed past him from their cliffside nests, Henry Emery must have wondered why he was there. John James Audubon knew why. America's premier ornithologist had sent Emery scaling the sandstone cliffs of one of the Magdalen Islands in the Gulf of St. Lawrence for the eggs of a black guillemot.

Audubon collected eggs as a routine part of his work. Many went to museums. Others—especially those of seabirds like the black guillemot and pigeon guillemot—went for breakfast, rewarding sailors who would risk their lives for science and the taste of fresh food. So it was that ship's captain Emery found himself lowered over a cliff in June of 1832, his mission to reach into darkened fissures, roust the nesting guillemots, and make off with their eggs.

The venture proved anything but routine. Emery was slammed into the cliff face by every gust of wind. Aroused birds shot past like bullets. When dislodged boulders came crashing down, he scrambled aside as sailors high above strained to hold on. An anxious Audubon later wrote: "You may imagine, good reader, how relieved I felt when I saw Mr. Emery drawn up, and once more standing on the bold eminence waving his hand as if to signal success." And if he was relieved, we may imagine how Emery felt.

Pigeon Guillemot

Recognition. 12–14 in. long. Slender and thin-billed. Breeding adult black; large wing patches white (crossed by black bar on pigeon guillemot); feet bright red. Winter adults mainly white; still show white wing patches.
Habitat. Inshore waters and coastal islands.
Nesting. Builds no nest; 1–2 eggs, dull white or greenish spotted with brown, hidden in crevice in rocks near shore. Incubation 3–4 weeks, by both sexes. Young downy; leave nest 5–6 weeks after hatching and swim out to sea with parents.
Food. Fish, mollusks, shrimp, and crabs.

Marbled Murrelet

Brachyramphus marmoratus

In winter, marbled murrelets in their black and white plumage usually feed close to shore.

The sound came at twilight: a sharp *keer, keer* cascading down from the tops of tall conifers that dripped with ocean fog. California loggers heard it. So did coastal Indians in British Columbia. It meant marbled murrelets were coming, returning as they did each summer evening to fill the night air with their calls. But why? Were these ocean-loving birds flying inland, sometimes dozens of miles, to build nests in the crowns of 200-foot trees?

Answering that question would mean solving one of North America's last great birding mysteries. The tireless Arthur Cleveland Bent tried to do it while compiling his multivolume *Life Histories of North American Birds*. Bent combed coastal forests for days in search of the marbled murrelet's nest, but with no luck. He then offered a reward to anybody who could show him one. Still nothing. Over the years, juvenile murrelets were occasionally sighted miles from the coast; but clues were not answers. Only in 1974—more than a century after the first attempts were made to find a nest—did a stroke of pure luck bring success. Climbing high in a Douglas fir, a tree trimmer in northern California accidently shook a murrelet chick from a nest 140 feet above ground. Chick and nest were studied, and experts rendered their verdict: the case of the mysterious murrelets had been solved at last.

Recognition. 9–10½ in. long. Chunky, with dark, pointed bill. Breeding adult dark brown mottled with rusty or gray. Winter adult black and white; dark crown extends below eye; wing patches white; underparts white.
Habitat. Bays and inshore waters.
Nesting. Nest (few ever found) a small platform of lichens and moss on limb of forest tree up to 140 ft. above ground; single egg pale green with brown spots. In Far North nests among rocks. Incubation and nestling periods unknown.
Food. Fish, mollusks, and crustaceans.

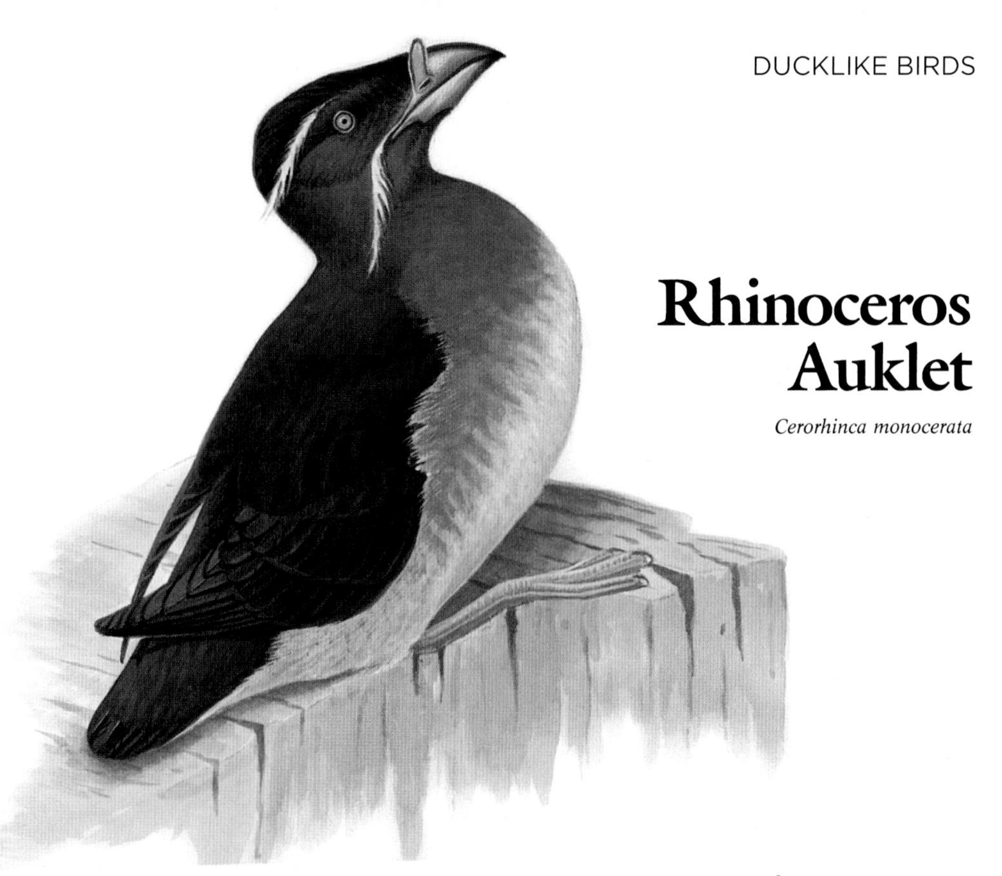

Rhinoceros Auklet

Cerorhinca monocerata

If birds could lay eggs in the water, this one would never touch land at all. As it is, the seagoing rhinoceros auklet reduces terrestrial contact to an absolute minimum. For much of the year this sardine-hunting diver seems to disdain even the sight of a shoreline, avoiding inland straits and channels in favor of bobbing on ocean swells. Not until spring—and the appearance of an odd little horn that sprouts at the base of its bill—does the bird relent. Island slopes beckon. Only there can a rhinoceros auklet dig its 8- to 20-foot long nesting burrows.

As if embarrassed to touch dry land, the auklet arrives under cover of darkness. Barking, growling, even shrieking like a parrot, this normally quiet bird sets about digging new burrows or renovating old ones. Single downy chicks soon hatch at the end of these passageways. Parents fly out to sea on nighttime feeding trips, return by dawn, and remain underground until dusk lures them seaward again.

The rhinoceros auklet's aversion to daylight is but one trait that distinguishes it from other members of the auk family. Some relatives—puffins and guillemots, for example—tend their young actively by day. Others, such as murres and razorbills, lay their eggs on rock ledges instead of inside burrows—differences that help related species avoid head-on competition for nesting sites and food.

The rhinoceros auklet loses its horn in winter, and its head plumes are much reduced.

Recognition. 14–15½ in. long. Breeding adult brownish, paler below; belly white; head with 2 rows of white plumes; bill reddish with pale knob at base. Winter adult similar, but lacks head plumes and horn on bill.
Habitat. Open ocean and coastal cliffs.
Nesting. Nest is a small mass of sticks and stems at end of burrow on wooded island. Single white egg, sometimes spotted with gray or lilac. Incubation period about 33 days, by both sexes. Young downy; leave nesting burrow 5–6 weeks after hatching and swim out to sea with parents.
Food. Fish and crustaceans.

Tufted Puffin

Fratercula cirrhata

Bringing food to their young, tufted puffins are often robbed by faster-moving kittiwakes.

As a deer sheds its antlers, so this bird sheds its remarkable beak. The divestiture takes place in late summer, after pairs of tufted puffins have fledged their chicks from burrows in the tops of island promontories. The breeding season is over. A life at sea awaits—nine months of riding out Pacific storms while diving for fish. Flashy displays serve no purpose here: no mates need convincing, no pair bonds need strengthening. The tufted puffin's brightly hued breeding mask has become excess baggage.

Layered during the spring and summer with sheathlike plates that transformed it into a massive, multicolored tomahawk, the puffin's bill now sheds no fewer than seven plates in separate molts. What remains is a dark winter beak, fully one-third smaller than before. Practicality has rendered a gaudy bird plain.

The puffin's repudiation of finery doesn't stop there. As winter approaches, it also molts the tufted eyebrowlike plumes that inspired more than one observer to call this bird the "old man of the sea." Gone, too, are the puffin's pure white facial feathers, replaced by blackish ones. Having traded flash for function, the tufted puffin patrols wintry offshore waters with a less showy but just as effective bill, its serrations allowing the deep-diving hunter to grip several fish sideways as it restlessly looks for more.

Recognition. 14½–15½ in. long. Breeding adult mainly black; face white, with long, curved, golden tuft behind each eye; bill large, bright orange. Winter adult much duller, with trace of tufts and smaller, dull orange bill.
Habitat. Inshore waters and coastal cliffs.
Nesting. Nest is a burrow in sandy bluff above beach or on top of sea island. Single egg white or pale blue, spotted and scrawled with lilac. Incubation and nestling periods unknown. Sometimes 2 broods a season; often nests in colonies.
Food. Mainly small fish.

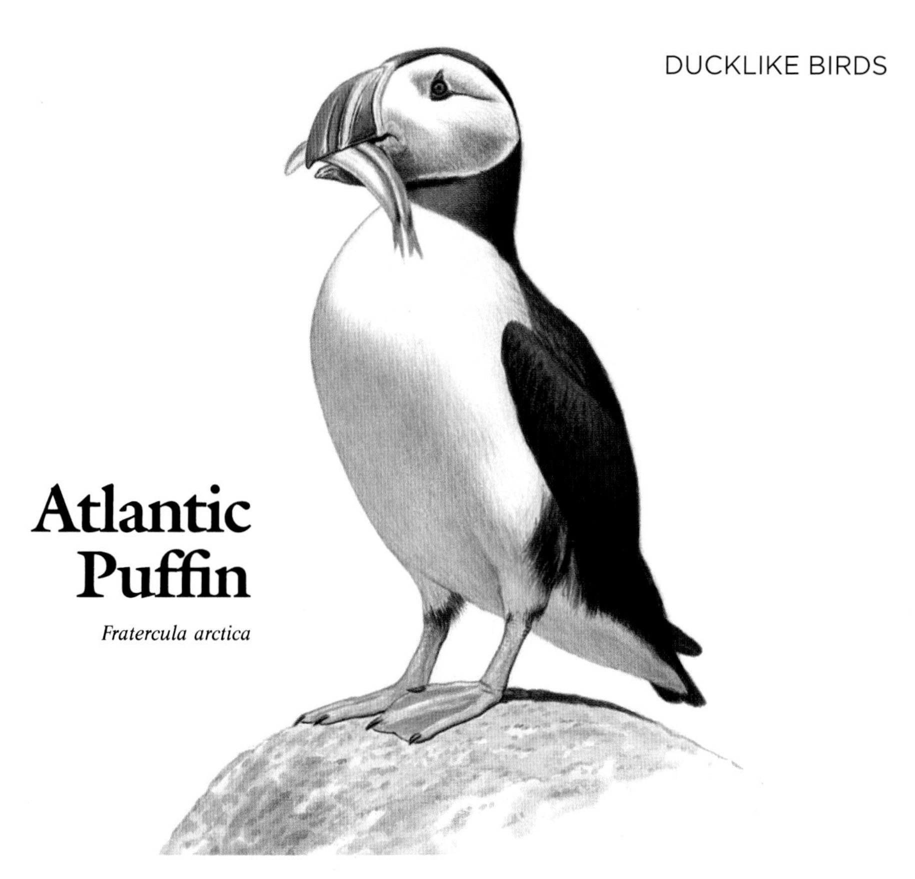

Atlantic Puffin

Fratercula arctica

Although puffins and penguins literally live poles apart, the two birds lead lives of uncanny similarity. Atlantic puffins dive for food in cold ocean waters. So do penguins. Puffins are confiding in the presence of humans, use their wings for underwater propulsion, and repel cold with dense waterproof feathers. Penguins do, too. The two birds even look alike. Sporting black and white tuxedos, they stroll with comical yet endearing clumsiness toward the water's edge—then knife through the sea like torpedoes.

Appearances notwithstanding, puffins and penguins aren't even distantly related. What has made them so alike is a need to survive in strikingly similar environments. Cold marine places, northern or southern, demand that a diving bird's body be densely feathered and streamlined.

For all the similarities, puffins and penguins do have their differences. While Atlantic puffins dig nest burrows in turf-covered cliff tops, Antarctic penguins lay their eggs out on the ice. Puffins can fly, too, unlike penguins—though they don't do it well. Splashing along the surface, their stubby wings flailing in a valiant attempt to take off, puffins succeed only some of the time. But if we are tempted to dismiss them as woeful aviators, we need only see them in flight where it counts—slicing through the ice-blue seawater—to marvel at their true virtuosity.

In a downy young puffin there is little sign yet of the breeding adult's ornate bill.

Recognition. 11½–13½ in. long. Chunky and large-billed. Black above; face and underparts white; bill triangular, red, yellow, and blue; legs orange-red. Winter adult much duller; face darker; bill smaller with little color.
Habitat. Open ocean; nests on rocky islands.
Nesting. Nest is a mass of grass, seaweed, and feathers in burrow in soil on islands. Single egg white, occasionally spotted with brown. Incubation period 5–6 weeks, by female only. Young downy; leave nest without parents about 7 weeks after hatching.
Food. Fish, mollusks, and crustaceans.

They seem as integral to the seaside as sunshine and salt air—the gulls, terns, and other large birds that soar, hover, skim the waves, or plunge below them along our coasts. While some of these "seabirds" are as much at home inland as at sea, others make an acrobatic art of ocean living, from low-flying shearwaters and skimmers to deep-diving boobies and gannets. Add the inimitable pelicans—comical when they stand still, surprisingly graceful in flight—and you have an array of birds without which our coastlines would be sadly incomplete.

Gull-like Birds

Mew Gull

Northern Fulmar

Fulmarus glacialis

Light-phase adult

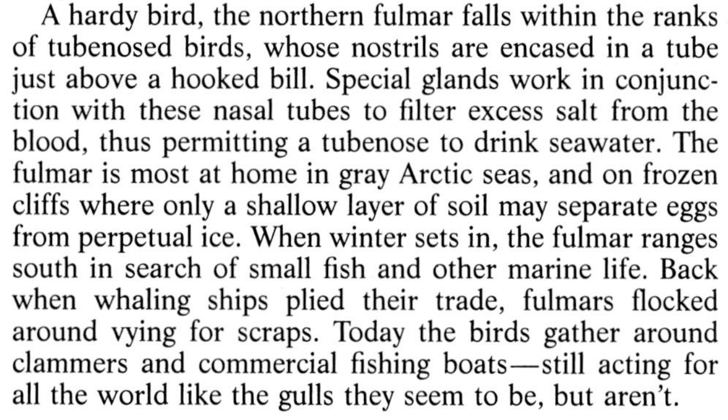

Gray wings, white head and tail—at first glance, the bird could be mistaken for a gull. Actually, the northern fulmar comes closer to a seafaring life than most "sea gulls"; they rarely range beyond sight of the coast, whereas the fulmar spends most of its time out on the ocean.

In fact, that may be just as well: examining fulmars is something best done at a distance. Disturbed, they vomit a vile, oily substance, the smell of which lingers almost as long as the memory of the encounter. Reek and risks notwithstanding, fulmar oil was once collected as a fuel for lamps and as a remedy for assorted ailments.

A hardy bird, the northern fulmar falls within the ranks of tubenosed birds, whose nostrils are encased in a tube just above a hooked bill. Special glands work in conjunction with these nasal tubes to filter excess salt from the blood, thus permitting a tubenose to drink seawater. The fulmar is most at home in gray Arctic seas, and on frozen cliffs where only a shallow layer of soil may separate eggs from perpetual ice. When winter sets in, the fulmar ranges south in search of small fish and other marine life. Back when whaling ships plied their trade, fulmars flocked around vying for scraps. Today the birds gather around clammers and commercial fishing boats—still acting for all the world like the gulls they seem to be, but aren't.

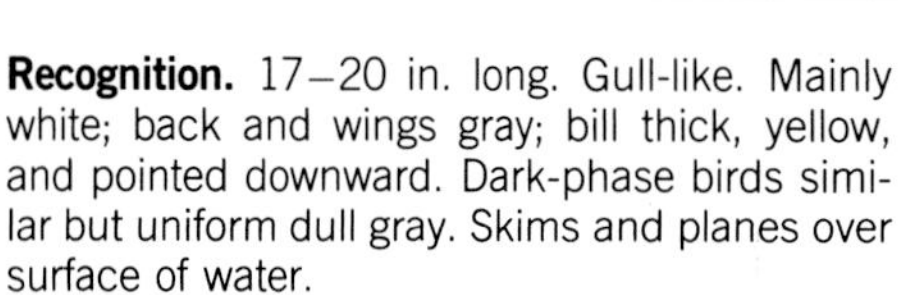

Recognition. 17–20 in. long. Gull-like. Mainly white; back and wings gray; bill thick, yellow, and pointed downward. Dark-phase birds similar but uniform dull gray. Skims and planes over surface of water.

Habitat. Open ocean, except when nesting.

Nesting. Nest is a shallow depression with a few stems or blades of grass, on a sea cliff or ledge. Single egg white. Incubation about 8 weeks, by both sexes. Young downy; leave nest about 7 weeks after hatching.

Food. Fish, mollusks, and crustaceans; refuse thrown from boats.

Leach's Storm-Petrel

Oceanodroma leucorhoa

Over a shimmering oil slick left in the wake of surfacing bait fish, several dozen small dark birds flutter and dance. "Mother Carey's chickens," some mariners call them. "Wilson's storm-petrels," say ornithologists. Among the flock is one larger, more angular petrel. Its flight brings to mind some stiff-winged, seagoing bat rather than a fluttering chicken. That one will be a Leach's storm-petrel.

Storm-petrels are birds shrouded in superstition. Seamen regard petrels gathering at the stern of a vessel as an omen of bad weather. The name "petrel" traces its origin to St. Peter and the belief that the fluttering birds walked on water—a talent any sailor might envy. The birds cannot really walk on water—though the dancing, foot-pattering flight of feeding birds does make it appear so.

Leach's, like other storm-petrels, nest in burrows, and during the breeding season they're active at night—the darker, the better. On the wooded islands where they nest, adults navigate toward the spruce-fir spires that stand over their burrows. Plummeting through the canopy, bouncing from branch to branch, they drop to the forest floor. Clambering over obstacles, each bird scrambles toward the one entrance among many leading to its own burrow. And within each cavity is one young petrel that, come September, will fledge and join its parents for a life at sea.

A young petrel remains in its burrow for more than two months before going to sea.

Recognition. 7½–9 in. long. Small, dark, and swallowlike. Rump usually white, sometimes dark on Pacific birds; tail notched. Skims and swoops over ocean on bent wings; flies like nighthawk, with deep wingbeats.
Habitat. Open ocean.
Nesting. Nest is a burrow dug by male in soil on offshore island, often under log. Single egg white. Incubation about 6 weeks, by both sexes. Young downy; leave nest 9–10 weeks after hatching. Nocturnal on nesting grounds. Usually nests in colonies.
Food. Fish, shrimp, and refuse from ships.

Sooty Shearwater

Puffinus griseus

In April and May, the great exodus reaches the coasts of North America. Just offshore, millions of chocolate-colored shearwaters pass in review before West Coast birders. Off the Atlantic Coast, the host of migrating birds numbers mere thousands. But unlike many birds heading north with the spring, the sooty shearwaters are not looking to nest. For these ocean travelers, nesting season is over. They are heading for the waters off Newfoundland and Alaska for a summer rich in fish. Only in August will the mass of birds tack south again.

It takes specialized skills to negotiate the distance between hemispheres, and that is precisely what shearwaters have—a special knack for long-range flight. Some birds muscle their way through the air on rowing wings. Others hitch rides on thermal air currents and updrafts, a trick called passive soaring. But shearwaters are also capable of dynamic soaring: their long, narrow wings tease lift out of the wind itself, climbing the gradient between slower layers of air near the water and faster-moving air above. Once aloft, gravity pulls them along in an effortless glide, allowing them to fly hour after hour on motionless wings. But when the air is still, even these supreme gliders cannot find lift. They can only sit idly on the water in great brown rafts, waiting like becalmed sailing ships for a rising breeze.

Recognition. 19–20 in. long. Slender and long-winged. Uniform dark brown, with whitish wing linings. Glides and planes over the waves; sometimes follows ships for food. Often seen in large flocks, especially on West Coast.
Habitat. Open ocean; nests in the Southern Hemisphere.

Nesting. Nest is a mass of grass at end of burrow on sea island. Single egg white. Incubation about 8 weeks, by both parents. Young downy; leave nest about 14 weeks after hatching. Nests in large colonies.
Food. Fish, squid, and crustaceans; sometimes takes refuse thrown from ships.

Greater Shearwater

Puffinus gravis

In April, on silent, almost motionless wings, the entire world population of greater shearwaters leaves its nesting burrows on three tiny South Atlantic islands. May and June will find the birds off the eastern seaboard of North America, feasting on vast schools of sand eels. By July, legions of greater shearwaters, several million strong, will have gathered off Newfoundland. Here, until September, they will forage in the rich waters of the Grand Banks and fight for the scraps tossed overboard by local fishermen.

Dark above, white below, and as silent as owls in falling snow, feeding shearwaters circle fishing trawlers like a host of feathered satellites. Though clumsy on land, shearwaters become the very picture of grace in flight, at one with the wind and the sea. Soaring just inches above the ocean, the hard-banking birds sometimes skim the surface with pointed wingtips—*shearing* the water. But when a morsel is tossed overboard, the artful ballet becomes a melee. Dozens of birds may vie for the scrap, plopping to the water's surface, lashing out with wings and bills as the prize is consumed piecemeal.

Once, seamen captured the birds by the thousands to use for food or fish bait. Today, shearwaters are still accidentally snared by fishermen's baited hooks—the price some must pay, perhaps, for the abundance enjoyed by the rest.

This species is one of the shearwaters that will dive after fish and squid.

Recognition. 18–20 in. long. Slender and long-winged. Brown above; cap dark, contrasting with white face, throat, and underparts. Glides and planes over the waves; follows ships for food.
Habitat. Open ocean; nests in the Southern Hemisphere.
Nesting. Nest is a mass of grass at end of burrow on remote oceanic island. Single egg white. Incubation about 8 weeks, by both parents. Young downy; leave nest about 12 weeks after hatching. Nests in colonies.
Food. Fish, squid, and crustaceans; refuse thrown from fishing boats.

Brown Booby

Sula leucogaster

The western brown booby's whitish head contrasts sharply with its eastern counterpart's.

Suppose that, after flipping through your field guide, you decide you'd like to see a brown booby. You admire the graceful shape, you want to see those wonderful webbed feet, or you simply think the name is amusing (Spanish sailors called them *bobo,* or "stupid," for their foolishly trusting nature on nesting islands). Now you do a little research: to see a brown booby in North America you'll probably have to sail to some offshore island, or you'll have to brave the desert heat of the Salton Sea in late summer, when a few boobies come to visit. Not an attractive choice—but how else can you get a brown booby for your North American life list?

Birding becomes more popular each year. To some it's a way to get outdoors, to others an esthetic delight. And to many it's a sport—a competitive sport. Myriad are the tales of birders braving any element, driving any distance, wading through any substance, and crawling over any obstacle to add a new bird to their life lists. But along the way they also see some wild shorelines, some fabulous landscapes, some fantastic plants, and some wonderful creatures. Above all, they have fun. Take away the life lists, the thrill of the chase, the elaborate planning—and the most serious birder is still, at bottom, someone who simply loves looking at birds.

Recognition. 26–28 in. long. Large, with pointed bill and pointed tail. Upperparts and breast dark brown; belly and wing linings white; dark breast and white belly sharply defined. Small yellowish throat pouch visible at close range. Young birds all dark brown.
Habitat. Tropical oceans.

Nesting. No nest; 1–3 pale bluish eggs laid on bare ground or rocks, often at edge of cliff where birds can easily launch into air. Incubation about 6 weeks, by both sexes. Young leave nest about 15 weeks after hatching.
Food. Fish, obtained by diving from air.

Northern Gannet

Morus bassanus

Not far from shore, well within the range of the unaided eye, a string of large seabirds moves by, rising and falling like a roller coaster on invisible rails. Suddenly one bird stalls, folds its wings, and plummets into the waves. A tower of spray marks the spot where the northern gannet has just caught its dinner. Birds dive from heights of more then 100 feet and may reach an equal depth beneath the surface. One feeding gannet attracts others. Soon the waters roil under a bombardment of birds, and the assault will continue until the beleaguered fish panic and retreat.

On land, this handsome seabird is more ungainly but hardly less spectacular. Nest cliffs towering high above the Gulf of St. Lawrence glow white beneath their blanket of birds. Gannet nests are impressive affairs, spaced along sandstone ledges, each just beyond the reach of its neighbor's bill. Despite their girth, gannet nests will receive only a single egg apiece. Young birds enjoy careful tending and all the fish they can eat—until fledging time draws near. In September, the room service stops and young gannets face a choice between hunger and the great emptiness that stretches beyond their ledge. After several days of reluctance and a series of false starts, each young gannet makes the great leap of faith, half flying, half falling to the water below and the beginning of its life at sea.

Gannets in pursuit of fish are famed for their spectacular dives, often visible a mile away.

Recognition. 35–40 in. long. Large, with pointed bill and pointed tail. Adults snow-white, with black wingtips; head tinged with yellow; bill gray. Young birds mottled brown, whitish below, becoming paler as they grow older. Flies over waves with steady, powerful wingbeats.
Habitat. Open ocean.

Nesting. Nest is a mass of seaweed and debris in dense colony on offshore island or isolated cliff. Eggs usually 1, pale chalky blue. Incubation about 6 weeks, by both sexes. Young leave nest 13–15 weeks after hatching.
Food. Fish, caught by diving from air.

American White Pelican

Pelecanus erythrorhynchos

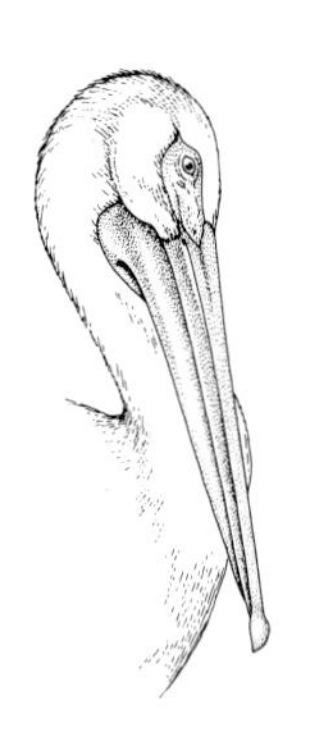

The vertical plate on this pelican's bill is shed soon after the nesting season.

Looking ludicrous on land with its oversize bill and feet, yet sublimely handsome in flight, this bird inspired two of the most delightful lines in modern verse: "A wonderful bird is the pelican, / His bill will hold more than his belican." And no poetic license has been taken here: the white pelican's bill can hold almost three gallons of water as it scoops for fish—more than twice the capacity of its stomach. And before the nine-week mark, a baby pelican's stomach will digest as much as 150 pounds of fish hauled to it by its tireless parents, who may fly 100 miles to find food for their youngsters.

But a good work ethic and family values are not these birds' only admirable traits. Nature offers few sights more serene than a flock of white pelicans on the wing—flying, gliding, or soaring high above the lakes where they breed, or along the warmer seacoasts to which they migrate in the fall. And they are master fishers as well, whether working singly or in small groups. An individual bird drops onto the water with a splash, submerges its huge bill, and scoops up a fish with startling speed. But white pelicans can also team up to form a floating semicircle facing a shoreline. With great wings throwing the water into spray, the pelicans systematically drive their fishy harvest into the shallows, where the feast will soon begin for all.

Recognition. 50–70 in. long. All white with black wingtips. Very large and bulky, with huge, flat yellow bill. Often seen in large flocks; frequently soars.

Habitat. Marshes, lakes, and coastal lagoons.

Nesting. Nest is often a large mound of soil and plant debris on island or in marsh; sometimes floating; occasionally no nest is built at all. Eggs 1–6, dull white. Incubation about 5 weeks, by both sexes. Young leave nest about 4 weeks after hatching.

Food. Fish, caught as bird swims on surface.

Brown Pelican

Pelecanus occidentalis

One of nature's memorable images—a line of brown pelicans advancing silently in the air, wingtip to wingtip, along a coastline at dawn or dusk—has endured only because a U.S. president was determined that it would. By 1903, so many brown pelicans had been killed by hunters and fishermen that Theodore Roosevelt set aside an island off Florida's east coast as a wildlife refuge for the bird. It was the first of many refuges now maintained worldwide, and the brown pelican is the true hero of this success story.

It is a noteworthy bird on several other counts as well. A spectacular diver, the brown pelican goes into a headfirst plunge of 60 or 70 feet, then emerges from the spray with a fish securely in the grasp of its large bill. Nothing is absolutely perfect anywhere, however, even for birds. Sometimes, just as a brown pelican is flipping the fish above its bill—so as to swallow it headfirst—an adventurous gull will dart by and snatch the fish in midair.

After every dive, a network of air sacs running through the pelican's skeleton and just under its skin helps lift the seven- or eight-pound bird to the ocean's surface, where it always reappears facing the wind for easier takeoff. In reality, though, no takeoff is likely to seem easy to this bird as it flails furiously with wings and feet to reach the speed it needs to climb skyward again.

Only a young brown pelican is really all brown; the adult always has white on its head.

Recognition. 42–54 in. long. Very large, with huge, flat bill. Adults gray-brown with white head and neck; back of head and neck dark rusty in nesting season. Young birds dull brown. Dives for fish from air; often seen perched on jetties and pilings.
Habitat. Coastal bays, lagoons, and estuaries.

Nesting. Nest is a mound of sticks, twigs, reeds, and plant debris 4–10 ft. above ground in bush or small tree; sometimes a large mound of mud and debris on ground. Eggs 2–3, whitish. Incubation about 4 weeks. Nests in colonies, usually on protected islands.
Food. Fish and some shrimp.

Parasitic Jaeger

Stercorarius parasiticus

Jaegers often follow birders on the tundra, snatching up chicks whose parents flee their nests.

The steel of this creature's spirit is tempered by two worlds—one of the land, the other at sea. *Jaeger* means "hunter" in German, and for much of its life the jaeger's hunting ground is the sea. It searches from the heights or rises from shadows of concealing troughs, a feathered harpy that snatches food from the mouths of other seabirds. Faster than gulls and more agile than terns, the jaeger harasses its victim until the overburdened bird is forced to drop its catch. Usually, the morsel belongs to the jaeger before it reaches the water.

In the endless sunlight of an Arctic summer day, the jaeger casts a shadow of terror across the lives of nesting birds. Coursing low over the tundra, jaegers spy out and feast upon eggs and nestlings. Lemmings, small rodents of the tundra, are also favorite prey. The spoils are carried back to the jaegers' own hungry nestlings—who, when grown, will likewise embark on lives of piracy and plunder.

The distribution of jaegers at sea is not firmly established, but in winter and during migration parasitic jaegers usually come much closer to land than the related pomarine and long-tailed jaegers. Still, it is the lucky observer who actually *sees* a jaeger, any jaeger. Most people are treated to little more than the glimpse of a sharp-edged shadow, cleaving a path through a distant cloud of gulls.

Recognition. 15–21 in. long. Brown above; cap blackish; underparts white with vague breast band; middle tail feathers longer and pointed; flash of white near wingtips in flight. Often pursues gulls and terns to rob them of food.

Habitat. Bays, ocean, and beaches; nests on Arctic tundra.

Nesting. Nest is an unlined depression in ground. Eggs 1–3, olive with brown spots and blotches. Incubation 24–28 days, by both sexes. Young downy; leave nest about 5 weeks after hatching.

Food. Fish stolen from other birds; rodents and small birds on tundra.

Bonaparte's Gull

Larus philadelphia

In a bird family notorious for its raucous cries, indiscriminate appetites, and often quarrelsome colony life, the Bonaparte's gull seems almost reserved by comparison. During the spring and summer it prefers to raise its family in the coniferous forests of Alaska and western Canada, well removed from the competitive feeding arenas of the seacoasts. The smallest, most delicately boned gull in the Western Hemisphere, the Bonaparte's exhibits some of the grace of terns in flight. It is also a relatively neat feeder, either skimming along the water's surface to snatch up small fish or else dropping right into the water, where it can readily pick out small crustaceans and marine worms.

Bonaparte's gulls usually begin their migrations toward winter quarters about August, making frequent stops en route to rest and feed in marshy terrain. Using flyways that will eventually attract some of them into the Caribbean, the Bonaparte's often follow the larger rivers toward the Atlantic and Pacific coasts. One river they obviously like (60,000 have been estimated there at one time) is the Niagara, where they ride the turbulent waters above and below the cataracts. The Bay of Fundy region sends additional thousands southward, giving bird watchers in the United States a chance to scan coastal waters for the well-mannered gull with a black head and bill.

Unlike other American gulls, the Bonaparte's nests in small trees, not in coastal colonies.

Recognition. 12–14 in. long. Breeding adult has black hood; bill black; back and wings pale gray, with bold white wedge on outer wing; wing linings white; legs red. Winter adult similar but head white, with small black spot behind eye. Young bird like winter adult, with dusky wing bands, black tail band.

Habitat. Lagoons, estuaries, and beaches; nests on northern lakes.

Nesting. Nest is a cup of twigs 5–20 ft. above ground in conifer. Eggs 2–3, buff with brown spots. Incubation about 24 days, by both sexes. Young downy; nestling period unknown.

Food. Insects, crustaceans, and marine worms.

Franklin's Gull

Larus pipixcan

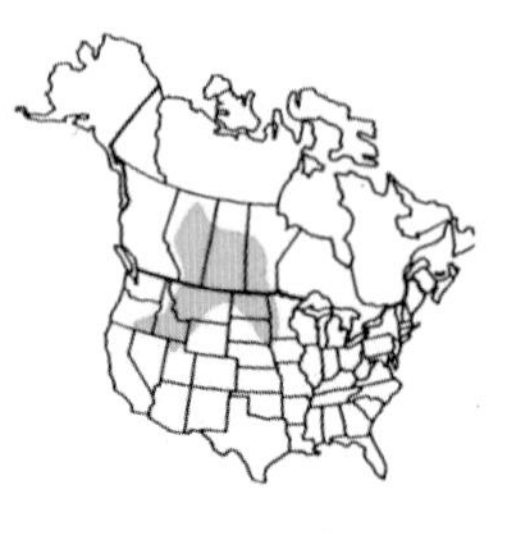

Flocks of Franklin's gulls follow plows, seizing insects driven from their hiding places in the soil.

Picture a large flock of gulls, standing on a beach or in a field—all ages, all colors, all mixed together. Someone once said that it's fun to identify gulls, but for a beginning birder this is no fun at all.

Making sense of such visual cacophony begins with understanding how gulls' plumages change. The Franklin's, for example, is a two-year gull, reaching its full adult plumage and readiness for breeding at two years of age. It is born in the summer, wears juvenile plumage briefly, then molts into a first-winter plumage. The next spring it molts into a first-summer plumage. In its second fall it molts into the adult winter plumage, and the spring after that into its breeding summer plumage. Thereafter it alternates normally between adult summer and winter plumages. So, for a two-year gull like the Franklin's, at any season one of two plumages must be watched for—first-year or adult.

The Bonaparte's gull also takes two years; others, like the ring-billed and Heermann's, are three-year gulls, while the California, herring, and great black-backed are four-year gulls. With each added year producing an extra set of summer and winter plumages that appear only once, even the experts sometimes grow bewildered. Mercifully, at least male and female gulls look alike: otherwise there would be twice as many plumages to keep track of.

Recognition. 13–15 in. long. Breeding adult has black hood with bold broken eye-ring; bill red; back and wings gray; wingtips blackish, crossed by white band; rear edge of wing has narrow white border. Young birds dark brown above, whitish below; head has dusky hood; bill dark; rump white; tail with broad black band.

Habitat. Inland marshes, lakes, and beaches.
Nesting. Nest is a cup of reeds and marsh grasses, in marsh. Eggs 2–3, olive with dark brown spots. Incubation about 3 weeks. Young downy; nestling period unknown.
Food. Insects and small fish.

Laughing Gull

Larus atricilla

From its dockside perch, the black-hooded herald draws back its head and hurls its challenge at winter: *Haah-ha-ha-ha-ha-ha.* Soon the welcome news is all over the coastal town. The "summer gull" has returned, and with it the start of another busy summer season.

From the Carolinas southward, the laughing gull is a year-round resident, though some may not know it for the same bird. In breeding plumage, the laughing gull's head is black, the bill rich and red. But in late summer, the dark head feathers are replaced by lighter ones until only a smudgy "earmuff" remains. (Many black-headed gulls follow this molting sequence—dark heads in summer and pale ones in winter—while white-headed species like the herring and California gulls do just the opposite.)

Rarely found far from the east coast, the laughing gull is no patron of landfills or garbage pits. A versatile and active feeder, it forages on everything from the eggs of horseshoe crabs to earthworms plucked from turned soil in the wake of tractors. Hundreds of laughing gulls may gather to snap up greenhead flies when a hatch of those vicious, biting insects is in progress. Gullets bulging, the birds return to their salt marsh colonies. Hungry nestlings waiting there feast on the captured insects—and coastal residents have one more reason to welcome the laughing gull's return.

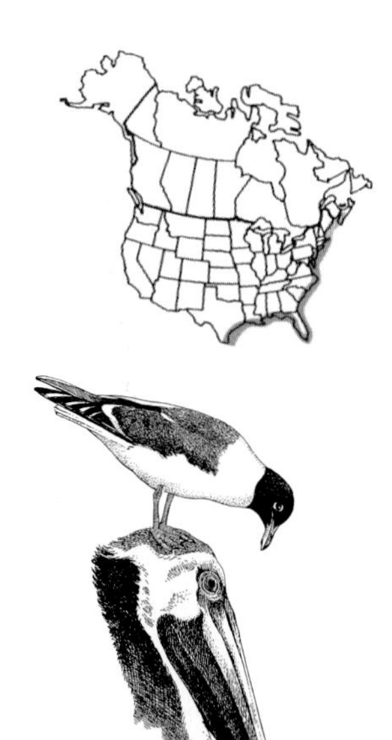

A laughing gull sometimes perches on a pelican's head in hopes of stealing a fish.

Recognition. 15–17 in. long. Breeding adult has black hood; bill red; back and wings gray; wingtips blackish, with no white band; rear edge of wing has narrow white border. Winter adults similar, but head white with gray mottling; bill blackish. Young birds dark brown above, whitish below; rump white; tail with broad black band.

Habitat. Beaches, salt marshes, and lagoons.
Nesting. Nest is a sturdy cup of reeds and marsh grasses, in salt marsh. Eggs 3–4, olive with dark brown spots. Incubation about 3 weeks. Young downy; nestling period unknown.
Food. Mainly fish, worms, insects, and refuse thrown from ships.

Heermann's Gull

Larus heermanni

Nesting on sweltering islands, Heermann's gulls must shield their eggs from the heat.

What a relief it is to lay eyes upon a Heermann's gull! Here is a gull that is easy to identify, its dark body handsomely set off by a blood-red bill and, in breeding garb, an elegantly white head. Even the chocolate-brown young are easily told from other gulls.

Both young and adult Heermann's clearly diverge from the "mainstream" appearance of our white-bodied (and sometimes black-headed) North American gulls. And they are exotic in habits as well as in looks. They breed in spring, primarily in northwestern Mexico—where the main task of the "incubating" parent is not to warm the eggs but to shade them from the blazing sun. After breeding, Heermann's gulls and their young disperse northward from Mexico across the western United States, reaching British Columbia by July and remaining along the Pacific coast until they move south to nest again.

No one knows how Heermann's gulls came to reverse the usual pattern of southward dispersal after breeding, but the food-rich West Coast is a natural destination for any gull breeding nearby. Nine other species of gulls "winter" there as well—eight of them migrants from farther north. In any event, the Heermann's contrary habits clearly meet its needs quite nicely, and its summer tourism provides a welcome sight on our western beaches.

Recognition. 18–21 in. long. Breeding adult mainly gray; head white; bill red; tail black. Winter adult similar, but head mottled with grayish brown. Young birds dark chocolate-brown with pinkish bill.
Habitat. Bays, harbors, and seacoasts.
Nesting. Nest is a mass of sticks and grass among rocks on offshore island. Eggs 2–3, cream-colored, spotted with brown and lilac. Incubation and nestling periods unknown. Nests in colonies.
Food. Fish, mollusks, shrimp and other crustaceans; also scavenges.

Mew Gull

Larus canus

While it's true that many members of the gull family look and act alike—swirling high above landfills, flocking noisily behind ferries and fishing boats—not all are pushy pierside beggars with a taste for garbage. The diminutive mew gull, for one, prefers to comb mud flats for worms and tread winter beaches for mollusks, rarely visiting the offal heaps where herring gulls, western gulls, and other birds scavenge for food.

Come spring, when flocks of mew gulls fly toward their inland breeding grounds, their behavior again strays from the expected. Instead of staking out the local dump, they flock to farmlands. They know it's planting time, and that deep-digging plows will unearth insect larvae in copious amounts. Because a number of these larvae feed on crops, the mew gulls (like California and Franklin's gulls on their breeding grounds) are warmly welcomed by the farmers.

Swooping low as the plow blade sweeps past, mew gulls hover buoyantly and pluck their food off the ground. They use similar flight skills on their coastal wintering grounds, dipping toward the salt water to grab sand eels with their bills. Mew gulls can drink the salt water, too, extracting the salt with special glands located above their eyes—glands that other gulls and seabirds also possess. Mew gulls may be different, but they're not different all of the time.

Recognition. 16–18 in. long. Mainly white; back and wings gray; wingtips black with large white spots; bill small and yellow; legs pinkish. Young birds mottled gray-brown; bill small, dark-tipped; legs dull pink.
Habitat. Seacoasts, marshy lakes, and rivers.
Nesting. Nest is a cup of seaweed, grass, and stems, on ground or in spruce tree. Eggs 1–4, buff or greenish, spotted with brown. Incubation 25–33 days, by both sexes. Young downy; leave nest soon after hatching; become independent at about 5 weeks. Often nests in colonies.
Food. Insects, earthworms, and crustaceans.

Ring-billed Gull

Larus delawarensis

It was once thought to be the most common gull in North America: John James Audubon himself called it "the great American gull." In the years since Audubon roamed the continent, however, the ring-billed gull seems to have surrendered some of its territory—and with it the status of being number one. For sheer breadth of distribution, at least, that rank can be claimed now by the herring gull. Thanks to its ability to eat almost anything, live almost anywhere, and offer a vigorous defense of the best nesting sites, the herring gull has become so ubiquitous that it is what most people mean when they mention a "sea gull."

The slightly smaller ring-bill does resemble the herring gull generally in build, coloration, and appearance in flight. But it is still recognizable almost immediately as a different bird. Its main identification mark, of course, is the black ring around the bill for which it is named. Keen-eyed birders also look for the yellow feet, whose color sets the ring-bill apart from the pink or flesh-colored underpinnings of *Larus argentatus.*

Farmers in Canada and the northern United States are glad to see the ring-bills return in spring to their breeding grounds. They are great insect-eaters on newly plowed fields—and for grasshoppers in particular, a visit by hungry ring-bills means an abrupt end to summer.

During the summer months, protein-rich grasshoppers are a staple of the ring-bill's diet.

Recognition. 18–20 in. long. Mainly white; back and wings gray; wingtips black with white spots; bill yellow with distinct black ring; legs yellow. Young birds mottled with pale gray and brown; tail white with narrow black band; bill dark-tipped; legs dull pink.
Habitat. Seacoasts, lakes, and city parks.

Nesting. Nest is a cup of grass and stems on ground. Eggs 2–4, pale brown, spotted with brown and lilac. Incubation about 25 days, by both sexes. Young downy; age at first flight unknown. Nests in large colonies.
Food. Insects and small fish; also scavenges.

California Gull

Larus californicus

In the heart of Salt Lake City, Utah, stands a monument to sea gulls that commemorates their contribution to the Mormons' settling of the surrounding area. To many visitors, such a landmark must seem somewhat puzzling.

The term "sea gull" is often a misleading one, for many gulls spend much of their lives far from the ocean. The California gull—itself somewhat misnamed—breeds primarily on islands in lakes of the interior plains of the West. This opportunistic feeder not only plunges after lake fish but also takes cutworms, grubs, small rodents, and other agricultural pests in freshly plowed fields—habits that make it welcome to farmers. Indeed, to the first Mormon settlers in Utah, the California gull seemed divinely sent.

In 1848 and again in 1855, plagues of grasshoppers broke out in Salt Lake Valley and threatened to destroy the settlers' crops, which would have meant almost certain starvation. But as one newspaper reported, "the gulls made war on them, and they have swept them clean." In 1913 the Mormons erected their monument to that event, and the California gull, almost certainly the species that came to the rescue, is honored now as the state bird of Utah.

California and ring-billed gulls often nest side by side on islands that dot inland lakes.

Recognition. 21–22 in. long. Mainly white; back and wings gray; wingtips black with white spots; bill yellow with red and black spots near tip; legs yellow-green. Young birds mottled with pale gray and brown; tail white with narrow black band; bill dark-tipped; legs pink.
Habitat. Seacoasts, lakes, and farmlands.

Nesting. Nest is a cup of stems and weeds on ground near water. Eggs 2–4, buff, speckled with gray and brown. Incubation 23–27 days, by both sexes. Young downy; leave nest soon after hatching; age at first flight unknown. Nests in colonies.
Food. Insects and rodents; also scavenges.

Herring Gull

Larus argentatus

Few writers have more eloquently noted the interdependence of life-forms along seashores than the famed naturalist Rachel Carson. She recognized the herring gull in particular as a bird superbly equipped for all weathers, storms, and seas. Typical was her vivid description of a rocky shoreline at high tide, with herring gulls dozing placidly in the sunshine, waiting for crabs and sea urchins to be more conveniently exposed by receding waters.

In many of its actions, the herring gull seems to know instinctively how to get the most done with the least energy—though there is plenty of that in reserve. Blessed with skeletal architecture almost perfectly designed for flight, it is not merely a strong forward flier; it can also hover forward or backward with hardly a flick of its wings. Rather than waste time and effort cracking open shellfish plucked from the shallows, the herring gull simply flies up over a driveway, road, or bridge, then drops its meal from a great enough height for it to crack open on impact.

The herring gull population on both sides of the Atlantic is rising fast, boosted by huge concentrations of urban garbage that make it easy for these gulls to find food. Inevitably, weaker species lose out in desired nesting areas, both along the shore and inland, where herring gulls may become permanent settlers as well.

A young herring gull pecks the red spot on the parent's bill to make it regurgitate food.

Recognition. 22–26 in. long. Mainly white; back and wings gray; wingtips black with white spots; bill yellow with red spot; legs pinkish. Young birds mottled with brown; tail with broad black band; legs pink.
Habitat. Seacoasts, lakes, and cities.
Nesting. Nest is a cup of grass, stems, and seaweed on ground. Eggs 2–3, bluish, greenish, or buff, spotted with brown. Incubation 25–33 days, by both sexes. Young downy; leave nest soon after hatching; first fly at about 7 weeks. Nests in colonies.
Food. Fish, small marine animals, eggs and young of other seabirds; also scavenges.

Western Gull

Larus occidentalis

To this sharp-eyed opportunist, human beings have long meant trouble-free food. In recent years, landfills have provided a cornucopia of refuse that helped gull numbers to soar. But even before piling food scraps so generously, humans set the table for gulls in other, less obvious ways.

By the mid-1800's, western gulls had learned to take note when lines of people appeared on offshore islands—egg hunters who methodically scared seabirds from their nests and gathered their valuable eggs. When lines of these "eggers" formed at one end of an island, western gull flocks convened overhead. Timing was critical. As the egg-hunting phalanx moved across the island with an uproar that sent seabirds flying, the gulls swooped down on vacated nests and cracked as many eggs as possible with quick, well-aimed raps from their bills. The hunters, arriving moments later, would pass by the damaged eggs for unbroken ones just ahead—leaving the gulls to gorge on the spoils.

Although egg hunting is a thing of the past, birders, fishermen, and picnickers must exercise caution around seabird colonies lest they provide the gulls with the same unwitting service. Entire rookeries of cormorants have been reported destroyed when people approached carelessly, parent birds were flushed from their nests—and flocks of ever-alert western gulls swooped down on cue.

Western gulls regularly feed on the eggs of seabirds in colonies along the Pacific coast.

Recognition. 24–27 in. long. Mainly white; back and wings blackish; wingtips with white spots; bill thick, yellow with red spot; legs pink. Young birds mottled with brown; tail white with broad black band; bill thick; legs pinkish.
Habitat. Seacoasts.
Nesting. Nest is a thick-walled cup of grass and stems, among rocks on sea cliff or island. Eggs 1–4, buff, spotted with brown. Incubation 24–29 days, by both sexes. Young downy; leave nest soon after hatching; first fly at about 7 weeks. Nests in colonies.
Food. Mainly fish; also scavenges.

Glaucous-winged Gull

Larus glaucescens

This bird quacks, trumpets, mourns, shouts, and clamors. If you listen, it might even coax. W. Leon Dawson listened. During two separate expeditions, in 1905 and 1907, the pioneering ornithologist perched on remote islands off the Washington coast and snooped on glaucous-winged gulls. The nesting birds didn't mind, so Dawson got close. What he heard was their strange mother tongue: a cacophony of mews, honks, and squeaks that might easily be dismissed as one great caterwauling. But each sound sent a message. Dawson hunkered down to puzzle it out.

They said *koo,* he discovered. The gulls also said *kawk* and *klook, klook.* Categorizing the various notes, Dawson gradually learned what each sound meant. A simple *koo,* he observed—shouted once or repeatedly—seemed to express disapproval. Harsh notes that accompanied these clackings would rise with the speaker's degree of alarm. There were friendly notes, too. *Oree-ah* served as a greeting, exchanged between mates when one returned from the sea. *Keer, keer* signaled pleasure, uttered en masse as gulls arched their necks and thrust their heads forward.

Sitting there transfixed, Dawson at one point scribbled nonstop for a week. His notes would describe a new language. Combining a keen ear, sharp eyes, ingenuity, and patience, he had broken the code.

Glaucous-winged and western gulls often interbreed; their young resemble herring gulls.

Recognition. 24–27 in. long. Mainly white; back and wings pale gray; wings edged with white; bill yellow with red spot; legs pink. Young birds mottled with pale gray-brown; head whitish; bill black; legs pinkish.
Habitat. Seacoasts.
Nesting. Nest is a cup of grass and seaweed, on bare ground or among rocks. Eggs 2–4, buff or olive, spotted with brown. Incubation about 26 days, by both sexes. Young downy; leave nest soon after hatching; first fly at 5–7 weeks. Nests in colonies.
Food. Mainly fish; also scavenges.

Great Black-backed Gull

Larus marinus

The largest and most majestic of our gull species, the great black-backed gulls soar and wheel in the sky high above their lesser relatives. They usurp the choicest sites in nesting colonies and perch on the most prominent posts. Their wingbeats are slower than those of other gulls, just enough to give their flight a touch of grandeur.

At home on their nests, however, great black-backs are like other gulls—watchful and caring parents, affectionate and attentive mates. On their feeding grounds (which for much of the year are the same coastal landfills their relatives patronize) they tend to be silent, but otherwise are unmistakably gull-like. The proliferation of garbage dumps has been a boon to many gull populations, especially the ubiquitous herring gull. And that has been bad news for various smaller seabirds, for whom sanctuaries have often been set aside not far from the same landfills. Just when the herring gulls need extra food for their own young each summer—there within easy reach are thousands of nests filled with eggs and hatchlings for them to feast on.

Nature does compensate, however. No sooner do the herring gulls set off on a raid than a flock of great black-backs swoops in to plunder the briefly unguarded nests. To mankind this sequence may seem cruel and almost cannibalistic. To nature it is simply population control.

Great black-backs prey on the young of smaller gulls, terns, and other birds.

Recognition. 28–31 in. long. Large. Mainly white; back and wings blackish; wingtips with white spots; bill yellow with red spot; legs pink. Young birds mottled with brown; head whitish; bill black; legs pinkish.
Habitat. Seacoasts, lakes, and cities.
Nesting. Nest is a cup of seaweed, grass, and moss, on ground. Eggs 2–3, olive or tan, spotted with brown. Incubation 26–30 days, by both sexes. Young downy; leave nest soon after hatching; first fly at 6–8 weeks. Sometimes nests in small colonies.
Food. Fish, carrion, eggs and young of other birds; also scavenges.

Black-legged Kittiwake

Rissa tridactyla

It is called the black-legged kittiwake, but this is one bird worthy of the name "sea gull." It lives its life, wild and free, among the winds and waters of the oceans, and comes to land only long enough to nest and fledge its young on the precipitous sides of an oceanside cliff. So tiny is its nest that the kittiwake must sit with its breast against the rocks and its tail sticking out into empty space. The little ones must learn to lie low and hold tight while tempests beat upon the nest of compacted seaweeds and mosses.

That nest sits snugly amid a cliffside colony of thousands of birds, which fly lightly and swiftly over the ocean waves in flocks of equally great numbers. They close in like gray clouds behind fishing fleets and pods of whales, not scavenging for handouts but seeking the tiny mollusks and crustaceans of the ocean's plankton as it is churned to the surface. The kittiwakes hover like terns, then plunge headfirst and swim underwater in pursuit of their slippery prey.

They drink the salt water, too; indeed, they will only drink salt water. Their bodies, like those of all truly oceanic birds, carry excess salt through the bloodstream to special nasal glands, then let it drip back into the sea. Like their food and drink, even their rest is provided by the sea. At nightfall the kittiwakes alight on the rolling waves, tuck their heads under their wings, and drift peacefully in sleep.

The second egg in a kittiwake's nest is often smaller, and the chick seldom reaches maturity.

Recognition. 16–18 in. long. Mainly white; back and wings pale gray; wingtips black, without white spots; bill yellow; legs black. Young birds mainly white; black band across nape; M-shaped blackish mark across wings and back; tail with narrow black tip.
Habitat. Open ocean.

Nesting. Nest is a sturdy cup of seaweed, moss, and mud, attached to narrow ledge on a northern sea cliff. Eggs 1–3, pale green or cream-colored, with brown spots. Incubation 23–28 days, by both sexes. Young downy; leave nest 35–55 days after hatching. Nests in colonies.
Food. Small fish and crustaceans.

Gull-billed Tern

Sterna nilotica

Two centuries ago, gull-billed terns by the hundreds of thousands nested in the salt marshes stretching from the New Jersey coast south to Virginia and beyond. Then the tiny eggs of these birds became springtime delicacies for the tables of New York society. At the same time their gracefully feathered bodies became just the thing to adorn the hats of fashionable ladies. Either sort of popularity would have threatened the gull-billed terns; this double-barreled demand spelled their doom.

Ironically, the birds' own nature conspired against them. Fearless in defending their nests, gull-billed terns attacked egg collectors by coming at them directly. The "eggers" in turn could easily shoot an attacking bird, thus adding to their catch for the day. Another trait made matters even worse. When a tern was shot and fell to the water, nearby terns rushed to hover over it, offering easy targets to the waiting guns. As more victims fell, more and more excited birds rushed in from surrounding marshes—and the slaughter ended only when the ammunition was exhausted.

North America's populations of gull-billed terns have never recovered from this torrent of destruction. Where they can be found at all, they exist today only as scattered pairs, usually nesting on the outer beaches—and only in the protective company of other, more numerous species.

A gull-billed tern's stout bill is ideal for a diet that includes frogs and crustaceans.

Recognition. 13–15½ in. long. Mainly white, with long, pointed wings and shallowly forked tail. Breeding adult has black cap; bill stout and black. Winter adult and immature have white head with dark smudge through eye.
Habitat. Salt marshes, beaches, and lagoons.
Nesting. Nest is a shallow depression lined with grass and bits of shell, in marsh. Eggs 2–5, buff, spotted with brown. Incubation about 23 days, by both sexes. Young downy; remain near nest after hatching; first fly at 4–5 weeks. Nests in colonies.
Food. Mainly insects caught in air; also frogs, crustaceans, and small fish.

Caspian Tern

Sterna caspia

The largest tern not only in North America, but in the world, the Caspian is often mistaken for a gull—no great surprise, since gulls and terns belong to the same family and might be expected to borrow an occasional trait from one another. But the Caspian tern shows so many gull-like characteristics it verges on impersonation.

There is the matter of the way it flies, for one. Terns fly low, flap their wings continuously, and hold their bills pointed down toward the water. The Caspian tern flies this way, but only when it's fishing. During a long flight it flies high above its fellows, points its bill straight ahead, holds its wings wide and still, and soars on the currents—like a gull. When a Caspian is fishing like a proper tern, it hovers, then dives into the water. Sometimes, though, it fishes like a gull instead: sitting on the water, it paddles with its rather small feet and snatches fish near the surface. It will also steal catches from other birds and rob other birds' nests of eggs and nestlings.

The least social of all terns, Caspians tend to travel singly or in small groups. They also keep their young in the nest by pecking and beating them until they crouch for shelter. But the severity is for good reason. As in gull colonies, if a youngster wanders to another nest, it will be treated as a predator—attacked, beaten to death, and eaten.

Caspian terns usually nest in colonies, but sometimes a pair will raise its young alone.

Recognition. 19–23 in. long. Large. Mainly white, with long, pointed wings and notched tail. Breeding adult has black cap with short crest; bill red; underside of wingtips blackish. Winter adult has white streaks on crown. Flight more gull-like than that of other terns.
Habitat. Lakes, rivers, beaches, and bays.

Nesting. Nest is a shallow scrape in sand, sometimes lined with grass. Eggs 1–4, pinkish buff, spotted with dark brown. Incubation 20–22 days, by both sexes. Young downy; remain near nest after hatching; first fly at 25–30 days. Nests in colonies.
Food. Mainly fish; also robs other birds' nests.

Royal Tern

Sterna maxima

No other birds in North America nest so closely together as do the royal terns—10,000 or 12,000 nests on a sandspit of an island, with such hairbreadth property lines that the feathers of one nesting bird almost, but not quite, touch the feathers of its immediate neighbors.

This is not communal nesting—each bird attends to its own nest—but in such crowded colonies, mating and egg laying usually occur so that all the chicks hatch within a few days of each other. This enhances the likelihood that most will survive; it also means that predators will be more easily driven away by massed attacks of the nesting birds.

When the colony's youngsters are able to run about on the sand, but can't yet fly, adult birds shepherd the whole crew—thousands of little royal terns—into one great group called a crèche. Despite its tremendous numbers, the crèche can be led around as a unit, away from marauding predators or into shelter from heat and storms. Freed from nest-guarding duties, parent birds fly back and forth from the sea bringing fish for their young, which they recognize by their voices. The youngsters in turn must push their way to the edge of the crèche and make their voices heard, for a parent bird will feed only its own chicks and no one else's.

Unlike most tern species, the royal tern seldom lays a clutch of more than one egg.

Recognition. 18–21 in. long. Large. Mainly white, with long, pointed wings and shallowly forked tail. Breeding adult has black cap with ragged crest; bill orange; underside of wingtip whitish. Nonbreeding adult and immature have white forehead. Flight graceful.
Habitat. Beaches and coastal lagoons.

Nesting. Nest is a simple depression in sand on beach. Eggs 1–2, pale buff or white, spotted with brown. Incubation about 30 days, by both sexes. Young downy; gather in large group after hatching; begin to fly at 25–30 days. Nests in large colonies.
Food. Mainly fish and squid.

Elegant Tern

Sterna elegans

Like almost all sea birds, elegant terns believe in strict loyalty to one mate. Although male and female may spend most of the year hundreds of miles apart, they will faithfully reunite on their nesting grounds each breeding season for as long as they live.

The male arrives first at the nesting colony and begins calling and displaying, in the stylized manner of the elegant tern, to every bird around him. He's not looking for someone new—but since elegant terns all look pretty much alike, he has to hear his mate's voice in order to recognize her. Voice recognition is the vital bond that links a mated pair, just as it links parents and their chicks—and it has to be renewed at the outset of every breeding season.

Even after they have found one another, the pair must overcome a certain standoffishness. This they do by performing their ritual displays until, at last, the female turns aside the threat of her bill. Gradually they draw closer; soon he is feeding her gifts of fresh-caught fish, and throughout the long nesting period they remain tirelessly attentive, doting on each other like young honeymooners.

Then the honeymoon ends. They fly off to their separate wintering sites, strangers once again. But next spring will bring voice recognition, courtship renewal, shared nesting duties, and the rearing of another family—together.

The nesting success of elegant terns depends on the supply of a single fish, the anchovy.

Recognition. 16–17 in. long. Large and slender. Mainly white, with long, pointed wings and shallowly forked tail. Breeding adult has black cap with ragged crest; bill slender and yellow. Winter adult has white forehead, with black extending forward through eyes. Flight graceful.
Habitat. Bays and inshore waters.

Nesting. Nest is a simple scrape in sand. Eggs 1–2, white or buff, spotted with brown and black. Incubation period unknown. Young downy; remain near nest after hatching; age at first flight unknown. Nests in colonies.
Food. Mainly small fish.

Sandwich Tern

Sterna sandvicensis

For Sandwich terns, royal terns, and a number of other species, colonial nesting all too often means a teeming encampment in which one nest may be no more than an outstretched wing away from another. Here is prime potential for chaos: surely there are too many occasions for error amid such pandemonium. But adult birds seem to have an amazing ability to recognize their own nests and eggs, reducing the chance that one might innocently usurp another's nest—and in the process kidnap its offspring.

The chicks of Sandwich and royal terns hatch at about the same time, and they are soon organized into a crèche that may number thousands of chicks of both species, with a few adults on guard duty. While it may look undisciplined, the crèche provides a secure day-care system that allows adults to go in search of food. Returning parents, who feed only their own offspring, seem to rely on a chick's call as a tracking device that makes identification possible in a veritable sea of look-alikes. Plumage variations among the chicks may also serve as recognition signals for parents.

As the mass of chicks advances toward the fledgling stage, their crèche moves closer to the water's edge. As it does, errant youngsters are kept in line and out of harm's way—the whole point of a system used successfully by water-loving birds for countless generations.

Like other large terns, this species has a white forehead and crown in winter.

Recognition. 14–16 in. long. Large and slender. Mainly white, with long, pointed wings and shallowly forked tail. Breeding adult has black cap and ragged crest; bill black with yellow tip. Winter adult and immature have white forehead. Flight graceful.
Habitat. Beaches, estuaries, and bays.

Nesting. Nest is a simple depression in sand. Eggs 1–3, buff or pinkish, usually spotted with brown and gray. Incubation 21–24 days, by both sexes. Young downy; gather in large groups; first fly at about 5 weeks. Nests in colonies, often with royal terns.
Food. Mainly fish; also squid and shrimp.

Forster's Tern

Forster's Tern *Sterna forsteri*
Common Tern *Sterna hirundo*

If a field guide's description of any bird includes three or more *buts, ifs, excepts, somewhats, seldoms,* or other such terms, a real identification challenge is at hand—one that can send would-be birders in search of easier hobbies. Forster's terns, in league with common, roseate, and Arctic terns, pose just such a problem. They all look the same.

But birding has a basic rule of bird identification: begin by eliminating, paring down to possibilities and probabilities. Of these four "problem terns," there is little chance that all would be in the same place at the same time—so one or more can usually be ruled out right away. In summer, for instance, a problem tern on the northern Gulf Coast is either a Forster's or a common, but in New England it's probably a common or a roseate. In winter, the Forster's is an odds-on favorite almost anywhere, while most reports of common terns should be viewed with polite skepticism.

Becoming a knowledgeable bird watcher requires practice and patience, of course, but total recall is not a prerequisite. A few good field guides will supply the facts; then, with the process of elimination, even the most baffling group of birds can quickly be cut down to size.

Recognition. 13–16½ in. long. Mainly white with forked tail. Breeding adults have black cap; bill orange with black tip on Forster's, red with black tip on common. Forster's paler above than common. Winter adults have white foreheads.
Habitat. Marshes, beaches, lakes, and rivers.
Nesting. Nest is a scrape in sand or depression among marsh plants; that of Forster's usually in marshes; that of common often on beaches. Eggs 2–4, olive or tan, spotted with brown. Incubation 21–26 days, by both sexes. Young downy; remain near nest after hatching; first fly at about 4 weeks. Nests in colonies.
Food. Mainly insects (Forster's) and fish.

Least Tern

Sterna antillarum

During much of the 19th century, when the wholesale killing of wild birds was accepted business practice, this graceful, diminutive tern was worth as much as twelve cents to suppliers of the millinery trade. The least tern, like some other tern species, was so highly prized as an adornment of ladies' hats that by 1913, when legal protection was enacted, its numbers had been decimated. It actually made a strong recovery in the next decade, but the struggle was hardly over. By the 1980's the least tern was again endangered across much of its range.

This time the cause was not hunting but habitat loss. The open, sandy beaches it favors for nesting have been claimed for development or recreational use. Competition from other species has reduced the least tern to poor-relation status on islands dominated by colonies of larger terns. Nesting on such sites as mainland beaches and flat roofs invites danger from predation and exposure.

In some areas, though, things are changing. On one stretch of Mississippi coast the least tern has proliferated, from dozens in 1972 to thousands in 1990. Along more than a mile of beachfront, bordered by a busy highway and crowded recreational beaches, a nesting area protected by local officials and conservationists has been dubbed, with understandable pride, "the least tern capital of the world."

Least terns usually patrol and hover over waterways with their bills pointed downward.

Recognition. 8½–9½ in. long. Small. Mainly white, with notched tail. Breeding adult has black cap and white forehead; bill yellow with black tip. Winter adult has crown streaked with white. Flight more fluttery than larger terns'.
Habitat. Marshes, seacoasts, lakes, and sandbars in rivers.

Nesting. Nest is a shallow scrape on beach or sandbar. Eggs 2–3, buff, spotted with dark brown. Incubation 20–22 days, by both sexes. Young downy; remain near nest after hatching; first fly at 15–17 days. Nests in small colonies, sometimes with other terns.
Food. Small fish and crustaceans.

Black Tern

Chlidonias niger

Black terns don't need solid ground or a beach, but build floating nests in marshes.

When the black tern made the Blue List, North American birders took note. The year was 1978, and the National Audubon Society had just published its annual listing of North American birds whose numbers were declining rapidly. Inaugurated in 1971, the Blue List already had been adopted as an important conservation tool by state and federal wildlife agencies. Based on field observations of Audubon members nationwide, it provided documentation—and an early warning—that birds like the black tern were in trouble and needed help fast.

The Blue List's basic method—enlisting a legion of lay observers to help promote bird conservation—had been applied before. In the early 1900's Audubon Society members returned from field trips in numerous states to report the widespread killing of shorebirds and wading birds to provide plumes for women's hats. These reports sparked legislation that helped save several species from extinction.

Such citizen participation continues. A growing number of states have enlisted volunteers to make detailed surveys of breeding birds in each portion of the state. The resulting data, compiled in breeding-bird atlases, allow officials to monitor local bird populations with great accuracy. The black tern's populations remain at low levels, but a key factor in their decline has been identified: the declining number of freshwater marshes. Government and private efforts are consequently being focused on the goal of preserving these wetlands, and with them the graceful black terns that have long been part of their special appeal.

Recognition. 9–10½ in. long. Breeding adult black with some white on upper surface of wings; bill black; tail grayish. Winter adult gray above; cap blackish; white below with dark smudges at sides of breast. Flight graceful.
Habitat. Marshes, lagoons, and lakes.
Nesting. Nest is a small hollow of dead plants, placed on floating marsh plants or muskrat lodge. Eggs 2–4, buff or olive, spotted with brown. Incubation about 22 days, by both sexes. Young downy; remain in nest 2 weeks after hatching; first fly at 3 weeks. Usually nests in small colonies.
Food. Flying insects, fish, and crustaceans.

Black Skimmer

Rynchops niger

Individuality is this bird's stock in trade—its bill, a finely tuned instrument, allows for a unique method of food-finding. Unlike all other birds, the black skimmer has a lower mandible that extends well beyond the upper, and is compressed into knifelike sharpness. As a skimmer cuts a path on shallow wingbeats, just above still water, its lower bill shears the surface. When it makes contact with a small fish, shrimp, or other morsel, the upper mandible snaps down on the prey quickly and firmly. Moreover, none but the versatile skimmer can narrow its pupils into catlike vertical slits, limiting glare and giving the entire operation an appearance of sheer effortlessness.

Flocks of skimmers often feed raucously together, plowing and replowing the water in follow-the-leader formations along sheltered coastlines and marshes. When windy days and roiled waters hamper their efforts, they take their show on the road—to quiet ponds and rivers. And when dozens, or hundreds, of skimmers close ranks and face into the wind, they look for all the world like a flock of low-slung Charlie Chaplins in crisp white shirts and shiny black coats. Sad, vaguely embattled, but unfailingly dignified.

The skimmer's lower bill offsets the effects of water friction by growing faster than the upper.

Recognition. 16–20 in. long. Adult has long red and black bill, with lower mandible longer than upper. Upperparts black, with white trailing edge on wing; underparts white. Young bird has upperparts mottled with brown; bill shorter.
Habitat. Beaches, inshore waters, and lagoons.
Nesting. Nest is a simple scrape in sand. Eggs 1–5, bluish or cream-colored, spotted with brown and gray. Incubation period unknown. Young downy; remain near nest after hatching; age at first flight unknown. Nests in colonies.
Food. Small fish and crustaceans.

Large Wading Birds

Our long-legged waders obviously have a knack for survival: some have ancestors in the fossil record dating back 50 million years. So it is no surprise that they and their relatives have managed to establish populations across much of the world. Be it a solitary night-heron surveying a moonlit shore, a bittern vanishing against a background of marsh grass, or a whooping crane unfolding spectacularly into flight—each of these stately birds has a bearing that marks it as one of the aristocrats of the avian realm.

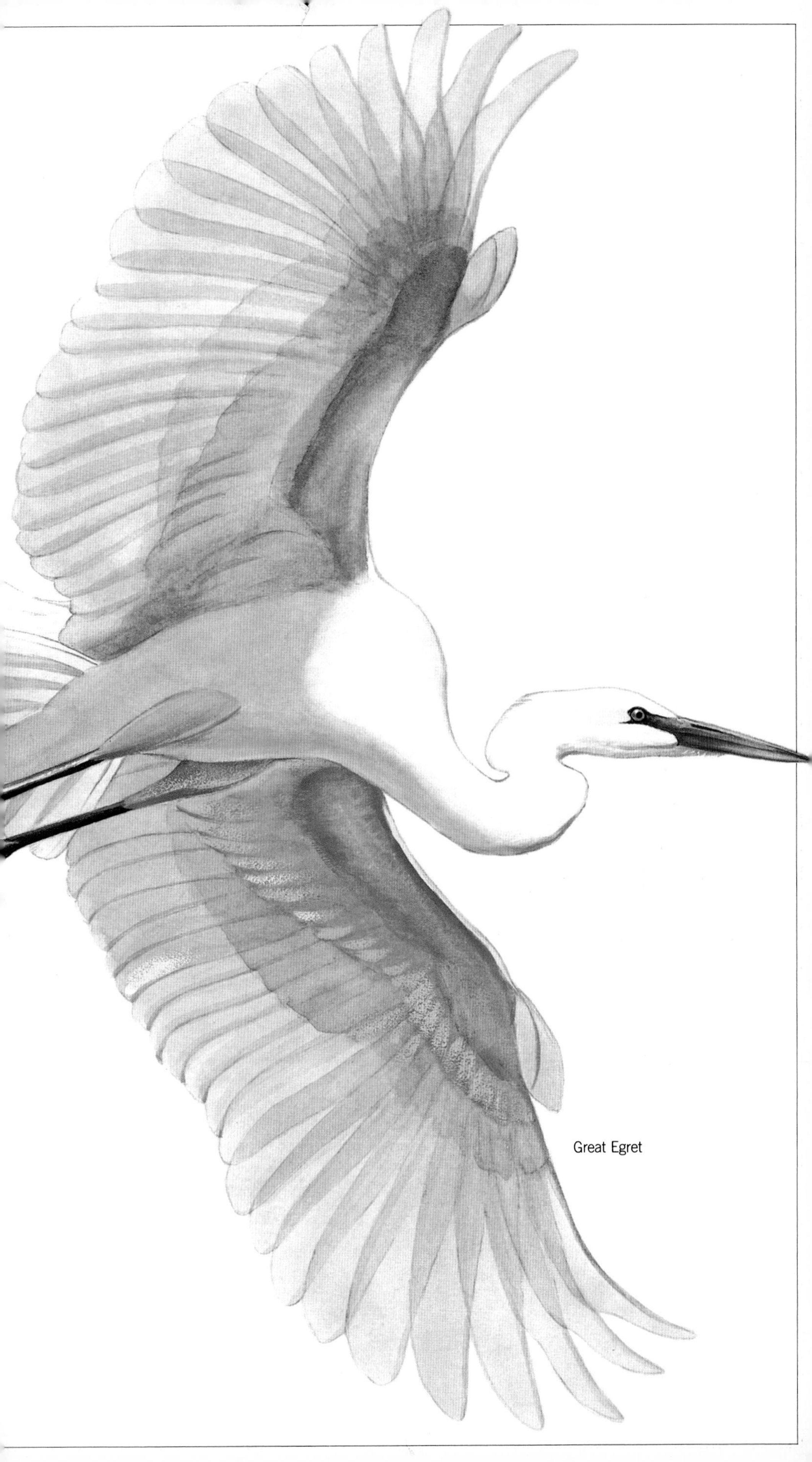

Great Egret

American Bittern

Botaurus lentiginosus

Unsuspecting people can walk within a few feet of a well-camouflaged American bittern.

Glide in a canoe through the world of marsh and fen, typical home of the American bittern. Tall plants with narrow green leaves stir gently, while red-winged blackbirds and marsh wrens sing from far back in the dense vegetation. Ahead, against a reedy background, a dead limb juts up; angular and broken, it shimmers with reflections from the shallow, sunlit water.

The canoe turns slightly to avoid a collision. Suddenly the dead limb springs to life, spreading broad, slightly pointed wings and flying away with a hoarse croak. Once again the American bittern has used amazing camouflage. Subtly colored, with stripes on its neck, throat, and breast that melt into the vertical lights and shadows of marsh plants, the bird points its bill skyward and draws its feathers in tight. It may remain motionless, taking on the form of a tree root or broken snag. Or, swaying slowly to mimic wind-stirred vegetation, it may simply make itself invisible. Nor is the American bittern easy to identify by sound. It calls usually between dusk and midnight, and its ventriloquial notes blend into the robust bullfrog chorus.

In March and April, the male bittern abandons his usual low profile in favor of elegant courtship displays that may be visible right out in the open. But the moment is fleeting. His venture into public life soon concludes, and the American bittern disappears again into his watery green world.

Recognition. 24–34 in. long. Dark brown and streaked. Neck with black stripe; wingtips dark in flight. Secretive; when alarmed, often freezes with neck taut and bill pointing straight up.
Habitat. Marshes and grassy lakeshores.
Nesting. Nest is a platform of reeds and grass close to water level in marsh. Eggs 3–5, buff or olive. Incubation about 29 days, by female only. Young downy; leave nest 6–7 weeks after hatching. Nests in isolated pairs.
Food. Mainly fish, frogs, small eels, water snakes, and insects.

Least Bittern

Ixobrychus exilis

The life of the least bittern is inexorably linked to that of our vanishing wetlands—cattail marshes are its first preference, but almost any freshwater area with dense plant cover will do. Though usually described as secretive and retiring, this small bird is also persistent, and it will continue to nest in marsh remnants in urban areas until they have been completely drained and filled.

The reason is not hard to understand. Eminently adapted to its environment, the least bittern is able to compress its sides so that it can move stealthily through dense stands of reeds and cattails. It also has long, flexible toes that enable it to climb and run through the reeds. Grasping as many as two or three stems in its feet, the bittern makes a quick transit somewhat like a squirrel running through the trees. Its eyes, positioned to peer around the sides of its beak, give the least bittern a head-on vision that helps it maneuver speedily through the marsh vegetation.

Surprised by an intruder, it strikes a camouflage pose similar to that of the American bittern, or else bursts into flight and then drops back into the concealing marsh. Inevitably, as wetlands dwindle, so will the bittern's population. Difficult as it is to see in the marsh, this smallest of our herons will be truly invisible when the marsh is gone.

The least bittern spends a good deal of time climbing high in the reeds of its marshy habitat.

Recognition. 11–14½ in. long. Small. Black on crown and back; neck and underparts buff; large wing patches buff and chestnut. Secretive; when alarmed, often freezes with neck taut and bill pointing straight up; climbs among reeds.
Habitat. Reedy and grassy marshes.
Nesting. Nest is a small platform of dried plant matter 8–14 in. above water in marsh. Eggs 2–7, pale blue or greenish. Incubation about 20 days, by both sexes. Young downy; leave nest about 25 days after hatching. Nests in isolated pairs or loose colonies.
Food. Mainly fish, frogs, and tadpoles.

Great Blue Heron

Ardea herodias

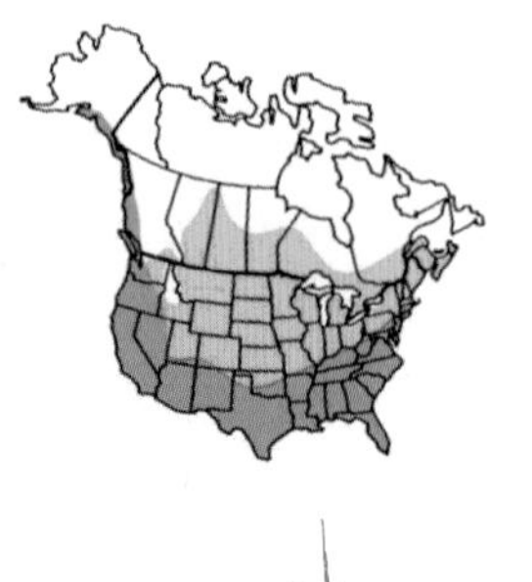

Found only in Florida, the great white heron is actually a color phase of the great blue.

Visitors to a great blue heron rookery should wear hard hats, just in case. One never knows when frog bones, chunks of rotted fish, or some other unsavory tidbits may come tumbling down when these birds feed their young.

Large rookeries may contain hundreds of bulky stick nests clustered high in tall trees. Gangly young birds snooze within their slightly hollowed interiors. Other juveniles teeter on nearby branches, squawking and flapping their wings. Overhead, adult birds appear, their huge wings beating slowly. They have been spearing frogs and fish in a distant watercourse. Their return with this food starts a riot. Sensing a meal in the offing, snoozing siblings begin to hop about on the nest edge, pushing, shoving, and squealing like hungry young pigs. A jostled bird sways out of control, then regains balance by flailing its wings.

The adult birds float downward, their legs extended like parachute strings. No sooner have they landed than they are besieged by a clamoring horde. Bills stab like swords as each youngster tries to wrest morsels from its parent's bill. In all the confusion, an occasional frog or fish part misses its mark and tumbles over the nest edge. The same can easily happen to portions of a snake or a young muskrat, which herons also eat. So don't forget that hard hat.

Recognition. 50–54 in. long. Very large. Adult mainly gray; head white with 2 black crown stripes; bill yellow. Young birds similar, but crown black. Great white heron (Florida) all-white phase with yellow bill and legs. Flies with neck folded onto back.

Habitat. Marshes, lakes, rivers, and shores.

Nesting. Nest is a platform of sticks in bushes or trees up to 100 ft. above ground, or on rocks or cliffs. Eggs 3–7, pale blue or greenish. Incubation about 28 days, by both sexes. Young downy; leave nest about 8 weeks after hatching. Usually nests in colonies.

Food. Fish, frogs, snakes, and small mammals.

Great Egret

Ardea alba

This sleek, snow-white wader was many things to many people in the late 1800's. Plume hunters called it long white, for the filamentous feathers that trailed down its back. Ladies thought it fashionable when they adorned their hats with these flowing nuptial plumes. Alarmed citizens, who saw egrets being shot on their breeding grounds, saw a species facing a death sentence.

Great egrets weren't the only birds to suffer when feathered hats came into fashion. During a stroll down Fifth Avenue in 1886, the ornithologist Frank M. Chapman counted feathers from terns, flickers, and nearly 40 other species atop the heads of New York's stylish women. Of all these adornments, egret plumes brought the highest price: $32 an ounce by the year 1900.

As the popularity of plumed hats grew, so did the outcry against them. The founders of America's young conservation movement decried the plume trade as inhumane. As they worked to enact laws protecting egrets and other plume birds, magazine articles and public lectures promoted more humane treatment of wildlife. Slowly attitudes changed, and by World War I breeding plumes on women's hats had all but disappeared. Bird lovers were able to breathe more easily. The great egret had won a reprieve.

No longer worn on hats, the plumes of the great egret can still be admired in the wild.

Recognition. 37–41 in. long. Large and white. Bill stout and yellow; legs and feet black. Has long, lacy plumes on back in nesting season. Flies with neck folded onto back.
Habitat. Marshes, lakes, lagoons, wooded swamps, and streams.
Nesting. Nest is a platform of sticks in bush or tree 20–40 ft. above ground or water. Eggs 1–6, pale blue or greenish. Incubation 23–26 days, by both sexes. Young downy; leave nest 6–7 weeks after hatching. Nests in colonies.
Food. Fish, frogs, water snakes, and insects.

Snowy Egret

Egretta thula

Some birds can drag their feet and still get things done. So it is with the snowy egret, a graceful wader that not only drags its feet but wiggles them, probes with them, and rakes them through the mud. Researchers analyzing the snowy egret's many methods of foraging with its feet have concluded that, among North American herons, this finely plumed hunter boasts an unequaled bag of tricks.

Foot stirring, their studies found, requires an egret to swish one foot through pond mire while keeping the other planted for balance. Because its feet gleam bright yellow, the sight of one flashing by often startles frogs and fish into motion. Once spied by the egret, they are promptly skewered on its long, pointed bill. Foot probing involves a different approach: the snowy egret sticks its foot deep into the mud and moves it about with purposeful strokes. Foot raking, in which an egret barely scratches its toes across the bottom, is a compromise of sorts. If the toes dug deeper, it would become foot probing; if they lifted up an inch, it would be foot stirring. Finally, there's foot paddling: the bird rams its foot into the mud and jiggles it up and down—basically a more fervent form of probing.

Of course, these fine distinctions mean nothing to the creatures being caught. What matters is that when this bird goes looking for food, it always puts its best foot forward.

Immature snowy egrets have yellow feet and a yellow stripe up the back of their legs.

Recognition. 22–26 in. long. Small and white. Neck very slim; bill thin and black; legs black; feet yellow. Has long plumes on back and head during nesting season.
Habitat. Marshes, wooded swamps, lagoons, lakes, and ponds.
Nesting. Nest is a platform of sticks on ground or in bush or tree 5–30 ft. above ground or water. Eggs 1–6, pale blue or greenish. Incubation 20–24 days, by both sexes. Young downy; leave nest at about 4 weeks. Nests in colonies.
Food. Mainly small fish, frogs, and snakes.

Cattle Egret

Bubulcus ibis

Mention "range expansion" to any learned observer of North American birdlife and it is likely to evoke an image of the cattle egret, a relatively graceless but eminently successful (and fascinating) member of the heron family.

Once limited to southern Europe, Africa, and Asia, this species is thought to have made an Atlantic Ocean crossing in the late 1870's, between Africa and Surinam, in South America. A natural wanderer, the egret found a niche in areas where cattle roamed or livestock were kept, and where it fed on insects stirred up by the animals' hooves. By 1916, with little competition from other birds to slow the process, it had spread north and west to Colombia.

The first North American breeding record of the cattle egret, from Florida in 1953, was followed by others from Texas and the Gulf Coast. From that point, nothing seemed out of reach. By 1971 cattle egrets were nesting as far north as Minnesota and as far west as California. By the mid-1980's, pioneers of the species had colonized more than half of the United States and parts of several Canadian provinces. In view of such a record, scarcely any place south of the Yukon seems exempt from brief visits by cattle egrets during their postbreeding wanderings—or even from potential settlers, restless as always, still testing the limits of habitat in this land of opportunity.

The rapid growth of the cattle egret's range was made possible by large grazing animals.

Recognition. 19–21 in. long. Small, stocky, and mainly white; bill short and yellow; legs pale. Breeding adults have buff on crown and breast. Often seen with livestock, catching insects flushed from grass.

Habitat. Marshes, swamps, fields, stock pens, and airstrips.

Nesting. Nest is a cup of sticks and twigs 5–12 ft. above ground or water. Eggs 2–6, pale blue. Incubation 22–26 days, by both sexes. Young downy; leave nest about 4 weeks after hatching. Nests in colonies.

Food. Mainly insects, earthworms, and frogs.

Tricolored Heron

Egretta tricolor

Waiting for prey, a motionless tricolored heron stretches its neck and stares at the water.

The pool of dark water, surrounded by vegetation and punctuated with the knobby knees of long-dead cypress trees, stands at the edge of a freshwater swamp. Even on bright days it remains in shadow, mysterious and forbidding. A visit in early spring may be rewarded by the music of northern parulas; in winter assurance can be found that bald eagles still nest in the area. But in late summer it's best to move on quickly, before hordes of insects descend or a careless step disturbs a lazy water moccasin.

But wait—behind the tangle of moss and vines, ghostly in the play of light and shadow, stands a tricolored heron. Its subtle coloring and soft contours are scarcely visible against the lush background. It does not move. It makes no sound. It is just there, as if rooted in time and place, barely foot-deep in the water.

Gradually it becomes evident that the heron is not moving, because it cannot. Without help it will die. An experienced visitor wades into the swamp, approaches the heron, tosses a jacket over its head, then extricates the bird from the morass of branches that has held it for hours, perhaps days. But it is safe now. With care, the heron will fatten up; its foot, mangled in a frantic bid for freedom, will heal. It will be returned to the swamp, where it belongs. And it will fly away without a moment's hesitation.

Recognition. 24–26 in. long. Slender. Mainly dark blue-gray, with white belly and rump. Frequently wades in deep water, with legs completely submerged.

Habitat. Marshes, lakes, and lagoons.

Nesting. Nest is a shallow cup of sticks on ground or up to 20 ft. above ground in bush or tree. Eggs 3–7, pale greenish blue. Incubation 21–25 days, by both sexes. Young downy; leave nest about 5 weeks after hatching. Nests in colonies, sometimes with other heron species.

Food. Mainly small fish, frogs, reptiles, worms, and insects.

Little Blue Heron

Egretta caerulea

Between the adolescent little blue heron (which is nearly all white) and the adult bird (which for all practical purposes is blue), there are some downright strange plumages that young little blues wear for brief periods before they grow up and learn how to dress.

In the beginning, the young birds are white—only their primary feathers are tipped in the same slaty blue as in basic-plumaged adults, and those tips aren't easy to see. Little blues wear this plumage during their first fall and most of the winter, when they may resemble snowy egrets with dirty feet. By February, they go into their tie-dyed phase, and time is the only cure. Through spring and often into summer, blue feathers gradually replace white ones as the birds molt into the so-called first nuptial pluage. But not all birds progress at the same rate; some of them, even if sexually mature, retain some white in their plumage into the first breeding season. Birds in any stage of this bicolored plumage are commonly referred to as calicoes and form a group unique among the herons.

The fickle-feather phase comes to an end by the time the little blue completes its molt into adult winter plumage. This plumage is definitive. There will still be seasonal trade-offs of old feathers for new, but the dress code leaves no room for doubt: calico is out and blue is in.

A young little blue heron can be puzzling to observers, since it is all-white like an egret.

Recognition. 25–29 in. long. Adult mainly slate-blue; head and neck dull purple; legs greenish; bill dull blue with blackish tip. Young birds white with dark bill.
Habitat. Southern marshes, lakes, lagoons, and wooded swamps.
Nesting. Nest is a cup or platform of sticks 2–40 ft. above ground in bush or tree. Eggs 3–6, pale blue or greenish. Incubation 20–23 days, by both sexes. Young downy; leave nest 6–7 weeks after hatching. Nests in colonies.
Food. Mainly fish, frogs, and reptiles.

Yellow-crowned Night-Heron

Nyctanassa violacea

More restricted and, over much of its range, less common than the black-crowned night-heron, the yellow-crown nevertheless offers would-be watchers one consoling advantage. It's active in daylight too.

But "active" may be a misleading word. "Out" is more like it. A roosting yellow-crown is as immobile as the limb on which it perches; when hunting, it may remain rigid for hours waiting for prey. Even the movements of a stalking yellow-crown seem agonizingly slow; but don't be deceived. The flashing strike from its coiled neck is as fast as a rattlesnake—or one of the water moccasins that abound in the night-heron's favorite southern haunts.

If the yellow-crown didn't already have a name, something like "crab plover" would suit a water bird whose diet is so heavily weighted in favor of crustaceans. And while it is indeed yellow-crowned, the name scarcely does justice to its subject. The black and white face seems lifted from an ancient Greek vase. The golden plumes sweep back along a bristling helmet of white-hot feathers. The bird's cloak is the color of pewter in moonlight, and the legs look amber-glazed. Surprised at close quarters, a roosting bird will often freeze rather than fly, studying the intruder with sleepy, half-opened eyes. In turn, the lucky intruder wins a rare opportunity to study a bird of uncommon beauty.

The stout bill of a yellow-crown is suited to the bird's hearty appetite for crabs and crayfish.

Recognition. 22–28 in. long. Adult mainly gray; head black with white crown and cheek patch, yellowish forehead; bill stout and black. Immature dark brown with whitish spots and streaks.
Habitat. Wooded swamps and marshes.
Nesting. Nest is a sturdy cup of sticks and twigs 15–50 ft. above ground. Eggs 2–8, pale bluish-green. Incubation 21–25 days, by both sexes. Young downy; leave nest about 25 days after hatching. Nests in isolated pairs; sometimes in small colonies.
Food. Mainly crayfish and crabs; also insects, frogs, and mollusks.

Black-crowned Night-Heron

Nycticorax nycticorax

From the shadows, a hunched, stolid-looking form menaces the riverbank. Atop its head, a pair of white plumes dance jauntily from a dark cap. Beneath the brim, unblinking eyes glow like fanned embers. What manner of creature is this? A smuggler? A river pirate? No, only the black-crowned night-heron. Approach it and the bird retreats with a frantic flapping of wings and a loud *squawk.*

In many places, this nocturnal heron is known as "the squawk," and that short, raucous call is as familiar as the sound of crickets in a suburban night. Even the bird's scientific name, *Nycticorax*—"night raven"—stems from its croaking cry. Every evening, well after most of the day-feeding herons and egrets have gone to roost, the night-herons emerge from rookeries and roost sites. All that the birds seem to require is a stand of protective vegetation, a measure of solitude, and proximity to water.

The nests themselves are hardly things of beauty—haphazard piles of sticks heaped upon supporting branches. Some are large and sturdy. But others are scant affairs that show unnerving amounts of daylight through them. Shaken by storm winds, nests, eggs, and flightless young often end up amid the debris that accumulates beneath the rookeries. Despite storms above, foxes below, and the multitudes of pestiferous crows that harass incubating adult birds, the squawk of the black-crowned night-heron continues to be one of America's welcome night sounds.

In flight, night-herons have a stockier appearance than other herons and egrets.

Recognition. 23–28 in. long. Stocky. Adult has black crown and back; wings pale gray; underparts white. Immature gray-brown with whitish spots and streaks.
Habitat. Rivers, wooded swamps, and marshes.
Nesting. Nest is a mass of reeds, sticks, or twigs, among reeds in marsh or up to 160 ft. above ground in tree. Eggs 1–6, pale bluish green. Incubation 24–26 days, by both sexes. Young downy; leave nest 6–7 weeks after hatching. Nests in colonies.
Food. Mainly fish, frogs, and crustaceans.

Green-backed Heron *Butorides striatus*

also known as Green Heron *Butorides virescens*

Patience is a natural virtue of most herons. Practicing a good-things-come-to-those-who-wait method of predation, they bring woe to a wide range of small fish, amphibians, reptiles, crayfish, insects, and other hapless prey with their still-hunting techniques.

From the slam-dunking great blue heron to the mannerly snowy egret, most herons bring dignity and grace to the hunt. As motionless as statues, necks outstretched, heads and bills pointing downward—only the rapid shifting of their eyes belies what passes for catatonia as they wait for edibles to come within striking distance.

The green-backed heron, the runt of the family, uses its short stature to achieve a kind of scrunch-and-crouch mastery. Neck tucked in and body leveled out in seeming defiance of gravity, it keeps a patient, low-profile watch over riverbank or mud flat—and makes up in results for what it lacks in style. The green-back's apparent transformation into a lifeless, loglike object helps lure one potential meal after another into its trap.

If patience occasionally wanes and the predator begins to walk and stalk, life in the marshes becomes fleeting indeed. In a riotous chase-and-nab ploy, the greenie creates a disturbance by stirring and raking the sediment with its feet. The prey has no choice but to flee, with the heron in hot pursuit. And if betting were permitted on these contests, birders would have no trouble picking the favorite.

When alarmed, the green-backed heron raises the spiky black feathers on its crown.

Recognition. 18–22 in. long. Small and dark. Back and wings blue-gray; crown black; neck purplish; legs orange-yellow. Immature similar but heavily streaked.
Habitat. Rivers, streams, ponds, moist woodlands, mud flats, and marshes.
Nesting. Nest is a platform of sticks up to 30 ft. above ground or water. Eggs 3–6, pale blue or greenish. Incubation 21–25 days, by both sexes. Young downy; leave nest after about 5 weeks. Nests in isolated pairs or small colonies.
Food. Fish, frogs, crustaceans, and insects.

Roseate Spoonbill

Ajaia ajaja

The tidal flats stretch for miles—green grass, silver pools, and scores of feeding egrets and white ibises. It is a model of harmony but for one discordant note: in the midst of it all, a single, brilliant splash of pink. No, it's not a mirage, or a flamingo—but one of our most exquisite marsh birds, the roseate spoonbill.

Incongruity seems to favor this bird. As the name implies, the spoonbill's mandibles are spatula-shaped. No other North American bird shares this curious (and very useful) adaptation. With its head beneath the surface, a feeding spoonbill swishes its sensitive bill from side to side, seining the water and mud for fish and other prey.

"It is much to be regretted," an ornithologist wrote of the roseate spoonbill in 1869, "that so many of the most beautiful water birds should be confined . . . to the southern extremity of our country." Perhaps so, but after the next 30 years the spoonbills were lucky to survive at all: thousands were slaughtered and their rose-blushed wings marketed as ladies' fans. Happily, with legal protection the birds have made a strong comeback and once again are dazzlingly visible along much of the Gulf Coast.

Downy young spoonbills show little sign of the distinctive bill they will sport as adults.

Recognition. 30–34 in. long. Bill long and spoon-shaped. Wings pink, with red on shoulder; tail orange; neck and body white; head naked and gray. Immature nearly all-white, including head, with pinkish tinge on wings. Flies with neck extended.

Habitat. Marshes, lagoons, and mangroves.

Nesting. Nest is a solid cup of sticks and twigs, 5–15 ft. above ground or water in bush or small tree. Eggs 1–4, white, spotted with brown. Incubation about 23 days, by both sexes. Young downy; leave nest 5–6 weeks after hatching. Nests in colonies.

Food. Fish, mollusks, shrimp, and insects.

Glossy Ibis

Glossy Ibis *Plegadis falcinellus*
White-faced Ibis *Plegadis chihi*

Glossy Ibis

White-faced Ibis

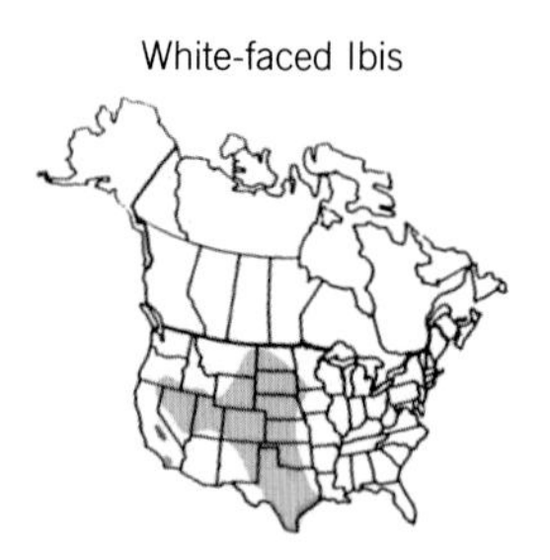

From a distance, under a cloudy sky, the birds look dark and lifeless—shadow images as stiff as the hieroglyphic likenesses of them carved on pyramid walls. But when the sun is unmasked and its warm rays touch the postured forms, these statues come alive. Bronze feathers flash with hidden hues; bold reds and subtle greens shimmer as the feeding flock works its way across the flats.

Fifty years ago, the glossy ibis was common enough but restricted to coastal marshes of the Deep South. Experts in the 1930's had scarcely ever seen one in New Jersey; yet by 1955 the bird had at least one established nesting site there, and 20 years later it was the most numerous colonial water bird nesting in the state. What had happened?

One theory views the bird as a relative newcomer to the New World and its northward push as a natural expansion of its range. Some speculate that ditching and draining, which destroyed so much of Florida's wetlands, forced the glossy ibis to move north; others believe that incursions by the white ibis pushed it out of its southern haunts.

In any event, along the Gulf Coast where Texas meets Louisiana, the range of the glossy ibis gives way to that of the white-faced ibis, a close relative with similar looks. Similar also in its expansive mood, the white-faced appears to be moving eastward at a steady rate.

Recognition. 19–26 in. long. Dark and glossy; bill gray and down-curved; legs gray. Breeding adult glossy has white border on bare facial skin. Breeding white-faced has broader band of white feathers bordering face. Nonbreeding birds very similar, but eyes are brown on glossy, red on white-faced. Flies with neck extended.

Habitat. Marshes, swamps, ponds, and farmland.
Nesting. Nest is a platform of sticks, on ground or up to 10 ft. above ground. Eggs 3–4, blue-green. Incubation about 3 weeks, by both sexes. Young downy; leave nest about 4 weeks after hatching. Nests in colonies.
Food. Mainly crayfish, insects, and frogs.

White Ibis

Eudocimus albus

Along the Tamiami Trail, the east-west road that cuts through the heart of the Everglades, cars speed toward their destinations. Overhead and unnoticed, a string of birds glides by. There is something dreamlike about these creatures. The wings and bodies gleam so white, they might have been spawned by clouds. The sickle-bills, painted faces, and hot-pink stockings look like the inspiration of a two-year-old artist working in lipstick. Wings set, the birds fairly float in the sultry Florida air, and when the flock settles to the ground, its slow, controlled descent has an almost underwater grace.

In coastal sections over much of the South, white ibises are common birds, northern emissaries from the tropics. Long before sunrise they leave their roosts, fanning out across marshes, swamps, and mangroves to feed. A flock will move across the shallow flats like farm harvesters in surgical gowns. The bright red mandibles reach ahead, probing the bottom like forceps. When a morsel is found, the bird throws back its head and swallows quickly, so as not to lose its place in line, or the tidbit to a neighbor.

In the evening, the birds return to their communal roosts—cloud-puff strings coming in from all directions. They disappear into the trees and the gathering dark until only the sound of them remains.

The young white ibis was once thought to be an altogether different species from the adult.

Recognition. 21½–27½ in. long. Adult white with black wingtips; bill red and down-curved. Immature brown on back and breast; belly white; bill brown to pink and down-curved. Flies with neck extended.
Habitat. Marshes, swamps, and mangroves.
Nesting. Nest is a flimsy platform of twigs 3–15 ft. above water in bush, sometimes in marsh grass; occasionally uses old nest of heron. Eggs 3–4, greenish, spotted with brown. Incubation 21–23 days, by both sexes. Young downy; leave nest after 4–5 weeks. Nests in colonies.
Food. Crustaceans, fish, and insects.

Wood Stork

Mycteria americana

Seen high overhead, North America's only stork is the very picture of beauty, poise, and grace. But as the distance falls away, so too does any illusion of beauty. At close quarters the gray, unfeathered neck appears grotesque, almost reptilian. Even so, when it is feeding, resting, or aloft, the wood stork moves with an air of dignity and courtliness. A feeding stork, wings spread and one exploratory foot thrust forward, seems engaged in some aquatic minuet. Soaring flocks, riding thermals aloft, fairly promenade within superheated columns of Florida air.

Unlike other southern wading birds, the wood stork was not gunned down for the table or the millinery trade. The birds were wary, unplumed, and therefore unprized. As late as 1954, one prominent ornithologist concluded that the bird "enjoys a rather static condition of well-being in Florida," and another wrote that "it is likely to survive for a long time in its native wilderness."

This turned out to be shortsighted prophecy. Birds last only as long as their habitat, and the swamps and wetlands that support wood storks and many other species have diminished. First came the diking of Lake Okeechobee, the source of the Everglades water. Then came ditching for mosquito control, then land development and periodic droughts. In 1984 the bird was identified as an endangered species by the federal government, and the flocks of the American stork have dwindled since then to a sad remnant.

Wood storks often nest among the very highest branches in the tallest bald cypress trees.

Recognition. 35–45 in. long. Tall, long-legged, and mainly white. Head naked and blackish; tail and rear half of wings black; bill long, stout, and down-curved. Young bird similar, but bill yellowish; head feathered and gray-brown.
Habitat. Swamps and marshes.
Nesting. Nest is a flimsy mass of sticks and twigs, 5–80 ft. above ground in tree. Eggs 3–4, whitish. Incubation 28–32 days, by both sexes. Young first fly 50–55 days after hatching. Nests in large colonies.
Food. Fish, frogs, reptiles, and insects.

Limpkin

Aramus guarauna

Though most Americans have never seen a limpkin, they have almost certainly heard it. Hollywood has long been enamored of the bird's loud, wailing cry and has dubbed it into the soundtrack of a host of African-jungle movies. Ironically, limpkins are not found in Africa at all, nor are any species closely related to them. This large wary bird of freshwater swamps, sole representative of the scientific family Aramidae, is strictly a New World bird. In the United States its range is limited to Florida, Georgia, and (by all appearances) Hollywood.

Skulking along a muddy riverbank, its head down and tail twitching, the bird resembles some overgrown rail. In flight, neck outstretched, wingbeats slow and measured, legs trailing behind, a limpkin looks for all the world like a small crane. But one limpkin feature is eminently its own: the bird's curious gait. Walking birds appear to favor one leg, to *limp*. This is the trait that gave the bird its name.

Long prized for its table-worthiness, the limpkin was nearly hunted to extinction before protective legislation gave it a chance to recover. Like the snail kite, the limpkin loves to eat freshwater apple snails. Outflanking the snail's defensive armor is a limpkin specialty. With the opening of the shell turned upward, the bird waits for the snail's "trap door" (or operculum) to open. Grabbing quickly, the bird tears the door off its hinges, then reaches for the exposed snail. With a flick of its bill the elusive prize is dislodged.

Apple snails are essential to the survival of limpkins, which rarely eat anything else.

Recognition. 23–28 in. long. Long-legged and long-billed. Brown, with white streaks and spots. Bill slightly curved. Loud, wailing call often heard at dusk and at night.
Habitat. Freshwater swamps and marshes.
Nesting. Nest is a a flat saucer of reeds 3–17 ft. above water in marsh vegetation or bush. Eggs 4–8, buff, spotted with brown and gray. Incubation by both sexes; period unknown. Young downy; leave nest soon after hatching; age at first flight unknown.
Food. Mainly large freshwater snails; also frogs, insects, and crayfish.

Sandhill Crane

Grus canadensis

Considered from any vantage point, the sandhill crane is clearly a bird to be reckoned with. Almost as if this mighty flier had a memory that stretched back over generations, when its ancestors were killed by the thousands as game birds, the sandhill is now one of the wariest birds in the American wilderness. And if cornered it can be a dangerous adversary, wielding its long, pointed bill with the speed and skill of a swordsman. Even the intrepid John James Audubon once plunged into a river up to his neck to escape the wrath of a crane with a broken wing, and in years past many a fine hunting dog was mortally wounded by an enraged parent bird protecting its young. Finally, the sandhill crane's wild courtship antics, which build up to the fervor of a tribal war dance, are among the most memorable sights in the avian kingdom.

During the summer, which it often spends in the marshes from California, Wisconsin, and Michigan northward, the sandhill takes its food chiefly from their waters. But it becomes more of a prairie feeder when it moves south for the winter. It was once a common migrant in the marshes and prairies of the Midwest, but people living there now seldom hear the wild call of the sandhill crane in flight—a vibrant, triumphant-sounding whoop that can be heard long after the great bird disappears into the clouds.

Unlike nestlings of other tall waders, downy young cranes can walk right after hatching.

Recognition. 34–48 in. long. Tall, long-necked, and long-legged. Mainly gray, with patch of bare red skin on crown. Flies with neck extended, not folded back as do herons; travels in noisy flocks.
Habitat. Marshes, prairies, and Arctic tundra.
Nesting. Nest is a large mound of plant matter in marsh or on dry ground. Eggs 1–3, buff or olive, spotted with reddish brown and lilac. Incubation 30–32 days, by both sexes. Young downy; leave nest soon after hatching; first fly at 10 weeks; remain with parents until following spring.
Food. Mainly grain, insects, and small animals.

Whooping Crane

Grus americana

Cranes are birds of the superlatives," one authority on the crane family has declared, and there is some merit to the claim, no matter how biased its source. To understand why, simply look at the whooping crane, a bird to inspire awe in the eye and heart of any beholder.

The whooping crane can fly so low that its wingtips brush the grasses and aquatic plants as it passes over. Yet it can also fly too high to be seen from the ground. The tallest and heaviest of North America's wading birds, it also owns one of the world's mightiest windpipes—a five-foot-long tube that produces its distinctive call. Before 19th-century hunters began their devastation, the deep, organlike tones of a thousand whooping cranes passing overhead were compared to the noise of some great army on the march.

By 1941 the total whooping crane population had fallen to just 11 birds, and many experts already considered it extinct. But rescue efforts have fostered at least a modest recovery, and continuing success would be especially welcome to anyone who has ever witnessed the whooping cranes' high-spirited mating dance. Typically, one crane will begin dancing, and before long others join in, often to impress prospective mates—but sometimes apparently just for the joy of being alive and of being a crane.

In northern Canada, whooping cranes like to nest in dense marshes and isolated bogs.

Recognition. 49–56 in. long. Tall, long-necked, and long-legged. Adult white, with black wingtips and bare red patches on crown and cheeks. Young bird similar, but tinged with rusty; lacks red patches on head.
Habitat. Marshes and river floodplains.
Nesting. Nest is a large mound of plant matter in marsh or on dry ground. Eggs 1–3, buff or greenish, spotted with reddish brown and lilac. Incubation 33–35 days, by both sexes. Young downy; leave nest soon after hatching; first fly at 14–18 weeks; remain with parents until following spring.
Food. Mainly fish, amphibians, and mollusks.

Shorebirds

With their slender bills and boundless reserves of energy, shorebirds are ideally suited to a life of foraging, whether on a sun-drenched beach or a muddy shoal, a marshy riverbank or an open field. Wandering over their chosen terrain, bills incessantly probing the sand or poking beneath rocks in search of food, they are sprightly, tireless bundles of determination. From the methodical sandpiper to the fast-darting plover, from the oystercatcher wedging its knife-blade beak into a mussel shell to the woodcock spiraling high above the treetops on a courtship flight at dusk, these birds are at once delightfully active and peacefully at home in their surroundings.

Sanderlings

American Oystercatcher

Haematopus palliatus

King Philip Came Over From Good Spain." So goes the mnemonic sentence countless students have learned to remember the hierarchy of plant and animal categories: kingdom, phylum, class, order, family, genus, species. But in the dynamic realm of nature, things are not always so precise. Our oystercatchers are an example—two birds very similar in lifestyle and, except for a few patches of white, in appearance. Their ranges overlap slightly, and hybrids are found from time to time. So, are they one species or two? Three ornithologists may produce three different answers. "They're good species," says the first. "Only subspecies," insists the second. "Members of a superspecies," asserts the third. Each represents a valid opinion on a difficult question: where do two birds with a common ancestor stand on the pathway of evolution?

The superspecies—a group of two or more species of common descent but with different ranges—is a refinement of the King Philip scheme. For some birds, such as the oystercatchers, the scientific jury is still out. Have they diverged so far that they cannot interbreed widely—making them different species—or are they just well-marked races of the same species? Putting them in a superspecies amounts to a sort of interim verdict, recognizing their close kinship without claiming to be the final word.

An oystercatcher doesn't "catch" oysters and mussels, but uses its bill to pry them open.

Recognition. 17–21 in. long. Large and stocky; boldly patterned in black or dark brown and white. Head and neck black; back dark brown; wing and tail patches and underparts white; bill long, bladelike, and orange-red; legs pale pink.
Habitat. Sandy beaches and mud flats.
Nesting. Nest is a hollow in sand, sometimes lined with pebbles and bits of dried plant matter. Eggs 2–4, greenish or buff, spotted with dark brown. Incubation 24–29 days, by both sexes. Young downy; leave nest soon after hatching; become independent at about 5 weeks.
Food. Mollusks, starfish, and marine worms.

Black Oystercatcher

Haematopus bachmani

Most birds stalk food that is difficult to catch but easy to eat. Oystercatchers take prey—mussels and other bivalves—that is easy to find but hard to consume. The question is, how does an oystercatcher get into these recalcitrant mollusks?

With the right tools, of course. Oystercatchers are equipped with long, stout, chisel-shaped bills that are ideal for loosening mussels from their footholds. Then they insert the compressed tips of their bills between the two half-shells like a fisherman's knife, severing the muscle that holds the bivalve closed. (Of course, some oystercatchers take a less surgical approach: they simply hammer with their bills at the sides of shells to crack them open.)

Oystercatchers are not the only birds with an appetite for well-protected prey. Various gulls carry mussels or clams high into the air and then drop them onto rocks or pavement below. Some birds need even more inventiveness to get at their food. The woodpecker finch of the Galápagos Islands uses a cactus spine to extract prey from holes and crevices, while the Egyptian vulture drops stones on the hard-shelled eggs of ostriches. But from Africa to the Galápagos to the coastlines of North America, all this ingenuity is focused on the same goal: to make the most of the earth's bountiful resources.

Surprisingly, oystercatchers are skilled swimmers even without the advantage of webbed feet.

Recognition. 16½–18½ in. long. Large and stocky; blackish brown, with no white in plumage; bill long, bladelike, and orange-red; legs pale pink.
Habitat. Rocky coasts and islands.
Nesting. Nest is a hollow in gravel, sometimes lined with pebbles or bits of shell and plant matter. Eggs 1–4, buff, spotted with dark brown. Incubation 24–29 days, by both sexes. Young downy; leave nest soon after hatching; become independent at about 5 weeks.
Food. Mussels, starfish, marine worms, crustaceans, and other marine animals.

Black-necked Stilt

Himantopus mexicanus

Long ago nicknamed the lawyer bird (for its persistent noisiness during the mating season), the black-necked stilt was, like many other wading birds, hunted nearly into extinction in the 19th century. Fortunately, lawmakers intervened in time to allow most of the threatened species to recover. Now the black-necked stilt is seen often in its favorite breeding areas in the South and West.

And it is a memorable sight, making its feeding forays into both fresh- and saltwater ponds atop the exceptionally long legs for which it is named—perhaps the longest in relation to body size of any bird's. The black-necked stilt is also a master at assuring its own air-conditioning during the breeding season. It keeps its ventral feathers wet by making as many as 100 trips daily from the nest into the nearest water, so that evaporation will help cool the parent, its eggs, and eventually its nestlings. Without such a cooling system, a female stilt could perish while spending an especially hot day motionless on her nest.

Breeding in a variety of locales—moist savannas, pond edges, marshes, and fields with a history of flooding—black-necked stilts may form colonies of about 40 pairs. When all are busy fishing, their long legs bent in shallow waters and needle-line bills poised for action, they form an almost dreamlike silhouette against sunset or dawn.

Just hatched, a young stilt already has the long legs that give this species its name.

Recognition. 13½–15½ in. long. Slender, with long, thin black bill and very long legs. Upperparts black; underparts white; tail white; legs orange-red.
Habitat. Marshes, mud flats, and shallow ponds and lakes.
Nesting. Nest is a shallow depression in ground, lined with grass, bits of shell, and other debris. Eggs 3–5, buff, spotted with dark brown or black. Incubation 22–26 days, by both sexes. Young downy; leave nest soon after hatching; become independent at 4–5 weeks. Nests in small colonies.
Food. Mainly aquatic insects and snails.

American Avocet

Recurvirostra americana

With its black and white–striped wings, bright eyes, and long upcurved bill, the American avocet is among the most beautiful and graceful of North America's shorebirds. Though its range east of the Mississippi River is limited, each spring birders throughout the West look forward to this summer visitor's arrival.

The avocet's bill is truly one of nature's masterpieces, both in its elegant design and in its utility as a hunting and fishing tool. The bird can use it with great speed and precision to catch insects on the wing or to nip them from the surface of a pond or marsh. Underwater another option is available. Dipping its head downward, the avocet sweeps the bill from side to side along the bottom, much like an old-time farmer scything a field. As the bill stirs up mud and sand, dislodging food and anything else that happens to be there, the water becomes so cloudy that it is impossible to see anything. But that poses no problem for the avocet. Its versatile bill is so sensitive to differences between the edible and the inedible that an avocet can make the right choice even though the food cannot always be clearly seen before it is consumed.

After the nesting season, the rusty head and neck of the American avocet turn whitish.

Recognition. 17–18½ in. long. Large, with long legs and thin, upturned black bill. Back boldly patterned in black and white; underparts white; legs and feet bluish gray. Head and neck rusty on breeding adult; whitish on nonbreeding adult.
Habitat. Shallow ponds, marshes, mud flats, and flooded fields.

Nesting. Nest is a shallow depression lined with bits of dry grass. Eggs usually 4, olive, spotted with brown. Incubation 22–29 days, by both sexes. Young downy; leave nest soon after hatching; are independent at about 5 weeks.
Food. Mainly aquatic insects.

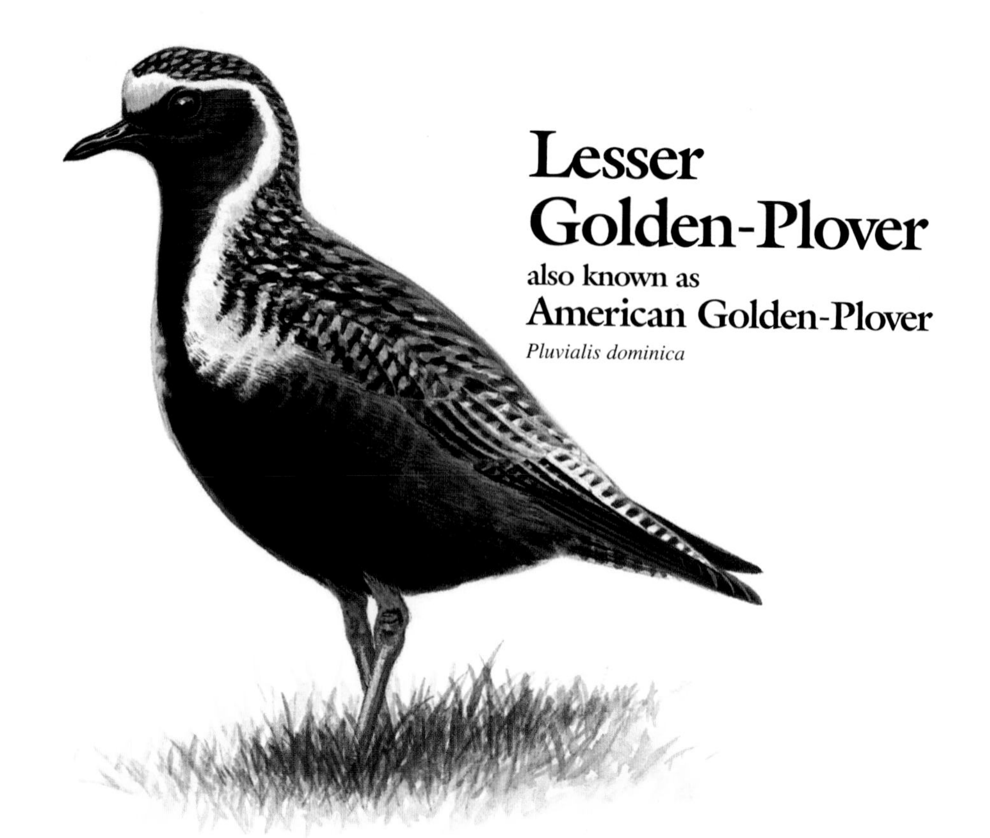

Lesser Golden-Plover

also known as

American Golden-Plover

Pluvialis dominica

It's not quite a foot long. Full-grown, it weighs all of six ounces. Nevertheless, this hardy bird hurtles its relatively small body into the air every year for two astounding migratory flights that together may carry it nearly 20,000 miles. In accomplishing these feats, the lesser golden-plover (which used to be called the American golden plover) will sometimes reach air speeds of 70 miles per hour as it cruises over vast stretches of open sea.

After nesting on the tundra behind North America's Arctic coast, the golden-plover population splits apart in August for the long southward migration. Some flocks fly east, then follow the Atlantic shoreline before breaking out over the ocean to destinations as distant as the pampas of Argentina. Others head for the Pacific, with far-off Australia as the goal. Few other species have the stamina to attempt voyages of such epic proportions.

A handsome creature, the golden-plover can be good eating, too. So much so that in the last century it was shot down in torrential numbers during migration—a staggering 41,000 were reportedly taken in a single day near New Orleans. Now protected by law, it is again proliferating, and when ocean gales blow in against the coastline they often carry exhausted plovers that can be seen resting briefly before continuing their incredible journeys.

Before their long migration to South America, golden-plovers fuel up on crowberries.

Recognition. 9–11 in. long. Upperparts all dark, without white rump or wing stripe, and no black patch under base of wing. Spring adult has black face and underparts; back mottled with brown and yellow. Fall adult dark above; underparts whitish; eyebrow whitish.
Habitat. Shores and prairies; nests on tundra.

Nesting. Nest is a hollow lined with moss and leaves. Eggs usually 4, buff, marked with black and brown. Incubation about 27 days, by both sexes. Young downy; leave nest soon after hatching; age at independence unknown.
Food. Mainly insects and crustaceans.

Black-bellied Plover

Pluvialis squatarola

For a shy bird, the black-bellied plover has managed to acquire an impressive collection of nicknames: *beetle-head, bottle-head, hollow-head,* and *grump* are a few unflattering samples. But only *black-breast* does any real justice to this trim, strong, handsome bird, whose habit of migrating in small, hard-to-hit flocks may be all that has kept hunters from blasting it into extinction.

The largest of North America's plovers, it breeds in the high Arctic and is an expert at foraging over wet sands exposed by an ebbing tide. Marine worms are a favorite delicacy, but the black-bellied plover also finds ample sustenance in meadows, salt marshes, and recently plowed fields. Moving over the ground in short, staccato runs of four or five yards, it strikes fast, eats quickly, and then scans the landscape for possible attackers.

Of the roughly 65 species of plovers found worldwide, this one is surely among the most cosmopolitan, wintering in places as far-flung as South Africa, India, Australia, Chile, and Argentina. But North Americans can also get a good look at plover flocks each May as the black-bellies migrate up the Mississippi Valley and both seacoasts on their way to the remote Arctic nesting grounds where they will bring forth yet another generation.

Recognition. 11–13½ in. long. Spring adult has black face and breast; undertail coverts white; upperparts mottled with black and white. Fall adult mottled gray above; underparts whitish. In flight has white rump, bold white wing stripe, and black patch under base of wing.
Habitat. Beaches and marshes; nests on tundra.

Nesting. Nest is a hollow in ground lined with moss. Eggs usually 4, greenish, gray, or tan, spotted with brown. Incubation about 26–27 days, by both sexes. Young downy; leave nest soon after hatching; become independent at 6–7 weeks.
Food. Small marine animals and insects.

Snowy Plover

Charadrius alexandrinus

Plovers and sandpipers present birders with special challenges to identification. Not only do these shorebirds come in many species, but their distributions overlap broadly, particularly during migration and on wintering grounds. Knowing plumages, habitats, and calls is critical to sorting the species out. And veteran birders get help from another clue as well: behavior.

While most sandpipers feed in groups, probing and searching the mud with their sensitive bills, plovers hunt by sight, chase down their prey, and dine alone. Running briefly, pausing, then running a little more and stopping abruptly to peck at the ground, they follow a run-pause-run-peck feeding pattern that distinguishes them even at a distance from other shorebirds.

Behavior can be the key to identifying other species as well. Chunky shorebirds feeding in shallow pools, their bills submerged and their heads pumping up and down like sewing-machine needles, are probably dowitchers. Birds acting similarly in a freshwater marsh are most likely snipes, and small shorebirds searching through pebbles and other beach debris are undoubtedly turnstones. On close inspection, a tail-bobber will almost certainly be a solitary sandpiper, while one that teeters will be a spotted sandpiper. Learning how shorebirds behave, as well as how they look and sound, furnishes the extra clues that turn many puzzles of identification into problems solved.

Perhaps because they eat more insects, snowies have thinner bills than other plovers.

Recognition. 6–7 in. long. Small, with thin black bill. Adult pale gray-brown above; face and underparts white; ear patch and mark at side of breast blackish. Immature similar, but ear patch pale, patch at side of breast pale gray-brown.
Habitat. Sandy beaches, salt ponds, and alkaline lakes.

Nesting. Nest is a scrape in sand lined with bits of shell. Eggs 2–3, buff, spotted with black and gray. Incubation about 4 weeks, by both sexes. Young downy; leave nest soon after hatching; become independent at about 1 month.
Food. Mainly insects and small crustaceans.

Piping Plover

Charadrius melodus

So melodious is the clear, whistled voice of the piping plover that the bird's popular and Latin names both refer to its musical ability. Its series of clear whistled peeps end with a bell-like *peep-lo,* which may be given as a separate and lovely two-note call. The notes seem to come from here, from there, from anywhere—except from where the sand-colored bird is actually standing, invisible on the sands of the upper beach.

The ornithologist Arthur Cleveland Bent argued that shorebirds' voices reflect their habitat as emphatically as their coloring does. Clear whistles and shrill cries belong to the birds of the shores and the mud flats. The gutteral calls of the snipe family fit right into their boggy homes among frogs and herons, and the reedy calls of the pectoral sandpipers seem a perfect voice for the salt marshes. Many shorebirds make sounds like small pebbles rolling together in the surf.

The vocal talents of shorebirds vary widely, yet the peeps, squawks, or whistles of each species accomplish what they are meant to do. With their voices shorebirds stake out nesting territory, provide the voice-recognition basic to mating, sound alarms, drive off intruders, and call and instruct their young. There are also times when shorebirds simply feel like adding their voices to the music of the natural world—and few contribute their share more melodiously than the piping plover of the upper dunes.

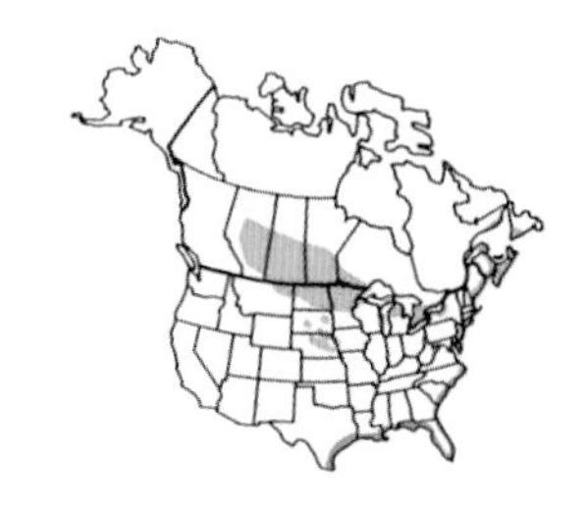

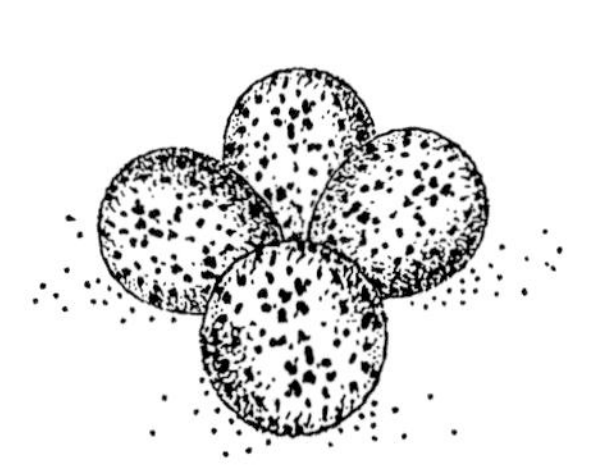

With their fine speckling, the eggs of a piping plover are well concealed on a sandy beach.

Recognition. 6–7 in. long. Breeding adult pale brown above; underparts white; single breast band and band across forehead black; bill orange with black tip; legs orange. Fall adult and immature duller; no band on breast or forehead; bill darker; legs yellowish.
Habitat. Sandy beaches.

Nesting. Nest is a hollow in sand, sometimes with a few shells or pebbles. Eggs usually 4, gray or buff, lightly spotted with black and lilac. Incubation 25–31 days, by both sexes. Young downy; leave nest soon after hatching; become independent at 3–4 weeks.
Food. Insects and small aquatic animals.

Semipalmated Plover

Charadrius semipalmatus

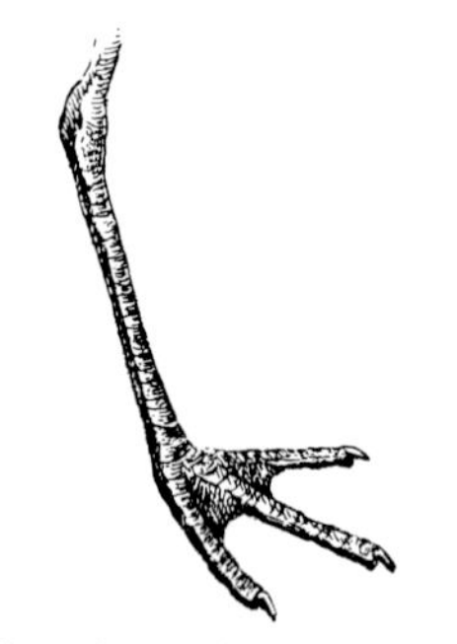

The rather technical-sounding name of this bird refers to the partial webs between its toes.

One of our most common plovers, the semipalmated is often seen roosting along beaches above the high-water mark while its feeding flats are covered by the tide. At first you may notice one or two as they begin to walk or run to keep a safe distance from you. Then they will seem to multiply, as if magically generated by the sand, until a dozen or more are keeping pace. How is it that you failed to spot all these boldly patterned birds immediately?

Semipalmated plovers (along with the killdeer and other closely related "ringed" plovers) are successful examples of what biologists call disruptive coloration. Their plumages feature bold patterns of black and white that effectively break their bodies up into discrete pieces, disrupting the birds' outlines and making them difficult to spot against a variegated background.

Disruptive coloration is only one means by which shorebirds make themselves inconspicuous. Many are colored to match their habitats—the American woodcock, for instance, whose upperparts are patterned to resemble the leaf litter on which it roosts and nests. Most are also countershaded, their dark upperparts gradually lightening to a white belly; this eliminates the sharp change in color that might catch an unwelcome eye. Even the most boldly patterned shorebirds, such as turnstones, expose their striking features only in flight and then subtly vanish again into the background when they land.

Recognition. 6½–8 in. long. Breeding adult dull brown above; underparts white; single breast band black; bill orange with black tip; legs orange. Fall adult and immature duller; bill darker; legs yellowish.

Habitat. Beaches, tidal flats, and lakeshores; nests on tundra.

Nesting. Nest is a hollow in ground, sometimes lined with grass or bits of shell. Eggs usually 4, buff, spotted with brown. Incubation 23–25 days, by both sexes. Young downy; leave nest soon after hatching; become independent at 3–4 weeks.

Food. Small marine animals and insects.

Killdeer

Charadrius vociferus

At least two days before killdeer babies peck their way out of their shells they are conversing in peeps and learning to understand the voices of their parents. When at last they kick the shells aside—usually within an hour or so of one another—they are wet and exhausted from the hours-long struggle. Their parents brood them to keep them warm and help them to rest. A few hours later their down is dry, their eyes are open, and they are following their parents from the nest, looking like mottled brown puffballs on toothpick stilts.

Soon they begin picking at the earth with their own small bills, finding out for themselves that they can eat the seeds and tiny insects that lie there. But danger is all around. At the slightest sign from their parents they close their dark eyes and freeze in place, not twitching a muscle, while their parents fly into the faces of cattle ambling toward them or try to lead more determined enemies away by pretending to be crippled. But if the parents signal them to flee, the chicks dash away, sometimes into a stream or pond where they swim away. If the predator is a hawk, they drop beneath the surface and swim safely under water.

Finally, after almost a month of learning about land and water, the youngsters grow flight feathers and join their fellows in flying and screaming happily above the pastures.

Although adult killdeers have two breast bands, the chicks have only one, like other plovers.

Recognition. 9–11 in. long. Large for a plover. Upperparts brown; underparts white; breast with 2 black bands; rump and base of tail rusty; wing stripe white. Identified by loud *kill-dee* call.
Habitat. Pastures, plowed fields, riverbanks, mud flats, and airports.
Nesting. Nest is a scrape in ground, lined with bits of grass, pebbles, and stems. Eggs 3–5, buff, spotted with brown. Incubation 24–28 days, by both sexes. Young downy; leave nest soon after hatching; become independent at about 25 days.
Food. Mainly insects, earthworms, and snails.

Greater Yellowlegs

Tringa melanoleuca

Land birds inhabit a three-dimensional world of trees, shrubs, and grasses. Shorebirds, in contrast, frequent the planes of mud flats, beaches, and marsh pools, some of which may seem inhospitable and unproductive as feeding areas. But look down into a marsh pool: a myriad of small fish and invertebrates can be seen moving around. Or reach into the muck of a tidal flat and pull up a handful: it is alive with small crustaceans and other tiny creatures. Rocky coastlines are dotted with mussel beds, and tiny shrimplike organisms abound in the surf-line sand.

Shorebirds have bills of diverse shapes and sizes to make the most of these riches. From short-billed sandpipers, which probe the mud's surface, to longer-billed curlews, which reach deep into the muck for worms, shorebirds boast a variety of bills that would put a mechanic's tool kit to shame. Knots, dunlins, godwits, whimbrels—each is designed for a particular niche, and together they take advantage of nearly every shoreline feeding opportunity.

Greater yellowlegs fall near the middle of this spectrum. Their long legs enable them to wade farther from shore, where instead of probing the mud they sweep the water for invertebrates or stab at small fish. At times their frenzied dashes verge on the comical, but they are only claiming a share of the shoreline's bounty that is uniquely their own.

Once nicknamed "telltales," yellowlegs take alarm quickly, scaring off other shorebirds.

Recognition. 14 in. long. Upperparts dark brown and streaked; underparts whitish with streaks on breast and belly; bill long, slim, and slightly upturned; legs long and yellow. Shows white rump in flight. Identified by loud, clear calls.
Habitat. Marshes, ponds, and mud flats; nests in northern bogs.

Nesting. Nest is a hollow in moss near water. Eggs usually 4, buff, spotted with brown. Incubation about 23 days, by female only. Young downy; leave nest soon after hatching; become independent at about 20 days.
Food. Mainly small fish, insects and their larvae, and snails.

Lesser Yellowlegs

Tringa flavipes

Except for a few southerly species, shorebirds appear to most of us merely as transients on their way north or south, part of a huge pulse of migration. Great hordes of sandpipers and plovers are scattered across mud flats and beaches, milling about in search of food or taking flight in mammoth flocks that almost seem to be a single organism as they turn synchronously through the air.

But in the Far North, where most of these birds breed, a different way of life awaits. There shorebirds are scattered far and wide across the boggy expanses that constitute much of their nesting territories. Gregarious no longer, they are busy with the duties of the brief summer.

A lesser yellowlegs swings through the sky in a pendulumlike display flight, swooping down in a graceful arc to perch atop a stunted spruce tree. Nearby, a male lesser golden-plover takes to the air, flopping about like a huge butterfly with slow, exaggerated wingbeats as he outlines his territory. A little farther off, a semipalmated sandpiper erupts into flight from behind a small hummock, chattering, twittering, and dashing wildly about the sky. At the peak of their plumage, the birds take to defending and advertising their territories with an almost manic zeal, compressing the charged drama of courtship and breeding into a "summer" that lasts only a few precious weeks.

The needlelike bill of the lesser yellowlegs is well adapted for snapping up small insects.

Recognition. 10–11 in. long. Smaller than greater yellowlegs. Upperparts dark brown and streaked; underparts whitish with streaks on breast and belly; bill long, slim, and straight; legs long and yellow. Shows white rump in flight.
Habitat. Marshes, ponds, and mud flats; nests in northern forests and on tundra.

Nesting. Nest is a hollow in ground, often far from water. Eggs usually 4, buff, spotted with dark brown. Incubation about 23 days, by female only. Young downy; leave nest soon after hatching; are independent at about 3 weeks.
Food. Mainly insects and crustaceans.

Solitary Sandpiper

Tringa solitaria

An oddball among its peers, the solitary sandpiper is not quite as much of a loner as its name might imply. But it does nest in the seclusion of Canadian and Alaskan coniferous forests in wet, wooded places that are not usual sandpiper habitat. And it also does the unthinkable for a sandpiper—it nests in trees. This sandpiper, however, does not build its own nest; rather, it lays its eggs in the abandoned nests of robins, grackles, or rusty blackbirds. As these may be anywhere up to 40 feet above the ground, the offspring face a formidable challenge.

Sandpiper chicks cannot fly for several weeks, but they run as soon as they are dry from the egg, and they all leave the nest before they're two days old. This is no problem for most sandpipers, which nest on the ground. But the chicks of the solitary sandpiper must step out of their high-rise nest and, like little tree ducks, beat their tiny wings frantically as they drop down to the forest floor below.

Following their parents, the chicks learn to wade in shallow water and to hunt aquatic animals of all sorts, from insects and crustaceans to tadpoles and even small frogs. They also learn to shake one foot rapidly under water, bringing creatures up out of the mud. Then, plunging their heads into the water, they bob and peck with great alacrity to catch their wiggling meal as it scatters.

On their nesting grounds, solitaries can be seen perched precariously in trees.

Recognition. 8–9 in. long. Dark olive-brown above; rump dark; throat and breast streaked with brown; head olive-brown with white eye-ring. Bobs its tail.

Habitat. Shores of ponds, wooded streams, marshes, and estuaries.

Nesting. Uses old bird's nest 3–40 ft. above ground. Eggs usually 4, greenish or buff, spotted with brown. Incubation about 24 days, by female only. Young downy; leave nest soon after hatching; tended by female only; age at independence unknown.

Food. Insects, spiders, crustaceans, and other small aquatic animals.

Willet

Catoptrophorus semipalmatus

Male willets react in a flash. And what they flash is their wings. Climbing high over a salt marsh or lake margin, they zoom down without warning, flashing signals with their wings as they knife toward the earth. To their mates, these black and white beacons help strengthen the growing bonds of the nesting season. To rival males who see them, the signals resemble No Trespassing signs.

Should a marsh hawk fly by, it too will receive a message. Frequently, several willets will join forces to mob the unwanted predator, flying close to its face while flashing their wings. These pulsing "threat colors" distract the hawk from surveying marsh grasses for nestlings and prompt it to seek less troublesome hunting grounds.

To protect their eggs and nestlings, willet hens can flash, too. If a raccoon or other predator approaches, a protective mother will feign injury, her bright wing markings and piteous calls guaranteeing that she will be noticed as she tries to lead the intruder away.

Finally, willet wings also flash after danger has passed. While flying in formation, the birds often signal each other, as if to identify themselves and, through a kind of avian camaraderie, to keep the flock in close ranks.

Once nesting is done, the willet exchanges its barred plumage for a plain gray winter coat.

Recognition. 14–17 in. long. Large. Mainly gray, streaked in summer; plain gray in winter. Wings with bold white stripe; rump white; bill long, straight, and heavy. Easily identified by loud *pill-will-willet* call.

Habitat. Marshes, beaches, and mud flats.

Nesting. Nest is an unlined scrape or a cup of grass. Eggs 4, olive, spotted with brown. Incubation 22–29 days, by both sexes. Young downy; leave nest soon after hatching; tended by female only; age at independence unknown.

Food. Insects, small marine animals, and seeds.

Spotted Sandpiper

Actitis macularia

No one suspected anything out of the ordinary in the lives of the little look-alike pairs of spotted sandpipers that nest throughout North America. Not until 1972, that is, when ornithologist Helen Hays burst into publication with the news that it is the female who returns first to the lakeside nesting areas, the female who fights with other females for a share of the shorefront property, and the female who ruffles her neck feathers and struts about among the males, choosing a mate. No one knows for certain who prepares the slight scrape in the earth that serves the pair as a nest, or who adds the bit of grass for lining. But it *is* the female who lays the eggs.

She usually lays four of them, and the male immediately takes over incubation while the female walks off for another week of romancing with any other available males. Sometimes it is only a fling, and she returns to her mate to take up her share of the nesting duties, but sometimes this little outing results in a second pair bond. Then this new pair builds a nest, the female lays four more eggs, the new male takes over the incubation, and the female again walks off for another week of flirtation. She may do this as many as four or five times before settling down to share nesting duties with her last mate—while all her former mates are left to cope with their youngsters on their own.

The "teeter-tail" is easy to spot in a crowd, thanks to the constant bobbing of its tail and rump.

Recognition. 7½–8 in. long. Breeding adult gray-brown above, with white wing stripe; breast and belly white with black spots; bill short, pink at base. Fall adult similar, but underparts plain white. Teeters constantly as it walks at edge of water. Flies close to water with stiff, rapid, fluttering wingbeats.

Habitat. Ponds and streams; seashores.

Nesting. Nest is a hollow hidden in grass. Eggs usually 4, greenish or buff, spotted with brown. Incubation 20–24 days, often by male only. Young downy; leave nest soon after hatching; become independent at 17–21 days.

Food. Insects, small fish, and crustaceans.

Upland Sandpiper

Bartramia longicauda

After the multitudes of North America's passenger pigeons had all been slaughtered for market, commercial hunters turned their guns on the innumerable upland sandpipers that flourished across our vast, rolling grasslands. By the early 1900's, with no restrictions on the carnage, scarcely an upland sandpiper could be found anywhere on the plains. The sweet-voiced birds—which, simply by feeding themselves, must have saved the country from many a scourge of grasshoppers, army worms, weevils, and cutworms—were almost annihilated.

Alarmed by the disappearance of so many birds, various states began to pass protective laws; but it was too late for some species and nearly so for others. Like all birds, upland sandpipers face many natural hazards—accidents, diseases, storms, and hungry predators, among others. In addition, they fly a difficult migration route of 7,000 to 8,000 miles—and are still hunted on their South American wintering grounds. Despite all this, the upland sandpipers have defied the odds and appear to be holding their own quite well across most of their large summer range.

An upland sandpiper pauses for just a moment as it alights, its wings gracefully raised.

Recognition. 11–12½ in. long. Small-headed, with long, slim neck. Upperparts scaled and streaked with brown; underparts pale, streaked and barred; wings dark-tipped; legs long and yellowish. Often seen perched on fence posts.
Habitat. Grasslands, meadows, and fields.
Nesting. Nest is a grass-lined cup on ground. Eggs usually 4, cream-colored or tan, spotted with reddish brown. Incubation 21–27 days, by both sexes. Young downy; leave nest soon after hatching; are independent at about 4 weeks.
Food. Mainly insects and grass seeds.

Whimbrel

Numenius phaeopus

Like many other shorebirds, the whimbrel nests on the Hudson Bay lowlands, a cold, boggy subarctic region where northern tundra meets stunted boreal evergreen forest. From a distance it might appear to offer little in the way of variety. But a study in the mid-1970's showed that whimbrels actually nest in three distinct habitats. One, called *hummock-bog,* is wet lowland marked by hummock mounds, an abundance of tundra dwarf birch, and patches of stunted black spruce and tamarack. The second, *sedge-meadow,* is also a low habitat, covered with sedges, grasses, and scattered shrubs. The third, *heath-tundra,* is dry, rolling upland with dense, low vegetation.

The study found that nest success was highest in the hummock-bogs, where 86 percent of the nests had at least one egg that survived the 3½-week incubation period. In the other two habitats only 54 percent had surviving eggs.

Why such a difference? Two factors may explain it. First, the density of nests is much greater in hummock-bogs, so there are more sitting birds to detect and chase predators. Second, the hummock-bog has better natural protection. Nests placed on the lee side of shrubs and larger hummocks are sheltered from cold winds and better hidden from egg-snatching gulls, ravens, and jaegers. As a realtor might say, location is everything. Whimbrel pairs who nest successfully in the hummock-bogs nearly always return; those who fail in the other habitats very often don't.

Weeks after the adults have left, unattended young whimbrels are still on the breeding grounds.

Recognition. 15½–19 in. long. Large, with down-curved bill. Brown and streaked; head striped with blackish brown; no white wing stripes or white rump.
Habitat. Marshes, prairies, shores, and mud flats; nests on tundra.
Nesting. Nest is a depression in ground. Eggs 3–5, olive, spotted with brown and lilac. Incubation about 28 days, by both sexes. Young downy; leave nest soon after hatching; become independent at 5–6 weeks.
Food. Insects, worms, crustaceans, and berries.

Long-billed Curlew

Numenius americanus

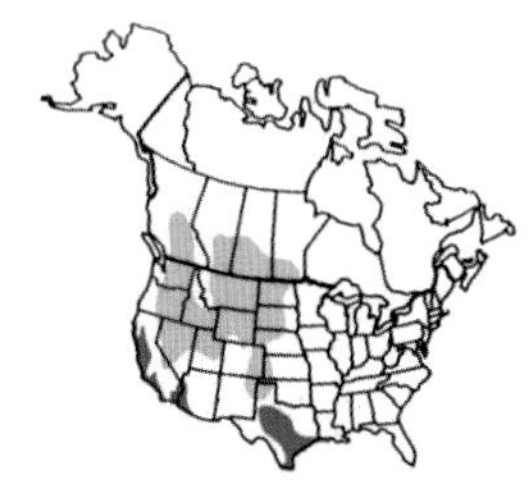

Most of us associate shorebirds with the beaches and estuaries that teem with a variety of "waders," as Europeans call them, during the migratory and winter seasons. So it may seem a bit odd that North America's largest shorebird is found chasing grasshoppers on a hot piece of Montana or Nevada pastureland. But long-billed curlews are quite at home on interior grasslands, having nested on prairies as far east as Illinois until settlers arrived with guns and plows.

Prairie breeding, however, is fraught with dangers. The curlew's huge, brown-blotched eggs are prized by badgers, coyotes, and long-tailed weasels. Some gopher snakes have been known to swallow the eggs whole. Curlews usually stay close to the nest, but their occasional absences invite predation by black-billed magpies and common ravens. And since most remaining grassland now serves as rangeland, trampling by sheep and cows takes a toll as well.

Even healthy chicks face a hazardous, earthbound five weeks before fledging, during which they must cope with summer heat, disease, and marauding hawks. The chances of survival are slim. A study in the rangeland of Idaho found that, on the average, two successfully laid four-egg nests together produced only one surviving fledgling.

The curlew's long bill can seek out prey buried more than six inches beneath the sand.

Recognition. 21–26 in. long. Large, with very long, down-curved bill. Warm brown and streaked above; underparts buff; wing linings cinnamon; head without stripes; no white wing stripes or white rump.

Habitat. Prairies and plains; salt marshes and beaches on migration and in winter.

Nesting. Nest is a grass-lined hollow in meadow. Eggs usually 4, whitish or buff, spotted with brown. Incubation 27–30 days, by both sexes. Young downy; leave nest soon after hatching; become independent at 32–45 days.

Food. Insects and small aquatic animals.

Marbled Godwit

Limosa fedoa

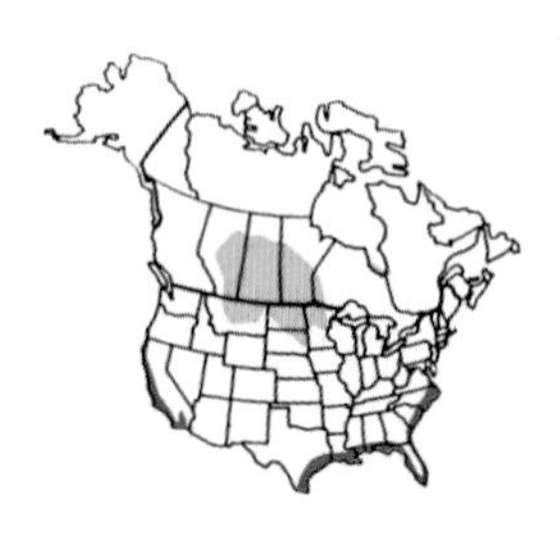

To photograph birds is to sweat and swat flies, to stand for long hours waist-deep in swamp water, or to crouch even longer in freezing makeshift blinds. Allan Cruickshank knew these tortures. As one of America's leading bird photographers, Cruickshank expected the worst.

Then along came a marbled godwit to make his life easy. Tawny like the grass that engulfed her, the mottled female was nearly invisible against the hollow of beaten-down grasses that held her four olive eggs. When Cruickshank approached, the bird scurried away. He built a blind and almost immediately the bird returned. Perhaps, he hoped, he would not need a blind. Perhaps this beautifully camouflaged shorebird—known for "sitting tight" while unknowing predators walked just inches away—would allow him to approach. The bird did that and more. To Cruickshank's amazement, his docile subject even allowed him to pick her up while he photographed the nest.

All that changed overnight, however. As soon as her eggs hatched, the godwit reacted hysterically to Cruickshank's presence, flying this way and that and berating him loudly before leading her young from the nest. Never again would the godwit allow Cruickshank's approach. But the man had his pictures—and in contrast to most such projects, getting them had literally been a walk in the park.

In its eagerness to find food, this big bird often plunges its entire head underwater.

Recognition. 16–20 in. long. Large, with very long, upturned bill; bill pink with dark tip. Upperparts mottled with brown; underparts cinnamon with dark bars; wing linings cinnamon or rusty.
Habitat. Plains and prairies; tidal flats on migration and in winter.
Nesting. Nest is a hollow concealed in grass. Eggs usually 4, greenish or olive, spotted with brown. Incubation about 3 weeks. Young downy; leave nest soon after hatching; become independent at about 21 days.
Food. Mainly mollusks, worms, crustaceans, and insects.

Wilson's Phalarope

Phalaropus tricolor

Where is it written that female birds must demurely sit by as males flaunt their gaudy courtship colors in spectacular displays? Not in any book describing Wilson's phalaropes. Flying in the face of conformity, these elegant shorebirds leave few sex roles unreversed.

Come springtime on their grassland breeding grounds, the larger, more brightly colored females court their future mates with high-flying aerial antics and musical water ballets in which they swim toward the males with their neck feathers flaring as they sing a strange chugging song. The males feign nonchalance until the amorous hens begin nudging and butting the targets of their affection. Sometimes two or three females will chase a single harried male as he flees from their fevered advances.

That male can run, but he can't hide—not for long, at any rate. One female or another eventually wins him, and mating takes place. Continuing to flout tradition, the male incubates the eggs his mate has deposited in a grass-lined hollow near the water's edge. When the nestlings hatch some 20 days later, it is again the male who rears them without help from the female. Some hens, in fact, abandon the nesting grounds before their young even hatch. Others form flocks and splash about in the shallows, gamboling without a care while dad keeps the youngsters in line.

Phalaropes obtain much of their food simply by plucking prey from the surface as they swim.

Recognition. 8½–9½ in. long. Bill needlelike. Breeding female has gray forehead and crown; black stripe through eye and down neck; rusty tinge on breast; back gray. Male duller. Winter birds gray above, underparts and rump white; line through eye blackish. Often swims; spins on water to stir up food.

Habitat. Marshes, lakes, and coastal bays.

Nesting. Nest is a grass-lined hollow on ground. Eggs usually 4, buff, heavily spotted with brown and gray. Incubation 16–21 days, by male only. Young downy; leave nest soon after hatching; tended by male only; age at first flight unknown.

Food. Insect larvae, shrimp, and seeds.

Black Turnstone

Arenaria melanocephala

Its bold wing pattern makes the black turnstone easiest to identify when it is in flight.

Most people have little experience with the eggs of wild birds. Judging from the few songbird eggs they may have seen, casual birders might assume that all wild birds' eggs are small. But shorebirds are a different story. Relative to the size of an adult, shorebird eggs are not merely large—they're huge. A clutch of four eggs in many cases outweighs the mother who laid them! The black turnstone is a perfect example: although a bit smaller than a robin, she lays eggs the size of a crow's.

The likely explanation for this formidable egg size lies in the way shorebirds develop. Born open-eyed, alert, and covered with down ("precocial," in the language of ornithology), shorebirds are ready within hours to leave the nest and feed themselves. A black turnstone chick must develop more fully in the egg than, say, a baby robin, which is hatched blind and helpless ("altricial") and is nurtured in the nest by its parents. As a result, the turnstone embryo requires a larger yolk—which provides energy and builds tissue—and added space in which the chick can grow to full precocity.

The size of a shorebird's eggs effectively limits these species to a standard clutch of four. The average adult can just about cover that many during incubation—and then only because the eggs are tear-drop–shaped ("pyriform"), which economizes on space by allowing them to fit snugly together like the wedges of a quartered pie.

Recognition. 9 in. long. Stocky and short-legged. Breeding adult black above and on breast, with white spot on face; belly white; legs dusky; bold pattern visible in flight. Winter adult similar but browner, without white spot on face.
Habitat. Rocky shores; nests on tundra.
Nesting. Nest is an unlined hollow in short grass. Eggs usually 4, buff, spotted with brown. Incubation about 22 days, by both sexes. Young downy; leave nest soon after hatching; age at independence unknown.
Food. Small marine animals.

Ruddy Turnstone

Arenaria interpres

Birds' common names are sometimes derived from their feeding strategies, but a curious habit of *Arenaria interpres* must have made a uniformly strong impression on early observers everywhere. Today that bird is called *Steinwälzer* in Germany, *tournepierre* in France, *revuelve-piedras* in Spain, and (to translate them all) *turnstone* in Britain. Both the ruddy and the black turnstone search for hidden prey by "turning"—inserting their short, pointed bills under rocks or other objects and flipping them with their strong neck muscles. Several will cooperate to turn a dead fish, or even to roll up mats of seaweed to find out what's underneath.

But no law says ruddy turnstones must turn something over to have a snack. They pick up insects and spiders, worms and snails, sea urchins, brittle stars, and crabs wherever they find them. They hammer limpets and barnacles open, and during spring migration dig for horseshoe crab eggs. They feast on sedge seeds when they arrive on their northern breeding grounds, and they fatten on crowberries before they leave. On beaches they have been known to savor sandwich bread and French fries.

There doesn't seem to be much these ruddies won't eat. They have even been seen feeding on eggs in tern colonies. Such behavior must be either new or rare, however, for the larger terns, not recognizing the turnstones as predators, just stood by calmly and watched them do it.

Turnstones not only turn over stones, but also dig deep holes in the sand in search of prey.

Recognition. 7½–9 in. long. Stocky and short-legged. In summer, rusty above with bold pattern in flight; head black and white; breast black; belly white; legs orange-red. In winter, brown above, whitish below, with dark breast patches.
Habitat. Beaches, rocky shores, and mud flats; nests on tundra.

Nesting. Nest is a hollow in ground, lined with stems. Eggs usually 4, olive, spotted with brown. Incubation 22–24 days, by both sexes. Young downy; leave nest soon after hatching; become independent at about 3 weeks.
Food. Small marine animals and insects.

Red Knot

Calidris canutus

Imagine a beach so densely packed with shorebirds that sunlight is blocked from the sand. If that beach were on Delaware Bay in mid-May, the shorebirds—numbering in the hundreds of thousands—would be red knots.

Separating New Jersey from Delaware on a coastline pressured by rapid suburban growth, Delaware Bay serves migrating knots as a crucial stepping-stone. En route to the Arctic from their South American wintering grounds, these chunky "beach robins" time their arrival to coincide with the breeding cycle of the bay's horseshoe crabs. Every May hordes of these ancient invertebrates haul themselves from the surf to deposit eggs by the millions in holes they have scooped in the sand. Knots, ruddy turnstones, and other shorebirds depend on these eggs to refuel their northward flights. The birds stay several weeks, gorging on the eggs from dawn until dusk to increase their body weight by 40 percent or more. Finally sated, they whirl off again in huge clouds toward the Arctic.

How many birds can convene on Delaware Bay? During the month of May alone, a million birds may comb the bay's egg-laden shores. On a given day during that period, more than 80 percent of North America's red knots—and half its ruddy turnstones—may join ranks there in one of the world's most awe-inspiring congregations of wildlife.

Outside horseshoe-crab season, knots usually travel in small parties and forage on beaches.

Recognition. 9½–10 in. long. Stocky, with straight, black bill. Breeding adult brown above with pale feather edges; face, breast, and belly rusty or pinkish; faint wing stripe. Winter adult gray above, whitish below.
Habitat. Mud flats, beaches; nests on tundra.
Nesting. Nest is a small hollow in ground. Eggs usually 4, buff, spotted with brown. Incubation 21–23 days, by both sexes. Young downy; leave nest soon after hatching; become independent at 18–20 days.
Food. Mainly mollusks, seeds, and insect larvae.

Sanderling

Calidris alba

To the thousands of vacationers thronging ocean beaches, sanderlings provide an endless source of interest and amusement. They are those incessantly feeding little birds who nimbly chase the frothy backwash down the sloping sands and just as nimbly race back along the perilous edges of incoming waves seething around them. Their little black three-toed feet twinkle on the beach as their thin black bills plunge into the surf, scissoring out the tiny crustaceans, worms, and mollusks washing about in the seas. All sandpipers appear to eat with great gusto, but none seem to be so forever famished as the sanderlings, which feed with such unflagging energy that it seems their fat little bodies should burst from the seafood stuffed inside.

Although they spend only 8 to 10 weeks on their tundra breeding grounds, most sandpipers are so eager to return to the ocean shores that they can barely wait the three or four weeks necessary for their young to fledge. Perhaps they grow hungry for the bountiful marine diet, or perhaps they are driven by an irresistible longing for the roar and surge of the tides that rule so much of their lives. Whatever the reason, some sandpipers simply desert their unfledged young. Sanderlings, however, do wait until their chicks can fly—not necessarily because of parental solicitude, but because their youngsters are capable of long flights after only 17 days of life.

Sanderlings chase the retreating waves to snatch tiny animals before the sand covers them up.

Recognition. 7–9 in. long. Bill short and black. Breeding adult rusty, mottled with black, above and on breast; belly white; bold white wing stripe. Winter adult grayish, with bold white wing stripe, black patch at shoulder.
Habitat. Beaches, mud flats, and sandbars; nests on tundra.

Nesting. Nest is a small hollow in ground, lined with moss and leaves. Eggs 3–4, olive, thinly spotted with brown. Incubation 24–31 days, by both sexes. Young downy; leave nest soon after hatching; become independent at 17 days.
Food. Insects and small marine animals.

Semipalmated Sandpiper

Semipalmated Sandpiper *Calidris pusilla*
Western Sandpiper *Calidris mauri*

Semipalmated Sandpiper

Western Sandpiper

When it comes to parenting, semipalmated sandpipers are among the most impatient of a notoriously short-fused clan. All sandpipers seem eager to flee their inland breeding grounds, but most manage to stay with their chicks at least until they can fly in some fashion—though some are not strong enough to migrate yet and so must follow later on their own. The semipalmated, however, abandons its chicks when they are no more than 14 days old and still flopping about with untested wings. Somehow the young survive and make their own way to South America, where this species spends most of the year.

Given the length of their migration, it is fortunate indeed that these plump shore-lovers are such fast fliers. A semipalmated actually holds the record for the fastest known long-distance bird flight. Four days after being banded in Massachusetts, this champion was in Guyana, 2,800 miles away. Sadly, its place in the record books carried a high price: the bird was shot by a hunter, who mailed the band number to the U.S. Fish and Wildlife Service.

In winter the western sandpiper looks so much like the semipalmated that they're virtual twins. But the western winters along U. S. coasts, so if birders in, say, California or Texas should spot a sandpiper that looks just like a semipalmated, they can safely add a western to their lists.

Recognition. 5½–7 in. long. Bill and legs black. Breeding adult gray-brown above with rusty feather edges; breast streaked with brown. Winter adult grayish above, whitish below. Most westerns have longer bills; summer birds have rusty on crown, cheeks, and sides of back.
Habitat. Seashores; nests on tundra.

Nesting. Nest is a grass-lined hollow. Eggs usually 4, buff or olive, spotted with reddish and gray-brown. Incubation 18–22 days, by both sexes. Young downy; leave nest soon after hatching; are independent at 14–19 days.
Food. Aquatic insects; small marine animals.

Least Sandpiper

Calidris minutilla

The tiniest members of the entire sandpiper family—no larger than sparrows—least sandpipers make their way along the coasts of the United States as they nest and feed during their twice-a-year migrations. But they are colored and marked so like the sands and surf of the beaches that they can scarcely be seen even when they move about. And they are colored and marked so like all the other pint-size sandpipers with whom they flock that it is nearly impossible to tell one species from another as they chase in and out with the waves. There are, however, a few clues.

As their name suggests, least sandpipers are generally smaller than any of the others. In addition, their legs are greenish yellow instead of the black that is common among their cousins. Semipalmated sandpipers can be distinguished (at very close range) by the short webbing between their toes. Female westerns have a longer, slightly downcurved bill. Baird's have buffier breasts; and in flight white-rumped sandpipers reveal the snowy rump feathers that earned them their name.

Only the most indefatigable birders, however, even attempt to separate and identify the members of these lively oceanside flocks in the field. The rest of us are satisfied just to call the whole collective crowd of sparrow-size, gray-brown, look-alike sandpipers "peeps."

The least sandpiper (top) tends to have a shorter, more slender bill than its semipalmated cousin.

Recognition. 5½–6½ in. long. Small, with short, slim bill, yellowish legs. Breeding adult brown above with rusty and buff feather edges; underparts whitish with streaks on breast. Winter birds duller above.

Habitat. Mud flats, marshes, and beaches; nests on tundra.

Nesting. Nest is a small hollow in ground, lined with grass or leaves. Eggs usually 4, buff, spotted with brown. Incubation 19–23 days, mainly by male. Young downy; leave nest soon after hatching; tended mainly by male; age at independence unknown.

Food. Mainly crustaceans and insects.

White-rumped Sandpiper

Calidris fuscicollis

Like several other members of the sandpiper family, white-rumps nest in the high latitudes of the Canadian Arctic during the long days of the Arctic summer. Then, as winter approaches in the north, they fly nearly 9,000 miles to the tip of South America, where they enjoy the extended daylight of summer there. As a result, white-rumps experience many more hours of sunlight each year than almost any other creature on earth.

Perhaps it is simply a yearning for light that drives these long-distance migrants, but this seems unlikely to be the whole explanation. To be sure, longer hours of daylight provide extra time in which to forage for food. Even so, the search for new habitats as the seasons change cannot in itself account for voyages of several thousand miles.

Bird migration in fact remains one of the great mysteries of the animal world. Especially intriguing is the subject of navigation. Experiments suggest that birds use the stars, the sun, polarized light, and even the earth's electromagnetic forces to guide them on their way. But how is all this information actually taken in and processed so that flocks fly straight and swiftly to their destinations? What could enable some young birds to set off without adults on a journey they have never made before? The questions are fascinating; and for now they outnumber the answers.

More than most other small sandpipers, white-rumps enjoy foraging at freshwater ponds.

Recognition. 6½–8 in. long. Larger than other small sandpipers; wings extend beyond tip of tail; rump white. Breeding adult rusty on crown and behind eye; sides spotted and streaked with black. Winter birds grayer, with whitish eyebrow.
Habitat. Mud flats, beaches, and prairies; nests on tundra.

Nesting. Nest is a small cup in ground. Eggs usually 4, buff or greenish, spotted with brown. Incubation about 22 days, by female only. Young downy; leave nest soon after hatching; become independent at 16–17 days.
Food. Mainly marine worms and insects.

Pectoral Sandpiper

Calidris melanotos

Courtship may be exciting, but this bird blows things out of proportion—literally. The huffing and puffing commences when pectoral sandpipers arrive on their Arctic breeding grounds, each bird beautiful in its own understated way but hardly stunning. Hidden beneath the males' throat and breast feathers, however, are the two sacs of tissue that turn these amorous suitors into real dandies.

When the time is right—which appears to be anytime a female comes within 50 yards and shows even the slightest interest—a lovestruck male will inflate his sacs and leap skyward. Filling his esophagus with air to create a ballooning throat pouch as big as his body, he rises 60 feet in the air, throws back his head, tilts down his tail, then glides earthward on stiff, outstretched wings, uttering a string of hollow, booming notes that sound like the whistle of air blown across the opening of a bottle.

Immediately on landing, the male rushes to the female, lifts his head high, and huffs up like a tiny tom turkey. He bows. He crouches. He struts back and forth while each throat pouch fills and empties like an overactive bellows. The female, however, hardly notices. Pectoral sandpipers are anything but monogamous, and she may have just mated with several suitors in rapid succession. This fellow may be flamboyant, but he's not the only man in her life.

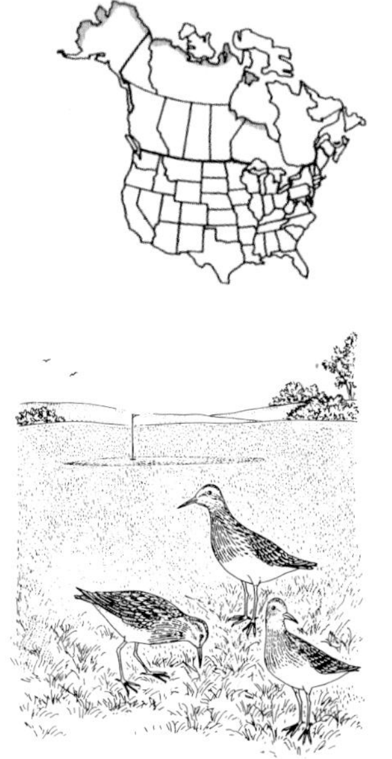

An affinity for open country often brings migrant pectoral sandpipers to golf courses

Recognition. 8–9 in. long. Large, with yellow legs. Upperparts brown with buff feather edges; breast dark brown and streaked, contrasting sharply with white belly.

Habitat. Marshes and grassy pools; nests on Arctic tundra.

Nesting. Nest is a cup of grass and leaves hidden on ground. Eggs usually 4, whitish or buff, spotted with brown. Incubation 21–23 days, by female. Young downy; leave nest soon after hatching; tended by female only; become independent at about 3 weeks.

Food. Mainly insects and small crabs.

Purple Sandpiper

Calidris maritima

Except for the summer months, when they breed on the Arctic tundra, purple sandpipers spend their lives on the rocky shores and islets of the northeast coast, pulling crustaceans, mollusks, and algae from the cracks and crevices in cliffs or probing into wet seaweed for small mussels and clams—which they appear to swallow shell and all.

Their daring can verge on the foolhardy as they creep along steep ledges slick with algae, the sea boiling and crashing just beneath them. But like other shorebirds, purple sandpipers can swim if they need to. At times they may even be seen adrift on a raft of seaweed, floating out to sea as they explore their makeshift vessel for the tiny delights of coastal dining that may be hidden there.

These sandpipers are not abundant everywhere in their range, but they do show an attachment to familiar haunts. Flocks of purples can be found, as the ornithologist John K. Terres noted, "feeding and resting in certain favored rocky places at the edge of the surf all winter, year after year."

Half a century ago purple sandpipers were rarely seen south of Long Island. But as people built mammoth stone breakwaters—ideal habitats for the purples—to protect the east coast's sandy beaches from erosion, these rock-loving birds extended their winter range nearly to Georgia, and may have increased their numbers in the bargain.

Stone breakwaters built along the east coast have extended the purple sandpipers' winter range.

Recognition. 8–9½ in. long. Stocky. Breeding adult brown-gray and streaked; legs yellowish and short; bill yellow at base. Bold white wing stripe visible in flight. Winter bird mainly dull gray; underparts whitish with brown streaks.
Habitat. Rocky shores; nests on tundra.
Nesting. Nest is a grass-lined hollow in ground. Eggs usually 4, greenish, spotted with brown. Incubation about 22 days, mainly by male. Young downy; leave nest soon after hatching; tended by male only; become independent at about 3 weeks.
Food. Mainly crustaceans and insects.

Rock Sandpiper

Calidris ptilocnemis

Seldom seen except along the Pacific coast's rocky shores, ledges, and reefs, the rock sandpiper feeds in the clefts and crannies of ancient, sea-broken stone. Although the birds nest in both the Asian and Alaskan Arctic, they all gather in Alaska before the fall migration. Some winter on the Aleutian Islands or the southern Alaskan coast; others head as far south as northern California.

Like most related species, the rock sandpiper nests in a depression in the mosses and lichens on the Arctic tundra, laying four eggs that are startlingly large compared to the size of the bird producing them. The male and female share the duties of incubation, and both care for the down-covered youngsters. Like all sandpiper babies, their chicks are able to run about and feed themselves almost as soon as they are dry.

We don't know when the little ones learn to fly, and we don't know whether their parents stay with them until that point or desert them earlier, as many sandpipers do. In any event, it is clear that little sandpipers teach themselves how to fly and that their parents have departed days before the chicks are strong enough to accompany them. The great mystery is how the fledglings know where the sandpiper flocks are gathering for the winter—and how they find their way there.

On the Pacific's rocky coastline, wintering rock sandpipers have no need of man-made habitat.

Recognition. 8–9 in. long. Stocky. Breeding adult has rusty upperparts; underparts whitish, with large black patch on lower breast; bill dark with pale base; legs short, yellow-green; bold white wing stripe visible in flight. Winter bird mainly dull gray.

Habitat. Rocky shores; nests on tundra.

Nesting. Nest is a small hollow in ground. Eggs usually 4, olive, spotted with brown. Incubation about 20 days, by both sexes. Young downy; leave nest soon after hatching; tended mainly by male; age at independence unknown.

Food. Mainly crustaceans, insects, and small mollusks.

Dunlin

Calidris alpina

The name was originally *dunling,* meaning little and dun-colored (a dull grayish brown). Then the *g* was dropped from the word for reasons lost to history, and the dunlings thus became dunlins. True to its name, whatever the spelling, the dunlin is a *little* sandpiper, about the size of a sanderling. Although in summer its back feathers are tinted with red, in winter these feathers are dun and seem flung over the bird like a cape.

The color contrast between each bird's back and underparts adds dramatic emphasis to the close-order flights performed by dunlin flocks that winter on our coasts. Flying so tightly packed that individuals seem scarcely a wingspan apart, the whole flock wheels and turns, rises and drops, in perfect synchronization, each bird showing first its dun-colored back and then its whitish belly.

Dunlins come to our shores in late summer but return to the north for nesting so early in spring that they often find the tundra pools and marshes still icebound. Sometimes they respond by reflocking and flying southward in reverse migrations that can last for days. The delay in courtship isn't a complete waste, however. It shortens the time for nesting, but it also gives the birds' voices more time to change, so that when courtship begins it is accompanied not by harsh wintry squawks but by richer, ringing songs.

The dunlin's spotted juvenile plumage is worn so briefly that few people ever see it.

Recognition. 6½–8 in. long. Bill long, black, and slightly drooped. Breeding adult rusty above; underparts whitish with large black patch in center of belly. Winter adult more uniform grayish brown above; underparts whitish.
Habitat. Mud flats and beaches; nests on Arctic and subarctic tundra.

Nesting. Nest is a cup of grass and leaves on ground. Eggs usually 4, greenish or buff, spotted with brown. Incubation about 22 days, by both sexes. Young downy; leave nest soon after hatching; become independent at 18–20 days.
Food. Small marine animals; insect larvae.

Surfbird

Aphriza virgata

This bird can scare you off a mountain. Suppose you're high on an Alaskan slope, hoping to spot Dall sheep, those sure-footed climbers that appear like ghosts out of the fog shrouding the peaks. Miles from the ocean, you're not thinking about sandpipers from the Pacific shore at all.

A Dall sheep trail winds upward through a rock slide. You take it, inching your way over the rubble as the trail narrows to the point of disappearing. Then it happens. You step forward and the rock seems to explode as a surfbird rockets upward, flying straight at your face. Wild shrieks and a mad flapping of wings bombard your senses. The bird seems to be everywhere. Finally, it flies away, calling loudly, its white rump and wings flashing white.

Surfbirds are so named for their winter habit of foraging along the rocky Pacific coast; but they nest in the Alaskan interior, again amid rocky slopes. How were you to know that your next step would have crushed four eggs an anxious parent was incubating in a camouflaged nest?

Although these birds depend primarily on protective coloring to save them from predators, they can be fiercely aggressive in defense of their nests and young. Rather than feign injury to lure an intruder away, they will fly directly at it, halting its progress and—with any luck—driving the startled trespasser straight out of their rocky domain.

Surfbirds in winter look as relentlessly gray as the rocks strewn along the shore.

Recognition. 9–9½ in. long. Stocky. Breeding adult mottled with black, white, and chestnut; breast and sides with black bars; rump and wing stripe white; legs greenish. Winter adult uniform gray on back, head, and breast; belly white.
Habitat. Rocky coasts; nests on tundra.
Nesting. Nest is a small hollow lined with moss and bits of grass. Eggs usually 4, buff, spotted with reddish brown. Incubation by both sexes; period unknown. Young downy; leave nest soon after hatching; age at independence unknown.
Food. Crustaceans, mollusks, and insects.

Short-billed Dowitcher *Limnodromus griseus*
Long-billed Dowitcher *Limnodromus scolopaceus*

Short-billed Dowitcher

Short-billed Dowitcher

Long-billed Dowitcher

Move over, Mr. Singer; here's a real sewing machine. It's the beak of the short-billed dowitcher, a four-inch needle that jabs mud flat ooze with incessant vertical strokes. There's a purpose to this fast, stitching motion: by liquefying the mud, it makes the dowitcher's task of grasping worms and crustaceans a great deal easier.

The dowitcher's hunting is also aided by a bill adaptation the bird shares with woodcocks and snipes. As these chunky nibblers probe mud flat shallows, they can bury their bills to the hilt and still open them to catch food. The trick lies in flexibility. Unlike the beaks of most birds, which are rigid from end to end, dowitcher bills have flexible tips that can separate slightly and snap closed.

Still, there's more to catching a worm than just grabbing it. First you must locate your prey down in the opaque slime. So dowitcher beaks also come equipped with a sense of touch, provided by tiny sense organs that lie buried in pits in their bony cores. This versatile type of beak serves not only the short-billed dowitcher but also its nearly identical relative, the long-billed dowitcher. And the latter, which is found more commonly in freshwater ponds, has at least one apparent advantage. Endowed with more of a good thing, as its name implies, the long-bill can probe even deeper for the bounty hidden below.

Recognition. 10½–12½ in. long. Long-billed with white rump and back. Spring adults brown above; rusty below. Fall birds gray above; whitish below. Long-billed has dense spots on throat; barred sides; some white on belly. Short-billed calls *tu-tu-tu,* long-billed, *keek!*
Habitat. Shores; nests on tundra.

Nesting. Nest is a hollow in ground, lined with twigs and grass. Eggs usually 4, buff, spotted with brown. Incubation about 21 days, by both sexes. Young downy; leave nest soon after hatching; age at independence unknown.
Food. Mainly insects, crustaceans, and snails.

Stilt Sandpiper

Calidris himantopus

Stilt sandpipers live up to their name. Their long, slender legs lift them well above most of their fellow sandpipers as they feed along the edges of ponds, marshes, and coastal pools. They are sometimes mistaken for lesser yellow-legs—in fact, they were once known rather ungraciously as mongrel or bastard yellowlegs. But what seems like a tricky distinction is relatively easy to make in the field. Stilt sandpipers have greenish-colored legs; yellowlegs, to put it simply, live up to *their* name, too.

Although stilt sandpipers travel in dense flocks of their own species, or occasionally in mixed flocks with other sandpipers, their most frequent feeding companions are the much larger and heavier dowitchers. And they feed just as dowitchers do, the whole mixed flock crowded together belly-deep in the water, bobbing their heads rapidly up and down in an unmistakable "sewing machine motion," as Roger Tory Peterson called it. Even in the relatively deep water where their long legs permit them to forage, stilt sandpipers plunge their heads down to the mud of the bottom to gather the worms, mollusks, larvae, and aquatic plants on which they thrive.

Did one species copy the other's manner of feeding? Or did each separately learn to feed with that rapid-fire bobbing of head and bill? Whatever the case, nature rewards methods that work, period—no matter how odd they may look through a pair of binoculars.

Frequent companions in flight, a stilt sandpiper (below) and a dowitcher are easy to tell apart.

Recognition. 7½–9 in. long. Slender and long-billed, with long, greenish legs and white rump. Breeding adult has rusty ear patch; underparts barred with dark brown. Winter bird uniform gray above, whitish below.

Habitat. Mud flats, marshes, and grassy pools; nests on tundra.

Nesting. Nest is an unlined hollow in ground. Eggs usually 4, greenish or olive, spotted with brown. Incubation 19–21 days, by both sexes. Young downy; leave nest soon after hatching; become independent at about 18 days.

Food. Mainly insect larvae and aquatic seeds.

American Woodcock

Scolopax minor

Variously known as the bogsucker, timberdoodle, hookumpake, and night peck, the American woodcock is a one-of-a-kind bird with the style of a true eccentric. It feeds, for instance, by thumping its feet vigorously on moist ground—probably to rouse any earthworms lurking below—then plunges its bill in up to the nostrils and pulls out a worm like a taut string of yarn.

But for sensational antics, the male woodcock reserves his best for the courtship ceremony. Every evening in spring, soon after sunset, he struts around the perimeter of his private dancing ground, *bzeeping* like a giant insect. Suddenly he takes off with a twittering trill (made by the whirring of his wings), circles high into the sky, and erupts with a burst of bubbling song. Then, still twittering, he descends in zigzagging swoops back to his dancing ground, where he mates with any woodhen attracted by his florid display. The dancing, singing, and swooping may go on for an hour or two and then be resumed in the predawn darkness of the next day. Or they may continue, without any sign of rest, the whole night long.

The show begins in mid-March and might continue to the end of June; but April is the liveliest month, since the female nests in May. And for any humans lucky enough to witness these rites of spring, the memory lasts a lifetime.

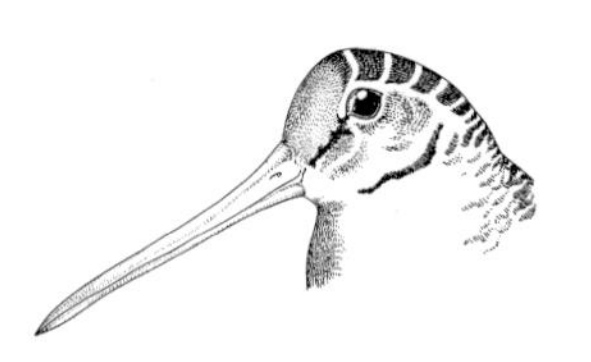

With eyes far back on its head, the woodcock can stay on guard even while probing for food.

Recognition. 10½–11½ in. long. Stocky and large-headed, with very long bill and rounded wings. Dead leaf pattern on upperparts; underparts rusty; crown with transverse black bars. Wings whistle in flight.
Habitat. Moist thickets and woods; brushy fields.
Nesting. Nest is a hollow in ground lined with leaves. Eggs usually 4, buff or cinnamon, spotted with brown and lilac. Incubation about 21 days, by female only. Young downy; leave nest soon after hatching; tended by female only; become independent at about 2 weeks.
Food. Mainly insects and earthworms.

Common Snipe

Gallinago gallinago

Henry David Thoreau squinted skyward but saw nothing save glowering dusk. A strange sound had drawn his attention—an eerie tremolo that faded like twilight, then rose again to haunt distant parts of the sky.

The 19th-century essayist knew this sound well. Straining his eyes in the near darkness, Walden Pond's most famous admirer watched as a long-beaked bird spiraled upward through the New England skies. Hidden in marsh grass below, a female snipe feigned indifference as her mate climbed in widening circles to a height approaching 300 feet, set his wings, spread his tail, and began to swoop. Then the strange sound commenced, a pulsating *who-who, who-who, who-who* that grew in intensity until the snipe pulled from his dive just a few feet above the ground.

For years a debate flourished over how the swooping male made that haunting sound. While most American scientists theorized that snipe wings made the noise, Europeans argued that the bird's outer tail feathers, held stiffly apart from the rest of his tail during dives, caused a vibrating hum as air rushed through and around them.

Subsequent field studies proved the Europeans correct. Thoreau had seen the bird "fanning the air like a spirit over some far meadow's bay." The snipe, in fact, was simply singing courtship songs—with his tail.

Air rushing through the parted tail feathers of a courting snipe creates an odd, rhythmic hum.

Recognition. 10½–11½ in. long. Very long, straight bill. Head striped; back brown with long whitish stripes; breast and sides spotted and barred with brown; belly white. In flight, shows rusty in tail. Flight call a harsh *scape!*
Habitat. Wet meadows, marshes, and bogs.
Nesting. Nest is a cup in ground, lined with grass and leaves. Eggs usually 4, buff or olive, spotted with brown. Incubation 18–20 days, by female only. Young downy; leave nest soon after hatching; are independent at about 20 days.
Food. Insects, crustaceans, and other small animals; also some seeds.

Parts of a Bird

Among the most helpful clues to bird identification are field marks—distinguishing features such as the size or shape of a bill, the length of a tail, or the colors and markings on different parts of the body. Shown here are some of the standard terms used in describing a bird's physical characteristics.

Index of Scientific Names

A

B

C

D

Q

R

S

T

U

V

W

X

Z

General Index

D

E

F

G

H

I

J

K

L

M

N

T

V

W

Credits

Paintings

Raymond Harris Ching: 20, 34, 36, 70, 130, 338, 345, 347, 348, 354, 358, 364, 386, 387, 481. **John Dawson**: 10, 12, 53, 61, 66, 92, 94, 99, 106, 123, 144, 154, 156, 166, 167, 175, 182, 184, 197, 207, 218, 222, 229, 233, 234, 236, 243, 250, 263, 264, 281, 282, 288, 304, 311, 314, 350, 449. **Walter Ferguson**: 1, 2, 68, 107, 109, 112, 113, 114, 116, 117, 118, 119, 120, 137, 138, 161 , 171, 172, 174, 179, 186, 191, 195, 200, 248, 249, 251, 259, 261, 265, 267, 271, 286, 295, 296, 301, 302, 305, 310, 312, 344, 353, 383, 391, 414, 415, 451, 468, 476. **Albert Earl Gilbert**: 8, 9, 11, 13, 14, 15, 16, 18, 19, 21, 22, 23, 25, 26, 27, 29, 31, 32, 33, 35, 38, 39, 40, 41 ,43, 45, 46, 47, 48, 49, 50, 51, 67, 76, 90, 93, 134, 135, 136, 181, 187, 189, 190, 193, 194, 198, 199, 203, 206, 210, 212, 214, 215, 217, 219, 237, 238, 239, 240, 241, 245, 258, 260, 269, 298, 300, 324, 332, 336, 339, 341, 343, 349, 363, 366, 367, 368, 371, 373, 374, 375, 376, 377, 388, 383, 385, 390, 403, 417, 419, 423, 424, 425, 429, 430, 431, 432, 433, 435, 436, 437,439, 440, 444, 445, 446, 452, 453, 457, 465, 466, 467, 469, 474, 479, 482, 491. **Cynthia J. House**: 73, 83, 89, 102, 103, 104, 105, 108, 121, 125, 126, 132, 133, 141, 146, 157, 158, 159, 160, 164, 165, 168, 169, 170, 173, 176, 180, 183, 205, 208, 209, 213, 220, 230, 244, 255, 256, 266, 283, 315, 321, 326, 327, 337, 361, 365, 369, 370, 372, 379, 380, 381, 384, 455, 461, 480. **H. Jon Janosik**: 80, 81, 85, 98, 110, 131, 226, 232, 247, 275, 299, 404, 422, 447, 448, 456, 459, 460, 462, 463, 464. **Ron Jenkins**: 276, 278, 316, 329. **Lawrence B. McQueen**: 128, 139, 142, 162, 196, 202, 211, 216, 221, 231, 235, 257, 277, 284, 285, 287, 290, 291, 292, 293, 392, 393, 394, 395, 396, 397, 398, 399, 400, 401, 405, 407, 408, 409, 410, 411, 412, 413. **Gary Moss**: 346. **John P. O'Neill**: 5, 153, 178, 280, 317, 320, 322, 389, 406, 416, 426, 434, 438, 441, 443, 478. **Hans Peeters**: 6, 17, 28, 30, 37, 42, 44, 60, 62, 65, 69, 71, 84, 87, 88, 91, 97, 122, 124, 127, 129, 143, 145, 185, 201, 227, 246, 333, 334, 335, 340, 342, 388, 418, 420, 421, 428, 454, 473, 475, 477. **H. Douglas Pratt**: 95, 96, 100, 101, 111, 115, 152, 242, 279, 289, 303, 306, 307, 313, 325, 351, 352, 357, 359, 382, 442, 470, 471, 472. **Tim Prutzer**: 74, 75, 77, 78, 79. **Chuck Ripper**: 24, 52, 54, 55, 56, 57, 58, 59, 63, 64, 82, 86, 140, 147, 148, 149, 150, 151, 155, 177, 224, 225, 228, 252, 253, 262, 268, 270, 272, 273, 274, 294, 297, 319, 328, 330, 331, 355, 356, 360, 362. **David Simon**: 308, 309, 318, 323. **John Cameron Yrizarry**: 163, 192, 223, 254, 402, 427.

Drawings

Amy Harold: 203–216. **Olena Kassian**: 43–54, 120–139, 214–230, 243–249, 254–256, 321–344, 417–433. **Cynthia J. Page**: 33–42, 140–159, 345–352. **Don Radovich**: 10–32, 55–97, 108–119, 160–180, 206, 209, 213, 216, 231–242, 245, 246, 248–253, 267, 268, 278–297, 300, 353–384, 403–416, 434–452, 454–464. **Dolores R. Santoliquido**: 98–107, 181–202, 257–277, 298–320, 385–402, 465–481.

Acknowledgments

The editors wish to thank Roger Tory Peterson, Thomas D. Nicholson, and Les Line for reviewing portions of this book. We are also grateful to Mary Beacom Bowers, Editor of *Bird Watcher's Digest,* and Kenneth J. Strom, Manager of the Lillian Annette Rowe Sanctuary, National Audubon Society, Gibbon, Nebraska, for their generous assistance.

Numerous sources were consulted in the preparation of this volume, but three are worthy of special mention: *The Audubon Society Encyclopedia of North American Birds,* by John K. Terres; *Life Histories of North American Birds,* by Arthur Cleveland Bent; and the *National Geographic Society Field Guide to the Birds of North America.*